Minimal Submanifolds and Geodesics

Minimal Submanifolds and Geodesics

Proceedings of the Japan-United States Seminar on Minimal Submanifolds, including Geodesics Tokyo, 1977

Edited by

MORIO OBATA
Keio University

1979
NORTH-HOLLAND PUBLISHING COMPANY
AMSTERDAM - NEW YORK - OXFORD

ISBN: 0 444 85327 8

Published in co-edition with
Kaigai Publications, Ltd.

Distributors for Japan.
Kaigai Publications, Ltd. Tokyo, Japan

Distributors outside Japan

For USA and Canada.
Elsevier/North-Holland Inc.
52 Vanderbilt Avenue
New York, NY 10017,
USA

For all other countries
North-Holland Publishing Company
Amsterdam · New York · Oxford

Printed in Japan

Preface

The Japan-United States Seminar on minimal submanifolds, including geodesics, was held at the National Education Hall in Tokyo from September 19 to 22, 1977 under the co-sponsorship of the Japan Society for the Promotion of Sciences and the National Science Foundation as a part of the Japan-U.S. Cooperative Science Program. The organizers were T. Otsuki of Tokyo Institute of Technology and S. S. Chern of the University of California at Berkeley. From Europe, M. Berger, the University of Paris VII, France, and W. Klingenberg, the University of Bonn, Germany, participated in the seminar.

These proceedings contain extended or corrected versions of the lectures delivered at the seminar and also contain several papers related to the subjects of the seminar, which were contributed later by participants.

Besides the papers in the proceedings the following two lectures were delivered at the seminar:

J. Simons: Gauge fields.
K. Uhlenbeck: Minimally immersed surfaces in Riemannian manifolds with symmetry.

The editor is grateful to Messrs. H. Kuroda and T. Mori of Kaigai Publications, Ltd. for their kind and efficient cooperation.

July 1978

Morio OBATA

Participants

(Index of the Photograph)

1. H. Omori, Tokyo Metropolitan University
2. M. Obata, Tokyo Metropolitan University (presently, Keio University)
3. J. Simons, State University of New York at Stony Brook
4. K. Uhlenbeck, University of Illinois at Chicago Circle
5. S. Sasaki, Science University of Tokyo
6. K. Yano, Tokyo Institute of Technology
7. W. Klingenberg, University of Bonn
8. M. F. Atiyah, University of Oxford
9. T. Otsuki, Tokyo Institute of Technology (presently, Science University of Tokyo)
10. S. S. Chern, University of California at Berkeley
11. R. Osserman, Stanford University
12. J. C. C. Nitsche, University of Minnesota
13. F. J. Almgren, Princeton University
14. R. Naka, Tokyo Institute of Technology
15. T. Takahashi, University of Tsukuba
16. R. Gulliver, University of Minnesota
17. D. A. Hoffman, University of Massachusetts
18. B. Smyth, University of Notre Dame
19. M. Berger, University of Paris VII
20. S. Kobayashi, University of California at Berkeley
21. S. Murakami, Osaka University
22. J. E. Taylor, Rutgers University
23. S. T. Yau, University of California at Berkeley
24. H. B. Lawson, Jr., University of California at Berkeley
25. M. L. Michelsohn, University of California at Berkeley
26. M. Hasegawa, Tokyo Institute of Technology
27. T. Ochiai, Osaka University (presently, University of Tokyo)
28. K. Kenmotsu, Tôhoku University
29. H. Takagi, Tôhoku University
30. H. Mori, Toyama University

31. Y. Shikata, Nagoya University
32. T. Sakai, Hokkaido University
33. M. Takeuchi, Osaka University
34. H. Nakagawa, Tokyo University of Agriculture and Technology
35. K. Shiohama, University of Tsukuba
36. H. Kitahara, Kanazawa University
37. R. Takagi, University of Tsukuba
38. T. Itoh, University of Tsukuba
39. M. Maeda, Yokohama National University
40. M. Tanaka, Tokai University
41. Y. Muto, Yokohama National University
42. K. Ii, Yamagata University
43. S. Tanno, Tôhoku University
44. Y. Ogawa, Ochanomizu Women's University
45. Y. Suyama, Tôhoku University
46. D. E. Blair, Michigan State University
47. K. Ogiue, Tokyo Metropolitan University
48. Y. Maeda, Tokyo Metropolitan University (presently, Keio University)
49. M. Itoh, University of Tsukuba
50. S. Nishikawa, University of Tokyo
51. Y. Matsuyama, Chuo University

42
43
44
48
49
50
51
28
29
30
31
32
33
45
19
46
47
34
35
36
37
38
39
27
15
16
17
18
20
21
22
23
24
40
41
26
14
25
5
6
7
8
9
10
11
12
13
1
2
3
4

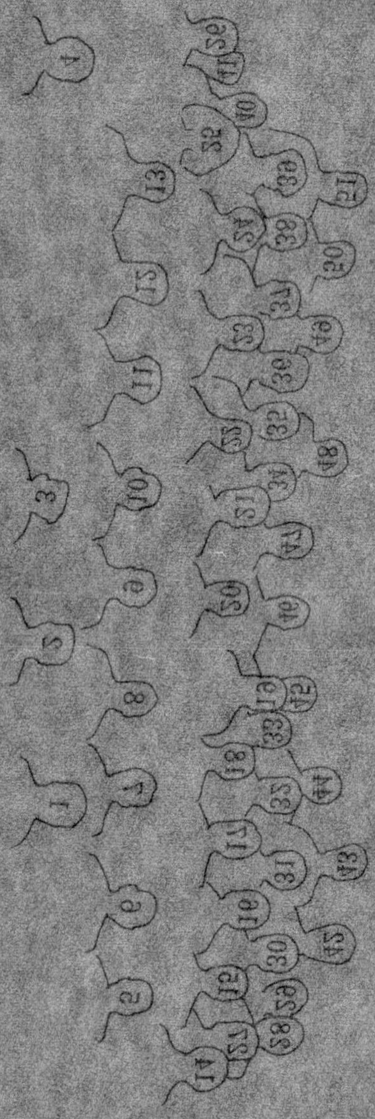

CONTENTS

Minimal Submanifolds and Geodesics
Kaigai Publications, Tokyo, 1978, 1–6

DIRICHLET'S PROBLEM FOR MULTIPLE VALUED FUNCTIONS AND THE REGULARITY OF MASS MINIMIZING INTEGRAL CURRENTS[1]

FREDERICK J. ALMGREN, JR.

This note is intended as a research announcement in general form of new regularity results for minimal surfaces [2].

The problem of least area (frequently called Plateau's problem) is really a collection of problems depending on the definitions one takes for surface, boundary of a surface, etc. In this note we will be talking exclusively about surfaces as *integral currents*, i.e. oriented surfaces with integer densities and integer topological multiplicities. In particular, we are studying the problem of least area in the context of *geometric measure theory* (the basic reference in the field is the treatise [3] with more recent results summarized in the colloquium lecture notes [5]). In this context we have the basic existence theorem

Existence Theorem. *Each $k-1$ dimensional integral cycle in $\boldsymbol{R}^n$ bounds a k dimensional integral current of least mass* [3 5.1.6].

In case $k = n-1$ such mass minimizing integral currents are highly regular; their singular sets, for example, are of Hausdorff dimension at most $k-7$ [4]. For $k \leqq n-2$ there are, however, many examples of mass minimizing integral currents having interior singular sets of dimension $k-2$. We will discuss the singular behavior of mass minimizing integral currents of codimension at least 2; our principal result is

Regularity Theorem. *A mass minimizing k dimensional integral current in $\boldsymbol{R}^n$ is regular almost everywhere with respect to k dimensional measure* [2 6.9].

For some time now this regularity question has been perhaps the most pressing fundamental question in geometric measure theory, and this theorem is the culmination of a research program of nearly a decade. In the following we will first give *examples* to suggest the substance and subtlety of the result followed by some *terminology* and a (sort of) *summary theorem* to indicate the overall flavor of the project and finally suggest an *outline of the general theory* which had to be developed to treat the problem.

[1] This research was supported in part by grants from the National Science Foundation.

Examples

As a source of examples we have the theorem

Example Theorem. *Each holomorphic variety in* $\boldsymbol{C}^n$ *is a mass minimizing (locally) integral current* [3 5.4.19].

To utilize this theorem we regard

$$\boldsymbol{C}^2 = \{(z, w)\} = \{(x_1, x_2, x_3, x_4)\} = \boldsymbol{R}^4 .$$

First example. The two dimensional mass minimizing integral current in $\boldsymbol{R}^4$ defined by the equation $w = \sqrt{z^3}$ ($w^2 = z^3$) has a singular point (i.e. a branch point) at $(0, 0)$.

Second example. The surface $w = \sqrt{(z - z_1)^3}$ has a branch point at $(z_1, 0)$.

Third example. The surface $w = \sqrt{\prod_{\nu=1}^{N} (z - z_\nu)^3}$ has branch points at $(z_1, 0), \cdots, (z_N, 0)$. In particular a 2-dimensional mass minimizing integral current in $\boldsymbol{R}^4$ can have isolated singular points at an arbitrarily large number of places.

Fourth example. Given a holomorphic function $f(z)$, the surface $w = f(z) + \sqrt{\prod_{\nu=1}^{N} (z - z_\nu)^3}$ has branch points at $(z_1, f(z_1)), \cdots, (z_N, f(z_N))$. A mass minimizing integral current thus can branch an arbitrarily large number of times off a minimal surface "center manifold".

Fifth example. In figure 1 is sketched a representation of a Cantor type set of positive 2 dimensional measure. Various centers of removed crosses are indicated at which one might seek to place a branch point of a mass minimizing integral current in the manner of the third example above. As the third example shows there are mass minimizing integral currents having any finite subset of these centers as branch points. If somehow one could construct a mass minimizing integral current S having each of these centers as a branch point (S needn't be a complex variety) then S would have a singular set of positive 2 dimensional measure (the singular set by definition is closed and the entire Cantor type set lies in the closure of the set of centers). The regularity theorem above says no such S exists.

Terminology and summary theorem

We will denote by $\boldsymbol{Q}$ the set of all unordered Q-tuples of points in $\boldsymbol{R}^n$ and by $\boldsymbol{B}(p, r)$ the closed m ball in $\boldsymbol{R}^m$ with center p and radius r. In case $F : \boldsymbol{B}(0, 1) \to \boldsymbol{Q}$ is reasonable we denote by $\mathbf{Area}\,(F\,; \boldsymbol{B}(0, r))$ the area of the graph of F (counting multiplicities as necessary) over $\boldsymbol{B}(0, r)$ and by $\mathbf{Dir}\,(F\,; \boldsymbol{B}(0, r))$ the Dirichlet integral of F, e.g. in case F locally separates into sheets $f_1, \cdots, f_Q$ then

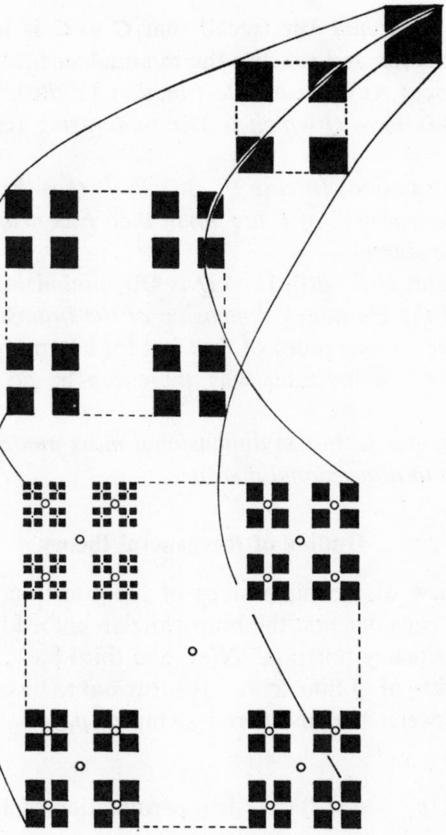

Figure 1

$$\textbf{Dir}\,(F\,;B(0,r))=\int_{B(0,r)}\sum\{(\partial f^k_\nu/\partial x_i)^2 : i=1,\cdots,m\,;\\ k=1,\cdots,n\,;\nu=1,\cdots,Q\}d\mathscr{L}^m\,.$$

Finally we set

$$\textbf{Exc}\,(F\,;B(0,r))=r^{-m}[\textbf{Area}\,(F\,;B(0,r))-Q\,\textbf{Area}\,(B(0,r))]\,.$$

As a consequence of the Taylor expansion of the area integrand in terms of first derivatives centered at 0 one makes the fundamental observation that

$$\textbf{Exc}\,(F\,;B(0,1))=2^{-1}\,\textbf{Dir}\,(F\,;B(0,1))+\mathscr{O}(|DF|^4)$$

which is crucial in proving the following theorem

Summary Theorem. (1) (Examples) *Holomorphic varieties uniquely* *minimize both* **Area** *and* **Dir** (recall that $C \to C$ is holomorphic if and only if it is harmonic and satisfies the minimal surface equation).

(2) (Existence) *Any reasonable function* $f : \partial B(0, 1) \to Q$ *extends to a function* $f : B(0, 1) \to Q$ *which is* **Dir** *minimizing* (essentially Rellich's theorem).

(3) (Approximation) *In case* $F : B(0, 1) \to Q$ *is* **Area** *minimizing and most of the "derivatives" of* F *are small then* F *is nearly* **Dir** *minimizing* (this implies (1) above).

(4) (Structure) *If* $F : B(0, 1) \to Q$ *is* **Dir** *minimizing then* F *is Hölder continuous and the Hausdorff dimension of the branch set of* F *is at most* $m - 2$ (this gives a new proof of that fact for holomorphic varieties; also the variety $w = \sqrt{z}$ indicates why there can be no interior derivative bounds on F).

(5) *The singular sets of m dimensional mass minimizing integral currents have zero m dimensional measure.*

Outline of the general theory

As an overview of the main steps of the construction of the present theory we now consider first the bilipschitzian embedding of Q into R^{PQ}, second the "frequency function" $N(r)$, and third basic approximations.

The embedding of Q into R^{PQ}. It turns out to be convenient to define the distance between the unordered Q-tuples $p_1, \cdots, p_Q$ and $q_1, \cdots, q_Q$ in R^n to be

$$\inf \{(\sum |p_i - q_{\sigma(i)}|^2)^{1/2} : \sigma \text{ is a permutation of } 1, \cdots, Q\} .$$

There is a more or less standard way of embedding Q into a Euclidean space by means of elementary symmetric functions (see in this regard the Appendix of [6]). For example, an unordered pair of points r_1, r_2 on the line R can be embedded in the uv-plane by the functions $u = r_1 + r_2$ and $v = r_1 r_2$ with image set given by the equation $4v \leqq u^2$. Unfortunately, neither this embedding nor its inverse is Lipschitzian (the problem is with v); this is not satisfactory for the delicate analysis of the present theory and a more suitable embedding is required. Such an embedding takes as embedding functions in the case above $u = \sup \{r_1, r_2\}$ and $v = \inf \{r_1, r_2\}$; this new map is clearly bilipschitzian onto its image defined by $u \geqq v$. Our general embedding function for unordered Q-tuples of points on the line is the function ξ which lists the points in decreasing order as an element of R^Q. It turns out that one can find lines $L_1, \cdots, L_P$ in R^n such that the mapping $\{\xi \circ L_\nu\}_\nu : Q \to R^{PQ}$ is a bilipschitzian embedding (here L_ν denotes orthogonal projection onto L_ν), and one has the following theorem

Theorem. (1) $\{\xi \circ L_\nu\}_\nu$ *is a bilipschitzian embedding of* Q *onto a polyhedral image* Q^* *in* R^{PQ} *which image is the union of linear subspaces of* R^{PQ} *intersected with half spaces.*

(2) Q^* *is a Lipschitz retract of* R^{PQ} *by a mapping* ρ (a general fact about a union of linear subspaces intersected with half spaces).

(3) *Lipschitz mappings of subsets of* R^m *into* Q *extend to be Lipschitz mappings on the whole of* R^m *into* Q (Kirszbraun's theorem [3 2.10.43]).

(4) *In case* $\{\xi \circ L_\nu\}_\nu \circ F$ *is differentiable at* x *then so is* F.

(5) *Mappings* F *into* Q *can be smoothed by convolution of the form*

$$\{\xi \circ L_\nu\}_\nu^{-1} \circ \rho \circ ([\{\xi \circ L_\nu\}_\nu \circ F] * \varphi_\varepsilon)$$

to produce general new types of comparison surfaces in geometric variational problems.

The frequency function $N(r)$. Suppose $F : B(0, 1) \to Q$ minimizes **Dir**. For each $0 < r \leq 1$ we define

$$N(r) = \frac{r\,\mathbf{Dir}\,(F : B(0, r))}{\int_{B(0,r)} F^2 d\mathscr{H}^{m-1}} ;$$

here $F^2 = \operatorname{dist}(F, (0, \cdots, 0))^2$. One checks that $N(r)$ is a dimensionless parameter which is invariant under homothetic changes in either domain or range. As particular cases, we have

Example 1. In case $m = 2$, $n = 1$, $Q = 1$ and $F(r, \theta) = a_n r^n \sin(n\theta)$ then $N(r) = n$ for each r.

Example 2. In case $m = 2$, $n = 1$, $Q = 1$ and

$$F(r, \theta) = \sum_{n=1}^{\infty} a_n r^n \sin(n\theta + \alpha_n)$$

then

$$N(1) = \frac{\sum_{n=1}^{\infty} n a_n^2}{\sum_{n=1}^{\infty} a_n^2} \geq \frac{\sum_{n=1}^{\infty} n(a_n/2^n)^2}{\sum_{n=1}^{\infty} (a_n/2^n)^2} = N(1/2) .$$

Example 3. In case $m = 2$, $n = 2$, $Q = 2$ and $F(z) = \sqrt{z}$ then $N(r) = 1/2$ for each r.

The general result is

Theorem. *Suppose* $F : B(0, 1) \to Q$ *minimizes* **Dir**.

(1) $N(r)$ *is monotonically nondecreasing as a function of* r (proved by estimation of dN/dr with use of first variation equalities).

(2) *The zeros of* F *do not have density one at* O (from which it follows, for example, that a harmonic function which is zero on a set of positive measure must be identically zero).

(3) *In case $N(r)$ is constant then F is homogeneous.*

(4) *F is homogeneous up to higher order at each of its zeros.*

(5) *F is Hölder continuous.*

(6) *The branch set of F is of Hausdorff dimension at most $m - 2$* (proved by downward induction on successive homogeneous approximations similar to the method of [4]).

Basic approximations. The main analytic estimates of the paper are indicated in the following theorem

Theorem. (1) *Any stationary integral varifold of small* **Exc** *can be approximated within mass of order* $\mathbf{Exc}^1$ *by a Q valued function having Lipschitz constant of order* 1 [1 6.2, 8.12].

(2) *Any mass minimizing integral current of small* **Exc** *can be approximated within mass of order* $\mathbf{Exc}^{1+\delta}$ *by a Lipschitz Q valued function* (the stronger estimate comes by virtue of convolution type comparison surfaces).

(3) *Any mass minimizing integral current of small* **Exc** *can be approximated off a suitable $C^{3,\alpha}$ center manifold by a Lipschitz Q valued function sufficiently nicely so that the $N(r)$ type analysis above applies and one can conclude that the singular set of the current cannot have positive measure.*

References

[1] W. K. Allard, *On the first variation of a varifold,* Ann. of Math. (2) **95** (1972), 417–491.

[2] F. J. Almgren, Jr., *Dirichlet's problem for multiple valued functions and the regularity of mass minimizing integral currents* (in preparation).

[3] H. Federer, *Geometric Measure Theory,* Die Grundlehren der math. Wissenschaften, Band 153, Springer-Verlag, Berlin, Heidelberg and New York, 1969.

[4] H. Federer, *The singular sets of area minimizing rectifiable currents with codimension one and of area minimizing flat chains modulo two with arbitrary codimension,* Bull. Amer. Math. Soc. **76** (1970), 767–771.

[5] H. Federer, *Colloquium Lectures on Geometric Measure Theory,* Bull. Amer. Math. Soc. (to appear).

[6] H. Whitney, *Complex Analytic Varieties,* Addison-Wesley Publishing Company, Reading, Mass., 1972.

PRINCETON UNIVERSITY
PRINCETON, NJ 08540
U.S.A.

Minimal Submanifolds and Geodesics
Kaigai Publications, Tokyo, 1978, 7–12

AUF WIEDERSEHENSMANNIGFALTIGKEITEN

MARCEL BERGER

Dans son livre [BL], Blaschke introduit la notion de "Wiedersehensfläche": il s'agit de surfaces de $\boldsymbol{R}^3$, ou plus généralement de variétés riemanniennes abstraites (M, g) de dimension 2, qui sont complètes et ont la propriété suivante: pour tout point m de M, toutes les géodésiques issues de m passent au bout du temps π par un même point m', différent de m et ceci en outre sans se rencontrer ni elles-mêmes ni entre elles.

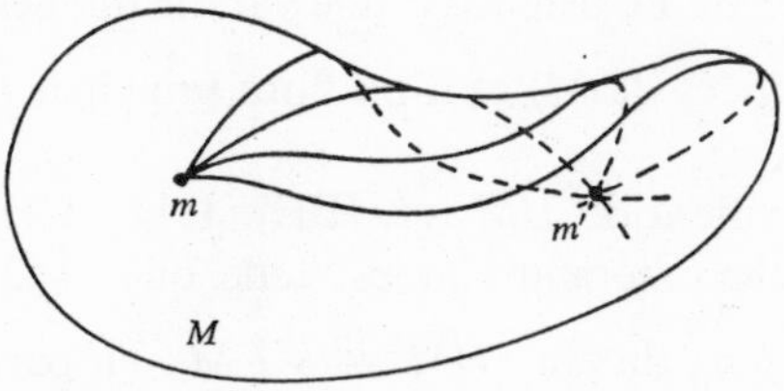

Fig. 1

Avec la notion, maintenant classique cf. [K-N], p. 97 ou [KI], p. 116, de cut-locus dans une variété riemannienne, on peut reformuler plus simplement la notion de Wiedersehensfläche en la suivante: "pour tout point m de M, son cut-locus est réduit à un point."

Blaschke conjectura dans [BL] que les seules Wiedersehensflächen sont les sphères, c'est à dire que si (M, g) est une Wiedersehensfläche, alors elle est isométrique à la sphère S^2 munie de sa métrique riemannienne *can*. Cette conjecture a été démontrée par L. W. Green en 1960 dans [G].

On peut de même en dimension quelconque donner la

Définition. Une variété riemannienne complète (M, g) est dite une *Wiedersehensmannigfaltigkeit* (W en abrégé) si le cut-locus de tout m de M est réduit à un point.

On a alors la généralisation suivante du théorème de Green

Théorème 1 ([BE], Appendice D). *Si (M, g) est une W et si sa dimension n est paire, alors elle est isométrique à la sphère canonique $(S^n$, can$)$* (voir aussi à la fin du texte).

Il n'est d'abord pas difficile de voir que l'hypothèse W entraîne que M est difféomorphe à S^n et qu'elle vérifie la *propriété SC* suivante: "toutes les géodésiques de (M, g) sont périodiques et de même plus petite période égale à 2π". Le théorème est alors la conjonction pure et simple des deux suivants

Théorème 2 ([W] ou [BE], 2.24). *Soit* (S^n, g) *une métrique riemannienne sur* S^n *qui vérifie SC. Si en outre n est pair, alors le volume* $\mathrm{Vol}(g)$ *de* (S^n, g) *est égal à celui, romains* $\beta(n)$, *de la sphère canonique* (S^n, can) (voir aussi à la fin du texte).

Théorème 3 ([BE], Appendice D). *Soit* (M, g) *une variété riemannienne, de dimension n quelconque et vérifiant W. Alors* $\mathrm{Vol}(g) \geqq \beta(n)$. *Si en outre* $\mathrm{Vol}(g) = \beta(n)$, *alors* (M, g) *est isométrique à* (S^n, can).

Schéma de démonstration du théorème 2.

Soit $UM \xrightarrow{p} M$ le fibré des vecteurs tangents de norme 1 de (M, g) et *désignons par* G $(t \mapsto G^t(\cdot))$ le flot géodésique sur UM. L'hypothèse SC entraîne que le cercle $S^1 = \boldsymbol{R}/2\pi\boldsymbol{Z}$ opère, via G, proprement et sans point fixe sur UM. D'où l'existence d'une variété quotient ΓM (la variété des géodésiques orientées de M) et d'un fibré principal $UM \xrightarrow{q} \Gamma M$ de groupe $S^1 = \boldsymbol{R}/2\pi\boldsymbol{Z}$.

La forme de Liouville α de UM (cf. [BE], 1.12) a les propriétés suivantes. Sa différentielle extérieure $d\alpha$ est telle que $(1/2\pi)d\alpha$ n'est autre que la forme de courbure du fibré $UM \xrightarrow{q} \Gamma M$; en particulier $(1/2\pi)d\alpha$ est basique, c'est à dire qu'il existe une 2-forme ω sur ΓM telle que $(1/2\pi)d\alpha = q^*\omega$. Ceci, joint au fait que l'intégrale $\int_\gamma (1/2\pi)\alpha$ vaut 1 sur chaque fibre γ (puisque γ est de longueur 2π), entraîne classiquement que la 2-forme ω représente la classe d'Euler du fibré $UM \xrightarrow{q} M$ via le théorème de Rham (cf. [G-H-V], p. 321). En particulier, puisque la dimension de ΓM est $2\,(n-1)$ et que ΓM est orientée naturellement, le réel $\int_{\Gamma M} \omega^{n-1}$ est bien défini et est un *entier*.

On utilise maintenant le fait que, pour toute métrique riemannienne sur S^n, le fibré unitaire UM est homéomorphe à la variété de Stiefel $\mathrm{St}_{n+1,2}$. Lorsque $n = 2p$, il se trouve que $\mathrm{St}_{2p+1,2}$ n'a comme cohomologie entière non nulle que $H^0 = \boldsymbol{Z}$, $H^{2p} = \boldsymbol{Z}_2$, $H^{4p-1} = \boldsymbol{Z}$. En appliquant alors la suite exacte de Gysin au fibré en cercles $UM \xrightarrow{q} \Gamma M$, on vérifie que l'entier mentionné plus haut vaut nécessairement 2.

On calcule alors le volume de UM de deux façons, en appliquant le théorème de Fubini à chacune des fibrations

$$UM \xrightarrow{p} M \quad \text{et} \quad UM \xrightarrow{q} \Gamma M$$

et on trouve bien ainsi que $\mathrm{Vol}(g) = \beta(n)$.

Schéma de démonstration du théorème 3.

Un point m_0 de M étant d'abord fixé, écrivons la mesure canonique dm de (M, g) en cordonnées sphériques de centre m_0:

$$(1) \qquad dm = f(u, r) dr \otimes d_{m_0} u ,$$

où r désigne la distance à l'origine, dr la mesure de Lebesgue sur $\boldsymbol{R}$ tandis que $d_{m_0}u$ est la mesure canonique de la sphère unité $U_{m_0}M$ de l'espace tangent $T_{m_0}M$ à M en m_0.

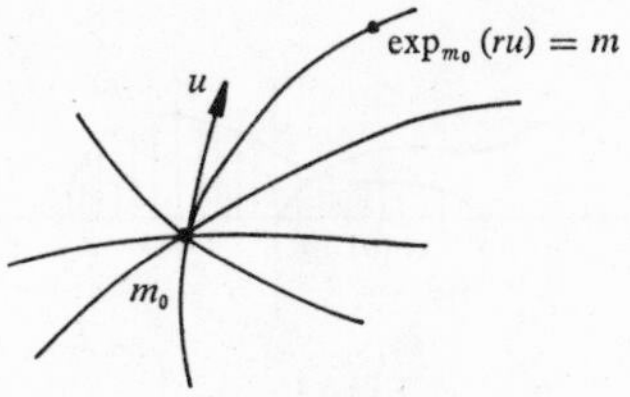

Fig. 2

La fonction $f(u, r)$ est bien définie par (1) ; elle mesure de combien dm diffère, au point $m = \exp_{m_0}(ru)$, de l'image par $\exp_{m_0}$ de la mesure de Lebesgue de l'espace euclidien $(T_{m_0}M, g_{m_0})$. Par exemple, pour tous u et r, on a:

$$\text{si } (M, g) \text{ est plate: } f(u, r) = r^{n-1} ,$$
$$\text{si } (M, g) = (S^n, can): f(u, r) = \sin^{n-r} r .$$

Si l'on fait maintenant varier m_0 dans M on obtient une fonction $f(u, r)$ définie sur $UM \times \boldsymbol{R}_+$.

L'hypothèse W entraîne entre autres que pour tout m de M la boule métrique $B(m, \pi)$ de centre m et de rayon π coïncide avec M donc en particulier

$$\operatorname{Vol}(g) = \int_0^\pi \int_{U_m M} f(u, r) dr \otimes d_m u .$$

En intégrant pour m parcourant M on a donc

$$\operatorname{Vol}^2(g) = \int_M \int_0^\pi \int_{U_m M} f(u, r) dm \otimes dr \otimes d_m u$$

que l'on peut réécrire, puisque $dm \otimes d_m u = du$ la mesure canonique de UM (cf. [BE], 1.123), comme

$$(2) \qquad \operatorname{Vol}^2(g) = \int_{UM} \int_0^\pi f(u, r) dr \otimes du .$$

On utilise maintenant la fibration $UM \xrightarrow{q} \Gamma M$ pour réarranger l'intégration de (2) en une intégration sur ΓM d'intégrales le long des géodésiques γ. On est ainsi conduit à étudier, pour tout u dans UM

$$\int_{r=0}^{r=\pi} \int_{s=0}^{s=\pi} f(G^s(u), r) dr \otimes ds \ ,$$

où G est toujours le flot géodésique et en utilisant le théorème de Liouville (cf. [BE], 1.125).

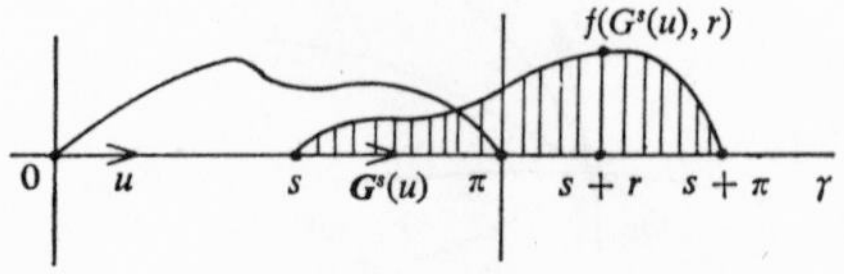

Fig. 3

L'intégrale double considérée représente l'intégrale en s de l'aire hachurée de la figure 3 lorsque s parcourt $[0, \pi]$. Le résultat désiré, pour obtenir le théorème 3 par la technique présédente, est que cette intégrale double est toujours supérieure ou égale à celle correspondant à la sphère canonique, c'est à dire que l'on désire que

$$(3) \qquad \int_{r \parallel 0}^{r=\pi} \int_{s=0}^{s=\pi} f(G^s(u), r) dr \otimes ds \geqslant \int_{r=0}^{r=\pi} \int_{s=0}^{s=\pi} \sin^{n-1} r \, dr \otimes ds.$$

Nous allons voir que tel est bien le cas.

On peut en effet écrire $f(u, r) = \det(A(r))$, où A est la résolvante de l'équation des champs de Jacobi le long de $\gamma : t \mapsto \exp(tu)$. Plus précisément, considérons l'équation différentielle

$$(4) \qquad Z''(r) + R(r) \circ Z(r) = 0$$

où $R(r)$ désigne l'endomorphisme de courbure $x \mapsto R(\dot{\gamma}(r), x)\dot{\gamma}(r)$ et $Z(r)$ un endomorphisme inconnu. Alors $A(r)$ est la solution de (4) qui satisfait aux conditions initiales $A(0) = 0$, $A'(0) = \mathrm{Id}$.

Le fait classique que toutes les solutions de (4) se déduisent de l'une d'entre elles (au moins dans un intervalle où celle-ci est non-singulière) entraîne facilement l'inégalité

$$(5) \qquad f^{1/n-1}(G^s(u), r) \geqslant f^{1/n-1}(u, s) f^{1/n-1}(u, s+r) \int_s^{s+r} \frac{dt}{f^{2/n-1}(u, t)}$$

valable pour tout s et r dans $[0, \pi]$ tels que en outre $s + r \leqslant \pi$.

L'inégalité (5) ne suffit certainement pas, en général, pour démontrer (3), puisque (5) n'apporte (à cause de la restriction $s + r \leqslant \pi$) d'information que sur la seule aire hachurée de la figure 4.

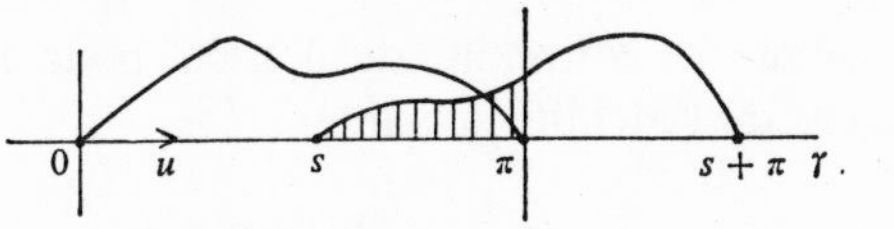

Fig. 4

Mais l'hypothèse W entraîne que $f(u, r + \pi) = f(u, r)$ pour tout u et r, ce qui implique l'inégalité des aires hachurées de la figure 5.

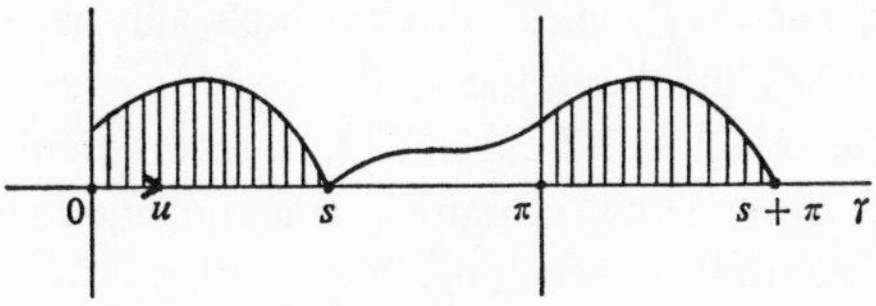

Fig. 5

On déduit finalement de ce qui précède que, pour une W, l'inégalité (3) est la conséquence de l'inégalité suivante

$$\int_{x=0}^{x=\pi}\int_{y=x}^{y=\pi}\int_{t=x}^{t=y} \frac{\varphi(x)\varphi(y)}{\varphi^2(t)} \sin^{n-2}(y-x)dtdydx$$
$$\geqslant \int_{x=0}^{x=\pi}\int_{y=x}^{y=\pi} \sin^{n-1}(y-x)dydx$$

valable pour toute fonction φ continue sur $[0, \pi]$ et strictement positive sur $]0, \pi[$. Cette inégalité a été démontrée par J. L. Kazdan qui a obtenu en fait le résultat tout à fait général suivant

Théorème 4 ([BE], Appendice E). *Pour toute fonction ρ continue sur $[0, \pi]$, positive ou nulle et telle que $\rho(\pi - t) = \rho(t)$ pour tout t, on a*

$$\int_{x=0}^{x=\pi}\int_{y=x}^{y=\pi}\int_{t=x}^{t=y} \frac{\varphi(x)\varphi(y)}{\varphi^2(t)} \rho(y-x)dtdydx$$
$$\geqslant \int_{x=0}^{x=\pi}\int_{y=x}^{y=\pi}\int_{t=x}^{t=y} \frac{\sin x \sin y}{\sin^2 t} \rho(y-x)dtdydx$$

pour toute fonction φ continue sur $[0, \pi]$ et strictement positive sur $]0, \pi[$. Si en outre $\rho(t) > 0$ pour au moins un t dans $]0, \pi[$, alors l'égalité ne peut être atteinte que si φ est proportionnelle à $t \mapsto \sin t$.

Schéma de démonstration du théorème 4.

L'idée de base est d'effectuer le changement de fonction

$$\varphi(t) = \sin t \, e^{\psi(t)} .$$

Alors, via l'inégalité de convexité de Jensen pour les intégrales, le théorème 4 résultera de l'identité suivante

$$\int_{x=0}^{x=\pi} \int_{y=x}^{y=\pi} \int_{t=x}^{t=y} [u(x) + u(y) - 2u(t)] \frac{\sin x \sin y}{\sin^2 t} \rho(y-x) dt dy dx = 0 ,$$

pour toute fonction u continue sur $[0, \pi]$ et toute fonction ρ satisfaisant aux hypothèses du théorème 4. La vérification de l'évanouissement précédent s'obtient alors sans trop de difficultés.

Remarque. Le cas de l'égalité, crucial bien sûr dans le théorème 3, provient du "en outre" du théorème 4.

Ajouté sur épreuves. Le théorème 2 a récemment été étendu par C.T.C. Yang aux dimensions impaires. En conséquence le théorème 1 est donc valable en toutes dimensions.

Bibliographie

[BE] Besse, Arthur L., Manifolds all of whose geodesics are closed, Ergebnisse Band n° 93, Springer.

[BL] Blaschke, Wihelm, Vorlesugen über Differentialgeometrie, 3rd Edition, Berlin, Springer 1930.

[G] Green, L., *Auf Wiedersehensflächen*, Annals of Math., **78** (1963), 289–299.

[G-H-V] Graeub, W., Halperin, S., Vanstone, R., Curvature, Connections and Homology, Vol. I, New York, Academic Press.

[KI] Kobayashi, S., *On Conjugate and Cut Loci*, in Studies in Global Geometry and Analysis, Edited by S. S. Chern, Studies in Mathematics, Vol. 4, Washington D.C., the Mathematical Association of America, 1976.

[K-N] Kobayashi, S., Nomizu, K., Foundations of Differential Geometry, Vol. 2, New York, Interscience Publishers, 1969.

[W] Weinstein, Alan, *On the volume of manifolds all of whose geodesics are closed*, J. Diff. Geometry **9** (1974) 513–517.

Laboratoire associé au C.N.R.S. n°212
Département de Mathématiques
Université Paris VII
2, place Jussieu
750005 Paris
France

Minimal Submanifolds and Geodesics
Kaigai Publications, Tokyo, 1978, 13–16

A GENERALIZATION OF THE HELICOID

D. E. BLAIR & J. R. VANSTONE

In [2] one of the authors gave a generalization of the catenoid characterizing minimal conformally flat hypersurfaces of Euclidean space. On the other hand it is well known that the only ruled surface in E^3 which is minimal is the helicoid. Our purpose here is to prove the following generalization.

Theorem. *Let M^n be a complete minimal hypersurface of E^{n+1} and suppose that M^n admits a codimension 1 foliation by Euclidean $(n-1)$-spaces. Then either M^n is totally geodesic or $M^n = M^2 \times E^{n-2}$ where M^2 is a helicoid in E^3.*

In contrast to the classical case this generalized helicoid is not locally isometric to the generalized catenoid of [2] which is the envelope of a particular locus of spheres, since such an isometric deformation would preserve the conformal flatness.

We denote by D the Riemannian connection on E^{n+1} and by g and ∇ the induced Riemannian metric and connection on M^n by the immersion $\iota: M^n \to E^{n+1}$. Suppose that the foliation is given by $u = 0$ for a 1-form u and let U be a unit vector field on M^n orthogonal to the leaves of the foliation. Since the foliation is by Euclidean spaces we can complete U to an orthonormal basis by vector fields $X_2, \cdots, X_n$ such that $D_{X_i}X_j = 0$ and hence, moreover, we may chcose local coordinates $(t, x^2, \cdots, x^n)$ such that $X_i = \partial/\partial x^i$ and $U = \alpha(\partial/\partial t)$ for some function α. Choosing, if necessary, the sign of t so that $\alpha > 0$, it follows that

$$[U, X_i] = -(X_i p)U$$

where $p = \ln \alpha$.

For simplicity we shall denote $X_i p$ by p_i. With respect to the above coordinates the metric is given by the matrix

$$\begin{bmatrix} 1/\alpha^2 & & & 0 \\ & 1 & & \\ & & \ddots & \\ 0 & & & 1 \end{bmatrix}.$$

Now computing the connection ∇ we have

$$\nabla_U U = \sum p_i X_i \,, \quad \nabla_{X_i} U = 0 \,, \quad \nabla_U X_i = -p_i U \,, \quad \nabla_{X_i} X_j = 0$$

where here and in the sequel sums are over the range of indices $2, \cdots, n$.

Again since M^n is foliated by Euclidean $(n-1)$-dimensional spaces, the Weingarten map A is given by

$$AU = rU + \sum q_i X_i \,, \qquad AX_i = q_i U \,.$$

From the Codazzi equation $(\nabla_U A)X_i - (\nabla_{X_i} A)U = 0$ we then obtain

$$X_i r - U q_i - r p_i = 0 \,, \qquad X_i q_j - p_i q_j - q_i p_j = 0$$

and hence

$$X_i q_j - X_j q_i = 0 \,.$$

Similarly the Gauss equation

$$g(R_{X_i U} U, X_j) = -g(AU, X_i) g(AU, X_j)$$

yields

$$X_i p_j = p_i p_j - q_i q_j \,.$$

These equations completely determine the dependence of the p_i and q_i on the x^i in a tubular neighborhood, T, of any leaf. For if ω is the 1-form defined by $\omega(U) = r$ and $\omega(X_i) = q_i$, then

$$d\omega(U, X_i) = \tfrac{1}{2}(U q_i - X_i r + r p_i) = 0 \,,$$
$$d\omega(X_i, X_j) = \tfrac{1}{2}(X_i q_j - X_j q_i) = 0 \,,$$

that is ω is closed. Therefore since the leaves of the foliation are simply-connected, $\omega = dq$ for some function q on T. Moreover, note that $X_i q = q_i$.

Now define a complex-valued function s on T by $s = p + iq$, then the Gauss-Codazzi equations yield

$$X_i X_j s = (X_i s)(X_j s)$$

or putting $S = e^{-s}$,

$$X_i X_j S = 0 \,.$$

The general solution of this system is

$$S = \tau_0 + \sum \tau_i x^i$$

where τ_0, τ_i are complex-valued functions of t. If all the τ_i vanish, $X_i s = 0$ and so the p_i and q_i vanish as well. The Weingarten map becomes $AU = rU$, $AX_i = 0$. Thus, if M^n is minimal, $\operatorname{tr} A = 0$ so that $r = 0$ giving the totally geodesic case of the theorem. Otherwise some $\tau_i \neq 0$ and, since e^s never vanishes, neither does S. Therefore the τ_i are linearly dependent over $\boldsymbol{R}$ and hence S must have the form

$$S = \tau_0 + \tau l(x)$$

where $\{\tau_0, \tau\}$ are linearly independent over $\boldsymbol{R}$ and depend only on t, while l is the real-valued linear form

$$l(x) = \sum l_i x^i .$$

Set $\sigma = \tau_0/\tau \notin \boldsymbol{R}$ and differentiate $e^{-(p+iq)} = \tau_0 + \tau l$, then

$$p_i = \frac{-l_i}{|l + \sigma|^2}(l + \operatorname{Re} \sigma) , \qquad q_i = \frac{l_i}{|l + \sigma|^2} \operatorname{Im} \sigma .$$

In particular

$$p_i = -\frac{(l + \operatorname{Re} \sigma)}{\operatorname{Im} \sigma} q_i$$

from which

$$\nabla_U U = \sum p_i X_i = -\frac{(l + \operatorname{Re} \sigma)}{\operatorname{Im} \sigma} \sum q_i X_i \tag{1}$$

and using the Codazzi equation for $X_i q_j$

$$\nabla_{\sum q_i X_i} \sum q_j X_j = -2\frac{(l + \operatorname{Re} \sigma)}{\operatorname{Im} \sigma} (\sum q_i^2) \sum q_j X_j . \tag{2}$$

Now suppose that M^n is minimal, then $r = 0$ and hence $Uq_i = 0$ from the Codazzi equation. Consider the 2-dimensional distribution spanned by U and $\sum q_i X_i$. Clearly this distribution is now integrable, for

$$[U, \sum q_i X_i] = \sum q_i \nabla_U X_i = -(\sum p_i q_i) U .$$

Moreover from this and equations (1) and (2), we see that the leaves M^2 of this foliation are totally geodesic in M^n. Similarly, if Y and Z are vector fields in the complementary distribution,

$$D_{\iota_* Y} \iota_* Z = \iota_* \nabla_Y Z , \qquad g(\nabla_Y Z, U) = 0$$

and writing $Y = \sum b_j X_j$

$$
\begin{aligned}
g(\nabla_Y Z, \sum q_i X_i) &= -g\Big(Z, \sum_{i,j} b_j(X_j q_i) X_i\Big) \\
&= -g\Big(Z, \sum_{i,j} b_j(p_j q_i + q_j p_i) X_i\Big) \\
&= 2\frac{(l + \operatorname{Re}\sigma)}{\operatorname{Im}\sigma} (\sum b_j q_j) g(Z, \sum q_i X_i) \\
&= 0 .
\end{aligned}
$$

Thus, the complementary distribution is also integrable with integral submanifolds which are totally geodesic in both M^n and E^{n+1}. It now follows that since M^n is complete, ι is a product immersion and hence M^2 lies in an E^3 (see e.g. [1] pp. 211–212). Since M^2 is totally geodesic in M^n, it is minimal in E^3. Since $M^n = M^2 \times E^{n-2}$ is foliated by $(n-1)$-dimensional Euclidean spaces, M^2 must be a ruled surface and hence a helicoid.

References

[1] Bishop, R. L. and R. J. Crittenden, Geometry of Manifolds, Academic Press, New York, 1964.

[2] Blair, D. E., *On a generalization of the catenoid,* Can. J. Math. **27** (1975) 231–236.

MICHIGAN STATE UNIVERSITY
EAST LANSING, MI 48824
U.S.A.
AND
UNIVERSITY OF TORONTO
TORONTO, ONTARIO
CANADA

Minimal Submanifolds and Geodesics
Kaigai Publications, Tokyo, 1978, 17–30

AFFINE MINIMAL HYPERSURFACES

SHIING-SHEN CHERN[1)]

Introduction

In this paper I wish to call attention to a class of variational problems which arise naturally in affine geometry. In recent years there have been important developments on minimal varieties and variational problems with more general integrands, such as Jean Taylor's crystalline integrands. These integrals involve first partial derivatives. In the affine differential geometry of hypersurfaces there is an invariant volume element which depends on second partial derivatives and the corresponding Euler-Lagrange equation is a partial differential equation of the fourth order. Because of their simple geometrical origin affine minimal hypersurfaces have many remarkable properties, some of which will be discussed below. On the other hand, fundamental problems, such as the analogues of Plateau and Bernstein problems, have not been touched.

1. Hypersurfaces in unimodular affine space (cf. [1], [2], [3], [5])

Let A^{n+1} be the unimodular affine space of dimension $n+1$, i.e., the space with the real coordinates $x^1, \cdots, x^{n+1}$, provided with the group G of transformations (called the "unimodular affine group")

$$x^{*\alpha} = \sum_{\beta} c^{\alpha}_{\beta} x^{\beta} + d^{\alpha}\,, \qquad 1 \leqslant \alpha, \beta, \gamma \leqslant n+1\,, \tag{1}$$

where

$$\det(c^{\alpha}_{\beta}) = +1\,. \tag{2}$$

In this space distance and angle have no meaning, but there is the notion of parallelism. Vectors

$$V = (v^1, \cdots, v^{n+1}) \tag{3}$$

are transformed according to the equations

$$v^{*\alpha} = \sum c^{\alpha}_{\beta} v^{\beta}\,. \tag{4}$$

1) Work done under partial support of NSF grant MCS 74-23180.

As a consequence of (2) the determinant of $n+1$ vectors, e.g.,

$$(V_1, \cdots, V_{n+1}) = \det(v^{\alpha}_{\beta}) , \tag{5}$$

where

$$V_{\beta} = (v^1_{\beta}, \cdots, v^{n+1}_{\beta}) , \tag{5a}$$

is an invariant.

A hypersurface consists of an n-dimensional manifold M and an immersion

$$x : M \to A^{n+1} . \tag{6}$$

We will study its properties invariant under the group (1), and we will apply the method of moving frames.

By a *unimodular affine frame*, or simply a *frame*, is meant a point x and $n+1$ vectors e_{α}, satisfying the condition

$$(e_1, \cdots, e_{n+1}) = 1 . \tag{7}$$

The importance of frames in affine geometry lies in the fact that there is exactly one unimodular affine transformation carrying one frame into another. In the space of all frames, which is the same as the group manifold G, we can write

$$\begin{aligned} dx &= \sum \omega^{\alpha} e_{\alpha} , \\ de_{\alpha} &= \sum \omega^{\beta}_{\alpha} e_{\beta} . \end{aligned} \tag{8}$$

The $\omega^{\alpha}, \omega^{\beta}_{\alpha}$ are the Maurer-Cartan forms of G. Differentiating (7) and using (8), we get

$$\sum_{\alpha} \omega^{\alpha}_{\alpha} = 0 . \tag{9}$$

Exterior differentiation of (8) gives the *structure equations* of A^{n+1} or the *Maurer-Cartan equations* of G:

$$\begin{aligned} d\omega^{\alpha} &= \sum_{\beta} \omega^{\beta} \wedge \omega^{\alpha}_{\beta} , \\ d\omega^{\beta}_{\alpha} &= \sum_{\gamma} \omega^{\gamma}_{\alpha} \wedge \omega^{\beta}_{\gamma} , \end{aligned} \tag{10}$$

We restrict to the submanifold of frames such that x lies on the hypersurface and $e_1, \cdots, e_n$ span the tangent hyperplane at x. Then

$$\omega^{n+1} = 0 \tag{11}$$

and the first equation of (10) gives

$$d\omega^{n+1} = \sum_i \omega^i \wedge \omega_i^{n+1} = 0 \,, \tag{12}$$

where, as later, we agree on the index range

$$1 \leqslant i, j, k \leqslant n \,. \tag{13}$$

By Cartan's lemma we have

$$\omega_i^{n+1} = \sum_k h_{ik}\omega^k \,, \tag{14}$$

where

$$h_{ik} = h_{ki} \,. \tag{15}$$

The quadratic differential form

$$II = \sum_i \omega^i \omega_i^{n+1} = \sum_{i,k} h_{ik}\omega^i\omega^k \tag{16}$$

is called the *second fundamental form* of M. In our terminology there is no first fundamental form.

Let $xe_1^* \cdots e_{n+1}^*$ be a frame satisfying also the condition that $e_1^*, \cdots, e_n^*$ span the tangent hyperplane to the hypersurface $x(M)$ at x. Then we have

$$\begin{aligned} e_i^* &= \sum a_i^k e_k \,, \\ e_{n+1}^* &= A^{-1}e_{n+1} + \sum a_{n+1}^i e_i \,, \qquad A = \det(a_i^k) \end{aligned} \tag{17}$$

or

$$\begin{aligned} e_i &= \sum b_i^k e_k^* \,, \\ e_{n+1} &= Ae_{n+1}^* + \sum_i b_{n+1}^i e_i^* \,, \end{aligned} \tag{17a}$$

where (b_i^k) is the inverse matrix of (a_i^k). Relative to the new frames $xe_1^* \cdots e_{n+1}^*$ let

$$\begin{aligned} dx &= \sum \omega^{*i} e_i^* \,, \\ de_\alpha^* &= \sum_\beta \omega_\alpha^{*\beta} e_\beta^* \,. \end{aligned} \tag{18}$$

Then we have

$$\begin{aligned} \omega^k &= \sum a_i^k \omega^{*i} \,, \\ \omega_i^{n+1} &= (e_1, \cdots, e_n, de_i) = A^{-1} \sum b_i^k \omega_k^{*n+1} \end{aligned} \tag{19}$$

and

$$\sum \omega^i \omega_i^{n+1} = A^{-1} \sum \omega^{*i} \omega_i^{*n+1} . \tag{20}$$

This leads to the important conclusion that the second fundamental form is defined by the hypersurface up to a non-zero factor. Its rank is an affine invariant. We will restrict ourselves to the study of *non-degenerate hypersurfaces* for which the rank of II is equal to n.

For a non-degenerate hypersurface we have thus

$$H \underset{\text{def}}{=} \det(h_{jk}) \neq 0 . \tag{21}$$

Under a change of frames (17) we have, by (20),

$$H^* = HA^{n+2} . \tag{22}$$

It follows that for even n the sign of H is an affine invariant.

Suppose M be oriented. We shall restrict to frames such that $e_1, \cdots, e_n$ define a coherent orientation. Then $A > 0$ and

$$|H^*|^{1/(n+2)} = |H|^{1/(n+2)} A . \tag{22a}$$

The normalized second fundamental form

$$\hat{II} = |H|^{-1/(n+2)} II \tag{23}$$

is affinely invariant. Since it is of rank n, it defines a pseudo-Riemannian structure on M, to which the methods of Riemannian geometry can be applied. In particular, M has the volume element

$$dV = |H|^{1/(n+2)} \omega^1 \wedge \cdots \wedge \omega^n . \tag{24}$$

If $\hat{II}$ is a definite form, M is locally convex.

Taking the exterior derivative of (14) and using (10), we get

$$\sum_k \left(dh_{ik} + h_{ik}\omega_{n+1}^{n+1} - \sum_j h_{ij}\omega_k^j - \sum_j h_{kj}\omega_i^j \right) \wedge \omega^k = 0 .$$

This gives

$$dh_{ik} + h_{ik}\omega_{n+1}^{n+1} - \sum_j h_{ij}\omega_k^j - \sum_j h_{kj}\omega_i^j = \sum_j h_{ikj}\omega^j , \tag{25}$$

where h_{ikj} is symmetric in all the indices i, k, j. Let (H^{ik}) be the adjoint matrix of (h_{ij}), so that

$$\sum_j H^{ij} h_{jk} = \delta_k^i H . \tag{26}$$

By (25) and (9)

$$dH = \sum_{i,j} H^{ij} dh_{ij} = -(n+2)H\omega_{n+1}^{n+1} + \sum_{i,j,k} H^{ij} h_{ijk} \omega^k . \tag{27}$$

Differentiating the second equation of (17), we get

$$de_{n+1}^* \equiv -A^{-2} dA e_{n+1} + A^{-1} \omega_{n+1}^{n+1} e_{n+1} + \sum_i a_{n+1}^i \omega_i^{n+1} e_{n+1}, \text{ mod } e_k .$$

It follows that

$$\omega_{n+1}^{*n+1} = \omega_{n+1}^{n+1} - d \log A + A \sum a_{n+1}^i \omega_i^{n+1} . \tag{28}$$

On the other hand, by (22a),

$$d \log |H^*| = d \log |H| + (n+2) d \log A . \tag{29}$$

Combination of (28) and (29) gives

$$\begin{aligned} &\omega_{n+1}^{*n+1} + \frac{1}{n+2} d \log |H^*| \\ &= \omega_{n+1}^{n+1} + \frac{1}{n+2} d \log |H| + A \sum a_{n+1}^i \omega_i^{n+1} . \end{aligned} \tag{30}$$

Since M is non-degenerate, we can choose a_{n+1}^i to annihilate this expression.

Suppose this be done, i.e., suppose

$$(n+2)\omega_{n+1}^{n+1} + d \log |H| = 0 , \tag{31}$$

or

$$\sum H^{ij} h_{ijk} = 0 . \tag{31a}$$

By (17) the vector e_{n+1} is defined up to a factor. The line through x in the direction of e_{n+1} is called the *affine normal* at x. The vector

$$\nu = |H|^{1/(n+2)} e_{n+1} , \tag{32}$$

affinely defined, is called the *affine normal vector*. The affine normal has interesting geometrical properties; we refer to [1] for a beautiful treatment.

Differentiating (31), we get

$$\sum_i \omega_{n+1}^i \wedge \omega_i^{n+1} = 0 , \tag{33}$$

which gives

$$\omega_{n+1}^i = \sum l^{ik} \omega_k^{n+1} , \tag{34}$$

where

$$l^{ik} = l^{ki} . \tag{35}$$

Equation (34) can be written

$$\omega_{n+1}^i = \sum l_k^i \omega^k , \tag{36}$$

where

$$l_k^i = \sum_j l^{ij} h_{jk} \tag{37}$$

The quadratic differential form

$$\begin{aligned} III = \sum \omega_{n+1}^i \omega_i^{n+1} &= \sum l^{ik} \omega_i^{n+1} \omega_k^{n+1} \\ &= \sum l^{ik} h_{ij} h_{kl} \omega^j \omega^l \end{aligned} \tag{38}$$

is called the *third fundamental form.*

With xe_{n+1} as the affine normal the allowable change of frame (17) becomes

$$\begin{aligned} e_i^* &= \sum a_i^k e_k , \\ e_{n+1}^* &= A^{-1} e_{n+1} , \qquad A = \det (a_i^k) , \end{aligned} \tag{39}$$

whence

$$\omega_i^{*n+1} = (e_1^*, \cdots, e_n^*, de_i^*) = A \sum a_i^k \omega_k^{n+1} . \tag{40}$$

On the other hand, by (39),

$$de_{n+1}^* \equiv A^{-1} de_{n+1} , \qquad \text{mod } e_{n+1} ,$$

while

$$\begin{aligned} de_{n+1}^* &\equiv \sum \omega_{n+1}^{*i} e_i^* , \qquad \text{mod } e_{n+1} , \\ de_{n+1} &\equiv \sum \omega_{n+1}^i e_i , \qquad \text{mod } e_{n+1} . \end{aligned}$$

It follows that

$$\omega_{n+1}^i = A \sum a_k^i \omega_{n+1}^{*k} . \tag{41}$$

By (40) and (41), we have

$$III = \sum \omega_{n+1}^i \omega_i^{n+1} = A \sum a_k^i \omega_{n+1}^{*k} \omega_i^{n+1} = \sum \omega_{n+1}^{*k} \omega_k^{*n+1} .$$

Therefore the third fundamental form is invariant under a change of frame keeping the affine normal xe_{n+1} fixed.

The elementary symmetric functions of the eigenvalues of III relative to $\hat{II}$ give the affine curvatures of M, which are scalar invariants.

In particular, the trace

$$L = \frac{1}{n} |H|^{1/(n+2)} \sum l_i^i \tag{42}$$

is called the *affine mean curvature*. A hypersurface with $L = 0$ is called *affine minimal*.

We consider the case when the hypersurface is given in the "Monge form":

$$x^{n+1} = F(x^1, \cdots, x^n) . \tag{43}$$

Let

$$x = (x^1, \cdots, x^n, F(x^1, \cdots, x^n)) . \tag{44}$$

The first equation of (8) holds, if we set

$$\begin{aligned} &\omega^i = dx^i , \\ &e_i = \Big(0, \cdots, 0, \underset{i\text{th}}{1}, 0, \cdots, 0, \frac{\partial F}{\partial x^i}\Big) . \end{aligned} \tag{45}$$

Putting

$$e_{n+1} = (0, \cdots, 0, 1) . \tag{46}$$

we get

$$\omega_i^{n+1} = \sum_j F_{ij} dx^j , \qquad F_{ij} = \frac{\partial^2 F}{\partial x^i \partial x^j} \tag{47}$$

and

$$II = \sum_{i,j} F_{ij} dx^i dx^j , \tag{48}$$

so that

$$h_{ij} = F_{ij} , \qquad H = \det (F_{ij}) . \tag{49}$$

We let (F^{ij}) be the inverse matrix of (F_{ik}), so that

$$\sum F_{ij} F^{jk} = \delta_i^k . \tag{50}$$

To find the affine normal we let $a_i^k = \delta_i^k$ in (17). Then it is along the vector e_{n+1}^* where, by (30), a_{n+1}^i is determined by

$$d\log|H| + (n+2)\sum a^i_{n+1}h_{ik}dx^k = 0\,.$$

It follows that the affine normal vector has the expression

$$(51)\qquad \nu = |H|^{1/(n+2)}\left\{e_{n+1} - \frac{1}{n+2}\sum_{i,k}\frac{\partial}{\partial x^i}(\log|H|)F^{ik}e_k\right\}.$$

2. Affine minimal hypersurfaces

Let D be a sufficiently small domain of M, with boundary ∂D. Its volume is

$$(52)\qquad V(D) = \int_D dV\,,$$

where dV is given by (24). We wish to compute the first variation $\delta V(D)$ under an infinitisimal displacement of D, with ∂D kept fixed. To express this situation analytically, let I be the interval $-\frac{1}{2} < t < \frac{1}{2}$. Let $f: M \times I \to A^{n+1}$ be a smooth mapping such that its restriction to $M \times t$, $t \in I$, is an immersion and that $f(m, 0) = x(m)$, $m \in M$. We consider a frame field $e_\alpha(m, t)$ over $M \times I$ such that for every $t \in I$, $e_i(m, t)$ are tangent vectors and $e_{n+1}(m, t)$ is along the affine normal to $f(M \times t)$ at (m, t).

The analytical development parallels that of § 1, in the sense that we are studying, instead of one hypersurface, a family of hypersurfaces. Pulling the forms ω^α, ω^β_α in the frame manifold back to $M \times I$, we have, since e_i span the tangent hyperplane at $f(m, t)$,

$$(53)\qquad \omega^{n+1} = a\,dt\,.$$

Its exterior differentiation gives

$$(54)\qquad \sum \omega^i \wedge \omega^{n+1}_i + dt \wedge (a\omega^{n+1}_{n+1} + da) = 0\,.$$

It follows that we can set

$$(55)\qquad \begin{aligned} \omega^{n+1}_i &= \sum h_{ij}\omega^j + h_i dt\,,\\ a\omega^{n+1}_{n+1} + da &= \sum h_i\omega^i + h\,dt\,, \end{aligned}$$

where

$$(56)\qquad h_{ij} = h_{ji}\,.$$

Exterior differentiation of (55_1) then gives

$$(57)\qquad \begin{aligned} &\sum (dh_{ij} - \sum h_{ik}\omega^k_j - \sum h_{jk}\omega^k_i + h_{ij}\omega^{n+1}_{n+1}) \wedge \omega^j\\ &\quad + \sum (dh_i - \sum h_k\omega^k_i + h_i\omega^{n+1}_{n+1} - a\sum h_{ij}\omega^j_{n+1}) \wedge dt = 0\,. \end{aligned}$$

Hence we can set

$$(58)\quad \begin{aligned} dh_{ij} &= \sum h_{ik}\omega_j^k + \sum h_{jk}\omega_i^k - h_{ij}\omega_{n+1}^{n+1} + \sum h_{ijk}\omega^k + p_{ij}dt\ , \\ dh_i &= \sum h_k\omega_i^k - h_i\omega_{n+1}^{n+1} + a \sum h_{ij}\omega_{n+1}^j + \sum p_{ij}\omega^j + q_i dt\ , \end{aligned}$$

where h_{ijk} is symmetric in all three indices, and

$$(59)\quad p_{ij} = p_{ji}\ .$$

We introduce H^{ik} as in (26), and find

$$(60)\quad \begin{aligned} dH &= \sum H^{ij}dh_{ij} \\ &= -(n+2)H\omega_{n+1}^{n+1} + \sum H^{ij}h_{ijk}\omega^k + \sum H^{ij}p_{ij}dt\ . \end{aligned}$$

As in § 1, by the change of frame (17), we can make

$$(61)\quad \sum H^{ij}h_{ijk} = 0\ .$$

Geometrically this means that e_{n+1} is in the direction of the affine normal to the hypersurface $f(M \times t)$ at $f(m, t)$. The resulting equation (60) can be written as

$$(62)\quad f^*\omega_{n+1}^{n+1} + \frac{1}{n+2} d\log|H| = bdt\ ,$$

where

$$(63)\quad b = \frac{1}{(n+2)H}\sum H^{ij}p_{ij}\ .$$

For later application we differentiate (26) and use (58), obtaining

$$(64)\quad \begin{aligned} dH^{ij} = &-\sum H^{ik}\omega_k^j - \sum H^{jk}\omega_k^i + H^{ij}(\omega_{n+1}^{n+1} + d\log|H|) \\ &-\frac{1}{H}\sum H^{ik}H^{jl}h_{klr}\omega^r - \frac{1}{H}\sum H^{ik}H^{jl}p_{kl}dt\ . \end{aligned}$$

We introduce

$$(65)\quad h^i = \sum H^{ij}h_j\ .$$

By (58) and (64), we find

$$(66)\quad \begin{aligned} dh^i = &-\sum h^k\omega_k^i + h^i d\log|H| + aH\omega_{n+1}^i \\ &-\frac{1}{H}\sum H^{ik}h^l h_{klr}\omega^r + \sum H^{ij}p_{jk}\omega^k \end{aligned}$$

$$-\frac{1}{H}\sum H^{ik}h^l p_{kl}dt + \sum H^{ij}q_j dt .$$

We use the structure equations (10), (9), and obtain

$$\begin{aligned} d(\omega^1 \wedge \cdots \wedge \omega^n) &= \sum_i (-1)^{i-1}\omega^1 \wedge \cdots \wedge \omega^{i-1} \wedge (\omega^i \wedge \omega^i_i + \omega^{n+1} \wedge \omega^i_{n+1}) \\ &\quad \wedge \omega^{i+1} \wedge \cdots \omega^n \\ &= \omega^{n+1}_{n+1} \wedge \omega^1 \wedge \cdots \wedge \omega^n \\ &\quad + \omega^{n+1} \wedge \sum_i \omega^1 \wedge \cdots \wedge \omega^{i-1} \wedge \omega^i_{n+1} \wedge \omega^{i+1} \wedge \cdots \wedge \omega^n . \end{aligned} \tag{67}$$

Pulling back under f, we get (cf. (42))

$$\begin{aligned} &|H|^{-1/(n+2)}d(|H|^{1/(n+2)}f^*(\omega^1 \wedge \cdots \wedge \omega^n)) \\ &= \left(f^*\omega^{n+1}_{n+1} + \frac{1}{n+2}d\log|H| + n|H|^{-1/(n+2)}Ladt\right) \\ &\quad \wedge f^*(\omega^1 \wedge \cdots \wedge \omega^n) \\ &= (b + n|H|^{-1/(n+2)}La)dt \wedge \tilde{\omega}^1 \wedge \cdots \wedge \tilde{\omega}^n , \end{aligned}$$

where $\tilde{\omega}^i$ is defined from $f^*\omega^i$ by "splitting off" the term in dt:

$$f^*\omega^i = \tilde{\omega}^i + a^i dt . \tag{68}$$

The operator d on $M \times I$ can also be decomposed as

$$d = d_M + dt\frac{\partial}{\partial t} . \tag{69}$$

Equating the terms in dt in the above equation, we get

$$\begin{aligned} &\frac{\partial}{\partial t}(|H|^{1/(n+2)}\tilde{\omega}^1 \wedge \cdots \wedge \tilde{\omega}^n) \\ &\quad + d_M\Big\{|H|^{1/(n+2)} \sum_i (-1)^i a^i \tilde{\omega}^1 \wedge \cdots \wedge \tilde{\omega}^{i-1} \\ &\qquad \wedge \tilde{\omega}^{i+1} \wedge \cdots \tilde{\omega}^n\Big\} \\ &= |H|^{1/(n+2)}(b + n|H|^{-1/(n+2)}La)\tilde{\omega}^1 \wedge \cdots \wedge \tilde{\omega}^n . \end{aligned} \tag{70}$$

On ∂D we have $a^i = 0$. Integrating over D and setting $t = 0$, we find the first variation of volume

$$
\begin{aligned}
V'(0) &= \frac{\partial}{\partial t}\int_D |H|^{1/(n+2)}\tilde{\omega}^1 \wedge \cdots \tilde{\omega}^n\Big|_{t=0} \\
&= \int_D (b + n|H|^{-1/(n+2)}La)dV\Big|_{t=0} .
\end{aligned} \tag{71}
$$

The last expression can be simplified, on account of the following lemma:

Lemma. *For* $t =$ const. *the form*

$$
\begin{aligned}
&\left(b + \frac{n}{n+2}|H|^{-1/(n+2)}La\right)dV \\
&= \left(b|H|^{1/(n+2)} + \frac{n}{n+2}La\right)\omega^1 \wedge \cdots \wedge \omega^n
\end{aligned} \tag{72}
$$

is exact and its integral over D for $t = 0$ *is zero.*

Proof. We introduce the form

$$
\begin{aligned}
\Omega &= \frac{1}{(n-1)!}\sum \varepsilon_{i_1\cdots i_n}h^{i_1}\omega^{i_2} \wedge \cdots \wedge \omega^{i_n} \\
&= \sum_i (-1)^{i-1}\omega^1 \wedge \cdots \wedge \omega^{i-1}h^i \wedge \omega^{i+1} \wedge \cdots \wedge \omega^n ,
\end{aligned} \tag{73}
$$

where $\varepsilon_{i_1\cdots i_n}$ is $+1$ or -1 according as $i_1, \cdots, i_n$ is an even or odd permutation of $1, \cdots, n$, and is otherwise zero. By using (10) and (66), we find

$$
\begin{aligned}
d\Omega &= \left(\omega_{n+1}^{n+1} + \frac{dH}{H}\right) \wedge \Omega \\
&\quad + \{a\sum l_i^i + (n+2)b\}H\omega^1 \wedge \cdots \wedge \omega^n .
\end{aligned} \tag{74}
$$

It follows that

$$
d(|H|^{-(n+1)/(n+2)}\Omega) = |H|^{-(n+1)/(n+2)}\left(d\Omega - \frac{n+1}{n+2}d\log|H| \wedge \Omega\right) .
$$

By (62) we have, for $t =$ const.,

$$
\begin{aligned}
&d(|H|^{-(n+1)/(n+2)}\Omega) \\
&= (\operatorname{sgn} H)\{a\sum l_i^i + (n+2)b\}|H|^{1/(n+2)}\omega^1 \wedge \cdots \wedge \omega^n ,
\end{aligned} \tag{75}
$$

We suppose the variation such that $h_i(m, 0) = 0$ for $m \in \partial D$. Hence the lemma follows.

It follows from (71) that

$$V'(0) = \frac{n(n+1)}{n+2} \int_D |H|^{-1/(n+2)} LadV \bigg|_{t=0} . \tag{71a}$$

If this is zero for arbitrary functions $a(m, t)$, $m \in D$, $t \in I$. satisfying $a(m, 0) = 0$, $h_i(m, 0) = 0$, $m \in \partial D$, we must have $L = 0$, i.e., M be an affine minimal hypersurface.

To better understand the analytical significance of an affine minimal hypersurface we suppose the latter be given in the Monge form (43). Then the affine normal vector ν is given by (51), with e_α defined by (45), (46). Computing mod ν, we find

$$\begin{aligned} |H|^{-1/(n+2)} d\nu \equiv & -\frac{1}{n+2} \sum_k d\Big(\sum_i (\log|H|)_i F^{ik}\Big) e_k \\ & -\frac{1}{n+2} \sum_{i,k} (\log|H|)_i F^{ik} dF_k e_{n+1} , \end{aligned}$$

the last term being congruent to

$$-\frac{1}{(n+2)^2} d\log|H| \sum_{i,k} (\log|H|)_i F^{ik} e_k , \qquad \text{mod } \nu ,$$

where the subscript i means partial differentiation with respect to x^i. From this it follows that the condition for the hypersurface (43) to be an affine minimal hypersurface is

$$(n+2) \sum_{i,k} ((\log|H|)_i F^{ik})_k + \sum_{i,k} F^{ik} (\log|H|)_i (\log|H|)_k = 0 , \tag{76}$$

which is a partial differential equation of the fourth order in the unknown function $F(x^1, \cdots, x^n)$.

Using the pseudo-Riemannian structure defined by $\hat{II}$ in (23), equation (76) can also be written

$$\Delta |H|^{-1/(n+2)} = 0^{(1)} , \tag{77}$$

where Δ is the Beltrami-Laplace operator relative to $\hat{II}$.

We have the following theorem:

Theorem. *There is no closed affine minimal hypersurface in A^{n+1}.*

Proof. We have the formula

$$d(x, \nu, \underbrace{dx, \cdots, dx}_{n-1}) = -(\nu, \underbrace{dx, \cdots, dx}_{n}) + (x, d\nu, \underbrace{dx, \cdots, dx}_{n-1}) , \tag{78}$$

where the determinant have entries which are vector-valued one-forms. By (32) and (31) we have

(1) This expression was communicated to me by S. Y. Cheng.

$$d\nu = |H|^{1/(n+2)}(\omega_{n+1}^1 e_1 + \cdots + \omega_{n+1}^n e_n) .$$

It follows that

$$\begin{aligned}(x, d\nu, dx, \cdots, dx) &= (n-1)!\,|H|^{1/(n+2)}(x, e_1, \cdots, e_n) \\ &\quad \times (\omega_{n+1}^1 \wedge \omega^2 \wedge \cdots \wedge \omega^n + \cdots \\ &\quad + \omega^1 \wedge \cdots \wedge \omega^{n-1} \wedge \omega_{n+1}^n) \\ &= (n-1)!\,|H|^{1/(n+2)}(x, e_1, \cdots, e_n) \\ &\quad \times (\textstyle\sum l_i^i)\omega^1 \wedge \cdots \wedge \omega^n ,\end{aligned} \tag{79}$$

which is zero if M is affine minimal. On the other hand,

$$\begin{aligned}(\nu, dx, \cdots, dx) &= n!\,(\nu, e_1, \cdots, e_n)\omega^1 \wedge \cdots \wedge \omega^n \\ &= (-1)^n n!\,|H|^{-1/(n+2)} dV .\end{aligned}$$

If M is a closed hypersurface without boundary, the integral over M of the left-hand side of (78) is zero. The same is therefore true of the integral of its right-hand side. By (79) the second term is zero if M is affine minimal. Hence

$$\int_M (\nu, dx, \cdots, dx) = 0 ,$$

which is clearly a contradiction.

3. Surfaces in A^3

Minimal surfaces in the three-dimensional affine space A^3 have many interesting properties. The classical account is [1]; cf. also [6]. For a recent result see [4].

On account of the great interest in classical minimal surfaces we wish to state the following theorem of G. Thomsen:

A minimal surface of the three-dimensional Euclidean space E^3 is at the same time an affine minimal surface, if and only if the image under the Gauss map of every asymptotic curve is a circle on the unit sphere.

I wish next to state a theorem on affine minimal surfaces proved recently by Chuu-lian Terng and myself [7]. It concerns with the analogue of Bäcklund's theorem in affine surface theory.

The classical Bäcklund's theorem, which is the basis of the theory of Bäcklund transformations and which has received much attention in recent studies of the soliton solutions of the sine-Gordon equation, says the following: *Let the surfaces M, M^* in E^3 be the focal surfaces of a line congruence, so that the lines of the congruence are the common tangent lines* *of M and M*. To every line let* $x \in M$, $x^* \in M^*$ *be the points of contact and* v, v^* *be the unit normal vectors to* M, M^* *at* x, x^* *respectively. If* $r = \text{dist}\,(x, x^*) = \text{const.}$ *and the angle* θ *between* v, v^* *is constant, then both M and M* have the Gaussian curvature* $-(\sin\theta/r)^2$.

The congruence in the theorem is a W-congruence, i.e., the asymptotic curves of M and M^* correspond under the mapping which sends x to x^*.

Our theorem is the following:

Let the surfaces M, M^* *in* A^3 *be the focal surfaces of a W-congruence, such that the affine normals to M and M* at corresponding points are parallel. Then both M and M* are affine minimal surfaces.*

Before concluding this paper I wish to state a conjecture:

Among affine minimal surfaces are the elliptic paraboloids

$$x^3 = (x^1)^2 + (x^2)^2\,, \tag{80}$$

whose affine normals are parallel. The following conjecture is an analogue to Bernstein's theorem: In A^3 consider the surface

$$x^3 = F(x^1, x^2) \tag{81}$$

defined for all x^1, x^2. If it consists entirely of elliptic points, and is affine minimal, then it is affinely equivalent to the paraboloid (80).

References

[1] W. Blaschke, Vorlesungen über Differentialgeometrie, II, Berlin, 1923.

[2] E. Calabi, *Improper affine hyperspheres of convex type and a generalization of a theorem by K. Jörgens,* Michigan Math. J. **5** (1958), 105–126.

[3] H. Flanders, *Local theory of affine hypersurfaces,* J. d'Analyse Math. **15** (1965), 353–387.

[4] E. Glässner, *Ein Affinanalogon zu den Scherkschen Minimalflächen,* Archiv der Math. **28** (1977), 436–439.

[5] H. W. Guggenheimer, Differential Geometry, New York, 1963.

[6] P. A. Schirokow and A. P. Schirokow, Affine Differentialgeometrie, Leipzig, 1962 (translated from Russian).

[7] S. S. Chern and Chuu-Lian Terng, *An analogue of Bäcklund's theorem in affine geometry,* to appear in Rocky Mountain J. Math.

University of California
Berkeley, CA 94720
U.S.A.

Minimal Submanifolds and Geodesics
Kaigai Publications, Tokyo, 1978, 31–41

REPRESENTATION OF SURFACES NEAR A BRANCHED MINIMAL SURFACE

ROBERT GULLIVER

In the study of the stability of an immersed minimal surface, it has been useful to exploit the representation of nearby surfaces by normal perturbation. That is, one assigns to each point on a nearby surface the closest point on the given surface, along with the distance and normal direction. The nearby surface now corresponds to a normal vector field along the original surface. With this representation, it has been possible to relate infinitesimal properties, such as the positive definiteness of the second variation of area, to questions of the isolated character of the surface; see, especially, [5]. In the case of a branched minimal surface, however, nearby surfaces in general may not be obtained by normal perturbation. In particular, the second variation of area for normal perturbations alone has only limited significance for the question of the area of the nearby surfaces.

The aim of the present paper is to develop a means for representing certain mappings near a given branched minimal surface x in three-dimensional euclidean space E^3, which include at least all nearby minimal surfaces, whether or not these have the same boundary curve as x. We shall assume that x extends as a minimal surface across the boundary of the parameter domain M, which is known to hold when x maps ∂M onto a system of real-analytic Jordan curves ([4]), and that x has only simple interior branch points. It will be seen that the representation involves a modified normal vector field, plus one of a finite-dimensional family of vector fields whose effect is to allow each branch point to move independently in its asymptotic tangent plane. In this perturbation, the branch points are not resolved. The validity of the representation for a nearby minimal surface $\tilde{x}$ depends on the presence of simple branch points of $\tilde{x}$ corresponding to those of x.

It should be pointed out that, since the representation is known to be valid only for nearby minimal surfaces, one cannot expect immediately to apply it toward results in which a given branched minimal surface is shown to be, for example, a relative minimum for area in a space of mappings having the same boundary. On the other hand, it is known that a branched minimal surface in E^3 can never provide a relative minimum for area, in the uniform topology, among surfaces with the same single Jordan curve as boundary ([6], [2], [3]). We hope, rather, for applications in which a given minimal surface is shown to be isolated or part of a one-dimensional manifold of minimal surfaces, along the lines of Nitsche's results for immersions in [6].

Definitions and conventions

For a Riemann surface M, a mapping $x: M \to E^3$ is called a *minimal surface* if x is harmonic,

$$x_{uu} + x_{vv} = 0 \tag{1}$$

and satisfies the conformality condition

$$|x_u|^2 - |x_v|^2 = x_u \cdot x_v = 0 \tag{2}$$

for any system of conformal coordinates (u, v) on M. Here we use subscripts u and v to denote partial derivatives. We write $x_w := (x_u - ix_v)/2$ for the complex gradient, and $Dx = (x_u, x_v)$ for the first-derivative transformation. $U_1 \cdot U_2$ denotes the inner product of two vectors U_1, U_2 in E^3, and $|U|$ is the length of the vector U. For a complex number $z \in C$, $|z|$ is its modulus. The norm $\|A\|$ of a linear transformation between two normed vector spaces is the supremum of $|A(U)|$ among vectors U with $|U| = 1$.

A minimal surface x is said to have a *branch point* at w_j if $Dx(w_j) = 0$. A branch point w_j is *simple* if $D^2x(w_j) \neq 0$. Note that x is an immersion at every point w which is not a branch point. In fact, either $x_u(w)$ or $x_v(w)$ is nonzero; the conformality condition (2) implies that they are both nonzero, and orthogonal. Thus $x_u(w)$ and $x_v(w)$ are independent. The oriented unit normal vector $N(w)$ is defined as usual when w is not a branch point. It is known, and will be shown following equation (4) below, that as w approaches a branch point w_j, $N(w)$ tends to a limit $N(w_j)$, which is called the *asymptotic normal vector* at w_j. The plane orthogonal to $N(w_j)$ is the *asymptotic tangent plane*.

For convenience, we shall deal only with the case where the parameter domain M is a domain in C; however, all considerations are of a local nature, and become valid after introduction of a local system of coordinates for a general Riemann surface M, perhaps with different constants. In defining norms for function spaces, we would use a fixed Riemannian metric on M where needed.

For $z \in C$ and $r > 0$, we write $B_r(z)$ for the open ball of radius r centered at z. For any set $K \subset C$, $\bar{K}$ is its closure and ∂K its topological boundary. $C^s(\bar{K}, \boldsymbol{R}^n)$ is the space of s times continuously differentiable functions from K into $\boldsymbol{R}^n$, with norm $\|x\|_{C^s(K,\boldsymbol{R}^n)}$ defined as a supremum over all partial derivatives of order up to s. The Lipschitz constant $[x]_K$ denotes the supremum of $|x(w_1) - x(w_2)|/|w_1 - w_2|$ over distinct $w_1, w_2 \in K$. The space of Lipschitz-continuous functions Lip $(K, \boldsymbol{R}^n)$ is defined with the norm max $([x]_K, \|x\|_{C^0(K,\boldsymbol{R}^n)})$, similarly the space Lip^s $(K, \boldsymbol{R}^n)$ of s times Lipschitz-continuously differentiable functions.

Statement of the Theorem

In order to define our representation of surfaces near a branched minimal surface $x: \Omega \to \boldsymbol{E}^3$, we shall first introduce a modified normal vector field n to x. We fix a smooth monotone decreasing function $\rho: \boldsymbol{R} \to \boldsymbol{R}$ such that $\rho(t) = 0$ for $t \geq 2$, and $\rho(t) = 1$ for $t \leq 3/2$. Let $\theta > 0$ be given so that for any two branch points w_i and w_j of x, $B_{2\theta}(w_i)$ and $B_{2\theta}(w_j)$ are disjoint. Let n be the following vector field along x: $n(w)\colon = N(w)$ whenever $w \notin \bigcup_j B_{2\theta}(w_j)$, and

$$n(w)\colon = N(w) + \rho(|w - w_j|/\theta)(N(w_j) - N(w))$$

when $w \in B_{2\theta}(w_j)$. Suppose the branch points of x are $w_1, \cdots, w_q$. We choose vector fields $U_1, \cdots, U_{2q}$ along x with the properties that $DU_k(w_j) = \Delta U_k(w_j) = 0$ for $1 \leq j \leq q$, $1 \leq k \leq 2q$; that $U_j(w_j)$, $U_{j+q}(w_j)$ form a basis for the asymptotic tangent plane to x at w_j; and that $U_k(w_j) = 0$ for $1 \leq j \leq q$ when $k \neq j, j+q$. We assume $U_k \in \mathrm{Lip}^2(\Omega_j \boldsymbol{E}^3)$.

Definition. For a function $\varphi \in$ Lip $(\Omega, \boldsymbol{R})$, a vector $\alpha = (\alpha^1, \cdots, \alpha^{2q}) \in \boldsymbol{R}^{2q}$ and $\gamma \in$ Lip (M, Ω), the mapping $\Phi(\varphi, \alpha, \gamma) \in$ Lip $(M, \boldsymbol{E}^3)$ is defined by $\Phi(\varphi, \alpha, \gamma) = (x + \varphi n + \sum_k \alpha^k U_k) \circ \gamma$.

It may be observed that the choice of the vector fields U_k is not unique. In fact, a different choice for the U_k away from branch points would merely require a corresponding change in φ and γ.

The effect of φ is to describe a modified normal perturbation, while γ introduces a change of coordinates. Thus our representation corresponds to normal perturbation plus perturbation along one of the $2q$-dimensional family of vector fields generated by $U_1, \cdots, U_{2q}$. The relevant study of the second variation of area will involve an operator on Lip $(M, \boldsymbol{R}) \times \boldsymbol{R}^{2q}$, since the change of coordinates does not affect area.

Theorem. *Let Ω be a domain in $\boldsymbol{C}$, M a bounded subdomain with $\bar{M} \subset \Omega$. Suppose $x: \Omega \to \boldsymbol{E}^3$ is a minimal surface having exactly q branch points, all simple and lying in M. For each $\varepsilon > 0$ there exists $\delta > 0$ such that if $\theta < \delta$ and if $\tilde{x}: M \to \boldsymbol{E}^3$ is a minimal surface with $\|\tilde{x} - x\|_{C^1(M,\boldsymbol{E}^3)} < \delta$, then there are $\varphi \in$ Lip $(M, \boldsymbol{R})$, $\alpha \in \boldsymbol{R}^{2q}$ and $\gamma \in$ Lip (M, Ω), all having norms less than ε, for which $\tilde{x} = \Phi(\varphi, \alpha, \gamma)$.*

The Weierstrass representation

We shall make use of the Weierstrass representation for minimal surfaces in $\boldsymbol{E}^3$, essential properties of which may be found in [7, pp. 63 ff.]. Let a system of orthogonal coordinates in $\boldsymbol{E}^3$ be given. To any minimal surface $x: M \to \boldsymbol{E}^3$ there correspond a holomorphic function $f: M \to \boldsymbol{C}$ and a meromorphic function $g: M \to \boldsymbol{C} \cup \{\infty\}$, such that

$$(3) \qquad 2x_w = f(1 - g^2, i + ig^2, 2g) \ .$$

The oriented unit normal vector to x is given in terms of g alone:

$$(4) \qquad N = (2\operatorname{Re}(g), 2\operatorname{Im}(g), |g|^2 - 1)/(1 + |g|^2) \ .$$

This equation states that the point $N(w)$ on the unit sphere and the point $g(w) \in \boldsymbol{C} \cup \{\infty\}$ correspond under stereographic projection. Of course, the normal vector in the sense of differential geometry is not defined at the branch points of x. But equation (4) now shows that N has a limiting value $N(w_j)$ at each branch point w_j, which is the asymptotic normal vector.

We shall need to know that the Weierstrass representation depends continuously on x. This need not be true in the presence of branch points of higher order. For example, if $f(w) = w^2$ and $g(w) = t/w$, then the minimal surface varies continuously in C^1 with the parameter t, while g varies discontinuously in $C^0(\Omega; \boldsymbol{C} \cup \{\infty\})$ at $t = 0$.

Lemma 1. *Suppose $x, \tilde{x}: \overline{M} \to \boldsymbol{E}^3$ are minimal surfaces of class $C^1(\overline{M}, \boldsymbol{E}^3)$, $\overline{M}$ compact. Let x and $\tilde{x}$ be represented by pairs of functions (f, g) and $(\tilde{f}, \tilde{g})$ respectively, in the Weierstrass representation. Suppose x has no branch points on ∂M, and only simple branch points in M. If $\tilde{x}$ is sufficiently close to x in $C^1(\overline{M}, \boldsymbol{E}^3)$, then $\tilde{f}$ is as close as desired to f in $C^0(\overline{M}, \boldsymbol{C})$, and $\tilde{g}$ is as close as desired to g in $C^0(\overline{M}, \boldsymbol{C} \cup \{\infty\})$.*

Remark. We shall use the metric d on $\boldsymbol{C} \cup \{\infty\}$ which corresponds to the standard metric on the unit sphere under stereographic projection. Thus $d(g(w), \tilde{g}(w))$ is the angle between the normal vectors $N(w)$ and $\tilde{N}(w)$.

Proof. Referring to equation (3), we see that $\tilde{f} - f = (\tilde{x}_w - x_w)^1 - i(\tilde{x}_w - x_w)^2$, where the superscripts refer to vector components. Thus, by making $\|\tilde{x} - x\|_{C^1}$ small, we force $\tilde{f}$ to be uniformly close to f on $\overline{M}$.

First consider a neighborhood $V \subset \overline{M}$ on which g is bounded and f is bounded away from zero: $|g(w)| \le C$, $|f(w)| \ge \beta > 0$ for all $w \in V$. Given $0 < \varepsilon < 1/2$, we may force $|\tilde{f}(w) - f(w)| < \varepsilon\beta$, as was just shown, and $|\tilde{f}(w)\tilde{g}(w) - f(w)g(w)| < \varepsilon\beta$ by consideration of the third component of equation (3). An application of the triangle inequality to $\tilde{f}(\tilde{g} - g) = (\tilde{f}\tilde{g} - fg) + (\tilde{f} - f)g$ leads to $|\tilde{g}(w) - g(w)| \le 2\varepsilon(1 + C)$, which is arbitrarily small. Thus $\tilde{g}$ is uniformly close to g in V.

Second, consider a neighborhood $W \subset \overline{M}$ in the closure of which x has no branch points and in which the values of the normal vector H lie in a hemisphere. Then after an appropriate rotation in E^3, the third component of $N(w)$ is always negative, which corresponds to $|g(w)| < 1$, for $w \in W$. Since $x_w(w) \neq 0$ for $w \in \overline{W}$, we see from equation (3) that $f(w) \neq 0$ and hence f is bounded away from zero. In particular, the conditions of the first case obtain, so that $\tilde{g}$ is arbitrarily close to g, uniformly in W. But the distance from $\tilde{g}(w)$ to $g(w)$ is invariant under rotations, and hence this shows that $\tilde{g}$ is arbitrarily close to g in $C^0(W, \boldsymbol{C} \cup \{\infty\})$.

Third, suppose w_j is a branch point with $g(w_j) \neq \infty$. Then there is a ball $B_{2\eta}(w_j) \subset M$ on which g is bounded, and $f(w) = 0$ only for $w = w_j$. The annulus $V = B_{2\eta}(w_j) - \bar{B}_\eta(w_j)$ satisfies the conditions of the first case, in particular $|f(w)| \geq \beta > 0$ for $w \in V$. Therefore $\tilde{g}$ is arbitrarily close to g, uniformly on V. Now w_j is a simple branch point of x, and as may be seen from equation (3), f has a simple zero at w_j. We require now $|f(w) - \tilde{f}(w)| < \beta$ on V. Then according to Rouche's theorem ([1], p. 152) $\tilde{f}$ has exactly one zero in $B_{2\eta}(w_j)$, which is simple. Again considering (3) and the differentiability of x_w, we see that $\tilde{g}$ is analytic in $B_{2\eta}(w_j)$. In fact, $\tilde{f}\tilde{g}^2$ must be analytic, so that any pole of $\tilde{g}$ would have to be a zero of $\tilde{f}$ of order at least 2. We may now apply the maximum principle to the analytic function $\tilde{g} - g$, to see that $\tilde{g}$ is arbitrarily close to g, uniformly on $B_{2\eta}(w_j)$.

Fourth, for any branch point $w_j \in M$, we may perform a rotation of coordinates in E^3 to obtain $g(w_j) \neq \infty$. Then there exists a ball $B_{2\eta}(w_j)$ as in the third case, such that $\tilde{g}$ is arbitrarily close to g in $C^0(B_{2\eta}(w_j), \boldsymbol{C} \cup \{\infty\})$.

Finally, $\overline{M}$ may be covered by a finite union of neighborhoods of the type W or $B_{2\eta}(w_j)$ considered in the second and fourth cases above, and the lemma is proved.

A lemma regarding square roots

If an analytic function has a simple zero at the origin, then it has an analytic square root. If the function $z: \boldsymbol{C} \to \boldsymbol{C}$ is only harmonic, then its square root need not be differentiable at the origin:

Lemma 2. *For positive numbers θ, λ and c, suppose $z \in C^1(B_\theta(0), \boldsymbol{C})$ is of the form*

$$z(w) = Aw^2 + \sigma(w) ,$$

where A is an invertible real-linear operator with $\|A^{-1}\| \leq 1/\lambda$, where

$\sigma(0) = 0$, $\|D\sigma(w)\| \leq c\,|w|^2$. *If* $\theta < \lambda/c$, *then there is a bilipschitz mapping* $\gamma: B_\theta(0) \to \boldsymbol{C}$, *with* $\gamma(0) = 0$, *unique up to sign, such that*

$$z(w) = [\gamma(w)]^2 .$$

Moreover, with an appropriate choice of sign, γ *varies continuously in* Lip $(B_\theta(0), \boldsymbol{C})$ *as* A *varies from the identity and* $w^{-2}\sigma(w)$ *varies in* Lip $(B_\theta(0), \boldsymbol{C})$.

Proof. We first consider the case $\sigma = 0$. Since the mapping which takes w to Aw^2 is two-to-one on a neighborhood of the origin, it has a well-defined square root $\gamma_1(w)$. Since γ_1 is positive homogeneous of degree one and smooth except at the origin, it is Lipschitz-continuous. This may be verified by comparing the values of γ_1 at two points w and w', where $|w| \leq |w'|$, via the intermediate point $|w'|\,w/|w|$. The Lipschitz-continuity of $\gamma_1^{-1}(w) = (A^{-1}w^2)^{1/2}$ is shown the same way.

In the general case, there is a C^1 diffeomorphism $F: U \to U'$, for some neighborhoods U and U' of 0, with $DF(0) = \text{id}$, such that

$$[F(w)]^2 = w^2 + A^{-1}\sigma(w)$$

for all w in U (proof of Lemma 2.2 in [2]). We claim that U may be chosen equal to $B_\theta(0)$. It may be computed that for $0 < |w| < \lambda/c$, $|w^{-2}A^{-1}\sigma(w)| < 1/3$, so that $1 + w^{-2}A^{-1}\sigma(w)$ has positive real part. Thus $F(w) = w(1 + w^{-2}A^{-1}\sigma(w))^{1/2}$ may be defined continuously on $B_\theta(0)$, choosing the square root with positive real part. Further, one may show $|w/F(w) - 1| < |(w/F(w))^2 - 1| < 1/2$, and then $|w/F(w) - 1| < 1/3$. This leads to $\|DF(w) - \text{id}\| \leq |w/F(w) - 1| + |w/F(w)|\,c\,|w|/2\lambda < 1$, which implies that F is a diffeomorphism on $B_\theta(0)$. Since $z(w) = A[F(w)]^2$, we may choose $\gamma = \gamma_1 \circ F$, which is therefore bilipschitz.

Now suppose that $\|A - \text{id}\|$ is small. For $\|A - \text{id}\| < 1/2$ we have $|\arg(w^{-1}Aw^2)| < \pi/6$, and we may choose γ_1 with $|\arg(w^{-1}\gamma_1(w))| < \pi/12$. With this choice of sign, $w^{-1}\gamma_1(w)$ is uniformly close to 1. We may compute that the first partial derivatives of $\gamma_1 - \text{id}$ are uniformly small, and therefore $\|\gamma_1 - \text{id}\|_{\text{Lip}\,(B_1(0), \boldsymbol{C})}$ is arbitrarily small. Meanwhile, given $\sigma_0 \in C^1(B_\theta(0), \boldsymbol{C})$, with $\|D\sigma_0(w)\| \leq c\,|w|^2$, consider σ close to σ_0 in the sense that $w^{-2}(\sigma(w) - \sigma_0(w))$ has small Lipschitz norm. Then $w^{-2}(A^{-1}\sigma(w) - \sigma_0(w))$ is small, and F is close to $\gamma_0(w) = w(1 + w^{-2}\sigma_0(w))^{1/2}$, in Lip $(B_\theta(0), \boldsymbol{C})$. Finally, $[\gamma - F] \leq [\gamma_1 - \text{id}][F]$ is arbitrarily small, so that γ is close to γ_0 in Lip $(B_\theta(0), \boldsymbol{C})$. This completes the proof.

Proof of the theorem: near branch points

In the course of the proof, we shall occasionally refer to "constants", meaning numbers which depend on x and an upper bound for δ.

Let $w_1, \cdots, w_q \in M$ denote the branch points of x. Let x and a nearby minimal surface $\tilde{x}$ be given in the Weierstrass representation (3) by pairs of functions (f, g) and $(\tilde{f}, \tilde{g})$, respectively, where f and $\tilde{f}$ are holomorphic, and where g and $\tilde{g}$ are meromorphic. We assume that orthogonal coordinates for E^3 have been chosen so that $N(w_j) \neq (0, 0, 1)$, and hence $g(w_j) \neq \infty$, for $1 \leq j \leq q$. Then, referring to equation (3), one sees that the presence of a simple branch point at w_j corresponds to $f(w_j) = 0$, $f'(w_j) \neq 0$. About each branch point w_j, consider a ball $B_{2\eta}(w_j)$ in which (as in the proof of Lemma 1) f has no zeroes other than w_j. Then $|f(w)| \geq \beta > 0$ for $w \in \partial B_\eta(w_j)$. Applying Lemma 1, we may choose δ small enough that $|\tilde{f}(w) - f(w)| < \beta$ on $\partial B_\eta(0)$. Then, according to Rouche's theorem $\tilde{f}$ has exactly one simple zero $\tilde{w}_j \in B_\eta(w_j)$. We require that η was chosen small enough that $g(w) \neq \infty$ for $w \in B_{2\mu}(w_j)$, and, applying Lemma 1, that δ is small enough that $\tilde{g}(w) \neq \infty$ on $B_\eta(w_j)$. Then $\tilde{x}$ has exactly one branch point $\tilde{w}_j$ in $B_\eta(w_j)$. Moreover, $\tilde{w}_j$ lies arbitrarily close to w_j when δ is taken small enough.

Since x and $\tilde{x}$ are harmonic, with $Dx(w_j) = D\tilde{x}(\tilde{w}_j) = 0$, we may write

$$\begin{aligned} x(w) &= x(w_j) + \operatorname{Re}\{a(w - w_j)^2\} + \sigma(w) \\ \tilde{x}(w) &= \tilde{x}(\tilde{w}_j) + \operatorname{Re}\{\tilde{a}(w - \tilde{w}_j)^2\} + \tilde{\sigma}(w)\,, \end{aligned} \tag{5}$$

where $a, \tilde{a} \in C^3$, $\sigma(w_j) = \tilde{\sigma}(\tilde{w}_j) = 0$, $\|D\sigma(w)\| \leq c|w - w_j|^2$ and $\|D\tilde{\sigma}(w)\| \leq \tilde{c}|w - \tilde{w}_j|^2$. The simple character of the branch point of x is expressed by $a \neq 0$. The conformality relation (2) implies that the real and imaginary parts of a are orthogonal and have the same length. Differentiating equation (5), we find that a is the limiting value of the vector-valued analytic function $(w - w_j)^{-1}x_w(w)$, as $w \to w_j$, and therefore

$$2\pi i\, a = \int_C (w - w_j)^{-1} x_w(w)\, dw$$

for any simple contour C around w_j. Similarly,

$$2\pi i\, \tilde{a} = \int_C (w - \tilde{w}_j)^{-1} \tilde{x}_w(w)\, dw\,.$$

For small δ, $\tilde{a}$ must be close to a, since $\tilde{x}_w$ is uniformly close to x_w and $\tilde{w}_j$ is close to w_j. The differentiated form of (5) is $x_w(w) = a(w - w_j) + \sigma_w(w)$, from which it follows that

$$2\pi i\, \sigma_w(w) = (w - w_j)^2 \int_C (z - w_j)^{-2}(z - w)^{-1} x_w(z)\, dz$$

with an analogous expression for $\tilde{\sigma}_w$ (p. 126 of [1]). Using again the proximity of $\tilde{x}_w$ to x_w, and of $\tilde{w}_j$ to w_j, we see that $\tilde{\sigma}$ is close to σ, particularly in the sense that

$$\zeta^{-2}(\sigma(w_j + \zeta) - \tilde{\sigma}(\tilde{w}_j + \zeta)) \tag{6}$$

has small Lipschitz norm, for ζ in some fixed neighborhood of 0.

Let $(\quad)^T$ denote the linear projection of E^3 onto C such that $N(w_j)^T = 0$, $(\operatorname{Re} a)^T = 1$ and $(\operatorname{Im} a)^T = -i$. Then $\zeta = \operatorname{Re}\{a\zeta\}^T$ identically. Since $\tilde{a}$ is close to a, the real-linear transformation $\tilde{A}: C \to C$ defined by $\tilde{A}\zeta = \operatorname{Re}\{\tilde{a}\zeta\}^T$ is close to the identity. We think of $(\quad)^T$ as the orthogonal projection onto the asymptotic tangent plane to x at w_j.

We may now determine the value of $\alpha \in R^{2q}$. The vectors $U_j(w_j)^T$ and $U_{j+q}(w_j)^T$ form a real basis for C. Therefore, there are unique real numbers α_j and α_{j+q} such that

$$\tilde{x}(\tilde{w}_j)^T - x(w_j)^T = \alpha_j U_j(w_j)^T + \alpha_{j+q} U_{j+q}(w_j)^T .$$

Thus the tangential projections at all q branch points are used to determine α. Observe that if δ is sufficiently small, than α is arbitrarily close to the origin in R^{2q}.

Now consider the mapping $y = x + \Sigma a_k U_k$. Since $U_k(w_j) = 0$ for $k \neq j, j + q$, we have

$$\tilde{x}(\tilde{w}_j)^T = y(w_j)^T .$$

We have assumed that $DU_k(w_j) = 0 = \Delta U_k(w_j)$, as well as $U_k \in \operatorname{Lip}^2(M, E^3)$, so that $Dy(w_j) = 0$ and $\Delta y(w) = O(|w - w_j|)$ as $w \to w_j$. Integrating twice, we may conclude that for some $b \in C^3$, y has the asymptotic form

$$y(w) = y(w_j) + \operatorname{Re}\{b(w - w_j)^2\} + \Sigma(w)$$

where $\Sigma(w_j) = 0$ and $\|D\Sigma(w)\| \leq C|w - w_j|^2$ for w in some neighborhood of w_j. If α is small, then b is close to a, so that the real-linear transformation $B: C \to C$ defined by $B\zeta := \operatorname{Re}\{b\zeta\}^T$ is close to the identity. Applying Lemma 2, we may conclude that for some value of δ, there are $\theta > 0$ and bilipschitz mappings $\tilde{\gamma}, \Gamma: B_{6\theta}(0) \to C$, with $\tilde{\gamma}(0) = \Gamma(0) = 0$, such that

$$y(w)^T = y(w_j)^T + (\Gamma(w - w_j))^2$$

for all $w \in B_{6\theta}(w_j)$, and

$$\tilde{x}(w)^T = \tilde{x}(\tilde{w}_j)^T + (\tilde{\gamma}(w - \tilde{w}_j))^2$$

for all $w \in B_{6\theta}(\tilde{w}_j)$. We require the Lipschitz constants $[\tilde{\gamma} - \mathrm{id}] < 1/3$ and $[\Gamma - \mathrm{id}] < 1/3$ on $B_{6\theta}(0)$, taking θ smaller if necessary. By construction, we have $\tilde{x}(\tilde{w}_j)^T = y(w_j)^T$. We define two bilipschitz mappings κ_+ and κ_- by

$$\kappa_\pm(w) = w_j + \Gamma^{-1}(\pm\tilde{\gamma}(w - \tilde{w}_j)) .$$

One may compute that both mappings $\kappa_\pm$ are defined on $B_{2\theta}(\tilde{w}_j)$, and that the Lipschitz constants $[\kappa_+ - \mathrm{id}] < 1$, $[\kappa_- + \mathrm{id}] < 1$. In particular, this implies that $\kappa_+(w) = \kappa_-(w)$ holds in $B_{2\theta}(\tilde{w}_j)$ only for $w = \tilde{w}_j$. Then for any other $w \in B_{2\theta}(\tilde{w}_j)$, there are exactly two solutions w' to the equation

$$\tilde{x}(w)^T = y(w')^T ,$$

and these are $w' = \kappa_\pm(w)$.

As δ tends to zero, $\tilde{A}$ and B tend to the identity, while $\tilde{\sigma}$ and Σ tend to σ in the sense of (6). According to Lemma 2, Γ and $\tilde{\gamma}$ tend toward a fixed bilipschitz mapping. At the same time, $\tilde{w}_j$ tends toward w_j, and therefore κ_+ approaches the identity in $\mathrm{Lip}\,(B_{2\theta}(\tilde{w}_j), \Omega)$.

Finally, as long as δ is sufficiently small, $|\tilde{w}_j - w_j| < \theta$, which implies $\overline{B_\theta(w_j)} \subset B_{2\theta}(\tilde{w}_j)$. We shall define the mapping γ required in the statement of the theorem by $\gamma(w) = \kappa_\pm(w)$ for $w \in B_\theta(w_j)$. Since $\tilde{x}(w)^T = y(\gamma(w))^T$ for $w \in B_\theta(w_j)$, there is a unique real number $\Psi(w)$ so that $\tilde{x}(w) = y(\gamma(w)) + \Psi(w)N(w_j)$. Upon taking the inner product in this equation with $N(w_j)$, we see that $\Psi \in \mathrm{Lip}\,(B_\theta(w_j))$. As $\delta \to 0$, $\tilde{x}$ and y each tend to x in the C^2 norm, and Ψ tends to zero in $\mathrm{Lip}\,(B_\theta(w_j))$. The function φ required for the theorem will later be defined by $\varphi = \Psi \circ \gamma^{-1}$.

Proof of the theorem: away from branch points

We shall now proceed to construct γ and Ψ on the remainder of M. Let K_θ denote the compact set $\overline{M} - \bigcup_{1 \le j \le q} B_\theta(w_j)$. The vector fields n, x_u and x_v are independent on K_θ. Therefore, for each $z \in K_\theta$, the mapping $X\colon \Omega \times \boldsymbol{R} \to \boldsymbol{E}^3$ defined by $X(w, t) = x(w) + tn(w)$ has rank 3 at $(z, 0)$. There is an open set $W(z) \Subset \Omega$, with $z \in W(z)$, and $t(z) > 0$ such that x is one-to-one and has full rank on $\overline{W(z)} \times [-t(z), t(z)]$. But these conditions remain under perturbation: there is a positive number $\mu(z)$ such that if $Q\colon \overline{W(z)} \to \boldsymbol{E}^3$ has C^1-norm less than $\mu(z)$, then the mapping $Y\colon \overline{W(z)} \times [-t(z), t(z)] \to \boldsymbol{E}^3$ defined by $Y(w, t) = X(w, t) + Q(w)$ is one-to-one with full rank. Since K_θ is compact, it is covered by a finite collection $\{W(z_1), \cdots, W(z_s)\}$, for some $z_1, \cdots, z_s \in K_\theta$. The covering $\{W(z_1), \cdots, W(z_s)\}$ has positive Lebesgue number, which we shall call

4ν. Let μ_0 be the smallest of the numbers $\mu(z_i)$, $1 \leq i \leq s$, and let t_0 be the smallest of the numbers $t(z_i)$. Then for any $Q \in C^1(\Omega, \boldsymbol{E}^3)$ with C^1-norm less than μ_0, the mapping $Y : \Omega \times \boldsymbol{R} \to \boldsymbol{E}^3$ defined as before is one-to-one and has full rank on $B_{4\nu}(z) \times [-t_0, t_0]$.

We are interested in $Q(w) = \Sigma \alpha_k U_k(w)$. For δ small enough, α is small and Q has C^1-norm less than μ_0. We now have that $Y(w, t) = x(w) + tn(w) + Q(w) = y(w) + tn(w)$ is one-to-one and has full rank on $B_{4\nu}(z) \times [-t_0, t_0]$, for any $z \in K_\theta$. There is a finite collection $\{B_\nu(\zeta_1), \cdots, B_\nu(\zeta_m)\}$ of balls of radius ν which covers K_θ. If α is sufficiently small, then the image $x(\overline{B_\nu(\zeta_k)})$ lies in the open set $Y(B_{2\nu}(\zeta_k) \times (-t_0, t_0))$, $1 \leq k \leq m$. If moreover δ is small, then $\|x - \tilde{x}\|_{C^1} < \delta$ forces the image $\tilde{x}(B_\nu(\zeta_k))$ to lie in $Y(B_{2\nu}(\zeta_k) \times (-t_0, t_0))$. We assume δ is small enough, and therefore α is small enough, to accomplish both of these ends for $1 \leq k \leq m$. Then to each $w \in B_\nu(\zeta_k)$ there correspond $w' \in B_{2\nu}(\zeta_k)$ and $t' \in (-t_0, t_0)$ such that $\tilde{x}(w) = Y(w', t')$. We claim that w' and t' are uniquely determined by the condition that $\tilde{x}(w) = Y(w', t')$ and by the estimates $|w - w'| < 3\nu$, $|t'| < t_0$. For suppose w'' and t'' satisfy the same conditions. Then $w \in B_\nu(\zeta_k)$ and $|w - w''| < 3\nu$ imply that $w'' \in B_{4\nu}(\zeta_k)$. That is, $(w'', t'') \in B_{4\nu}(\zeta_k) \times (-t_0, t_0)$ as well as (w', t'), while $Y(w', t') = Y(w'', t'')$, implying $(w'', t'') = (w', t')$. We may now define $\gamma(w) = w'$, $\Psi(w) = t'$; since K_θ is covered by $\{B_\nu(\zeta_1), \cdots, B_\nu(\zeta_m)\}$, γ and Ψ are defined on all of K_θ and are continuous. We have, for all $w \in K_\theta$,

$$\tilde{x}(w) = x(\gamma(w)) + \Sigma \alpha^i U_i(\gamma(w)) + \Psi(w) n(\gamma(w)) \; . \tag{7}$$

Moreover, the inverse function theorem shows that the restriction of Y to $B_{2\nu}(\zeta_k) \times (-t_0, t_0)$ has an inverse of class C^2, so that γ and Ψ are of class C^2 on $B_\nu(\zeta_k)$. This shows that $\gamma \in C^2(K_\theta, \Omega)$ and $\Psi \in C^2(K_\theta, \boldsymbol{R})$.

Finally, by leaving ν fixed we may examine the behavior of γ and Ψ on K_θ as δ tends to 0. The restriction of Y to $B_{2\nu}(\zeta_k) \times (-t_0, t_0)$ approaches X in the C^2 norm, and therefore Y^{-1} tends toward X^{-1}. Meanwhile $\tilde{x}$ tends to x in $C^1(K_\theta, \boldsymbol{E}^3)$. Therefore γ tends to the identity in $C^1(K_\theta, \Omega)$ and Ψ tends to zero in $C^1(K_\theta, \boldsymbol{R})$. We shall henceforth assume δ is sufficiently small that the Lipschitz constant $[\gamma - \mathrm{id}] < 1$.

We now wish to show that our definitions for γ make it continuous on $\partial B_\theta(w_j)$, $1 \leq j \leq q$, that is, on the boundary between the two sets where γ has been defined. For $w \in \overline{B_\theta(w_j)}$, we have $n(w) = N(w_j)$ by definition of the vector field n. It follows from equation (7) that $\tilde{x}(w)^T = y(\gamma(w))^T$ for $w \in B_\theta(w_j)$, where the projection $(\quad)^T$ is as constructed in determining γ on $\overline{B_\theta(w_j)}$. This implies that for each $w \in \partial B_\theta(w_j)$, either $\gamma(w) = \kappa_+(w)$ or $\gamma(w) = \kappa_-(w)$. Now $\kappa_+(w) = \kappa_-(w)$ only for $w = \tilde{w}_j$, which is not on $\partial B_\theta(w_j)$. Since γ is continuous on $\partial B_\theta(w_j)$, either $\gamma = \kappa_+$ or $\gamma = \kappa_-$

must hold everywhere on $\partial B_\theta(w_j)$. But $[\gamma - \mathrm{id}] < 1$ and $[\kappa_- + \mathrm{id}] < 1$ both hold on $\partial B_\theta(w_j)$; considering two distinct points on $\partial B_\theta(w_j)$, one sees that $\gamma = \kappa_-$ is impossible. Therefore $\gamma = \kappa_+$ on $\partial B_\theta(w_j)$, and by similar considerations, on a neighborhood of $\partial B_\theta(w_j)$. We conclude that γ is Lipschitz continuous on all of $\overline{M}$.

Defining $\varphi = \Psi \circ \gamma^{-1}$, we have shown that $\tilde{x} = \Phi(\varphi, \alpha, \gamma)$, as was to be proved.

References

[1] Ahlfors, L., Complex Analysis, 2nd ed. McGraw-Hill, New York, 1966.
[2] Gulliver, R., *Regularity of minimizing surfaces of prescribed mean curvature,* Ann. of Math. **97** (1973), 275–305.
[3] Gulliver, R., *Branched immersions of surfaces and reduction of topological type,* II. Math. Ann. **230** (1977), 25–48.
[4] Lewy, H., *On the boundary behavior of minimal surfaces,* Proc. Nat. Acad. Sci. USA **37** (1951), 103–110.
[5] Nitsche, J. C. C., *A new uniqueness theorem for minimal surfaces,* Arch. Rat. Mech. Anal. **52** (1973), 319–329.
[6] Nitsche, J. C. C., *Contours bounding at most finitely many solutions of Plateau's problem,* Preprint.
[7] Osserman, R., A Survey of Minimal Surfaces. Van Nostrand, New York, 1969.
[8] Osserman, R., *A proof of the regularity everywhere of the classical solution to Plateau's problem.* Ann. of Math. **19** (1970), 550–569.

UNIVERSITY OF MINNESOTA
MINNEAPOLIS, MN 55455
U.S.A.

Minimal Submanifolds and Geodesics
Kaigai Publications, Tokyo, 1978, 43–59

GEOMETRIES ASSOCIATED TO THE GROUP SU_n AND VARIETIES OF MINIMAL SUBMANIFOLDS ARISING FROM THE CAYLEY ARITHMETIC

REESE HARVEY & H. BLAINE LAWSON, JR.

§0. Foreword. This article is a report on part of a major investigation undertaken by the authors two years ago. The original motivation came from the study of minimal varieties in euclidean space. The objective was to find systems of partial differential equations, analogous to the Cauchy-Riemann equations, whose solutions represent absolutely area minimizing subvarieties of $\boldsymbol{R}^n$. The investigation has lead to a number of interesting results, and the work is still in progress.

One of the intriguing aspects of this study has been the discovery of natural geometries which strictly contain the usual complex geometries. This phenomenon is particularly interesting in dimension eight where there exists an enormous collection of area minimizing 4-folds associated to the group Spin_7. This geometry and its associated system of partial differential equations are discussed in § 5.

§1. Generalities. It is customary in differential geometry to define a geometric structure on a manifold by a *distinguished atlas of coordinate charts*, that is, by an atlas whose coordinate transformations lie in a particular pseudogroup of local diffeomorphisms of $\boldsymbol{R}^n$. However one could alternatively define a geometric structure on a manifold by a *distinguished family of subvarieties*. This second point of view entails certain interesting philosophical distinctions from the first and is closer in spirit to the classical geometries. Moreover, it has certain technical advantages. For example, it provides a natural concept of topological conjugacy of structures. It also broadens the class of structures available on a manifold. In fact, one of the main points of this article will be to show that there exist natural and interesting geometries of this second type which cannot be realized from the first point of view.

The geometries we shall be concerned with can all be defined in the following general way. Suppose X is a C^1 manifold and denote by $G_p(X)$ the bundle of oriented tangent p-planes on X. Let $\mathscr{G}$ be any subset of $G_p(X)$. Then $\mathscr{G}$ determines a family of subvarieties of X as follows. An oriented C^1 submanifold of dimension p in X is called a *$\mathscr{G}$-manifold* if all of its oriented tangent planes lie in $\mathscr{G}$. More generally, a locally integral current S of dimension p in X is called a *$\mathscr{G}$-variety* if $\|S\|$-almost all of its oriented tangent planes lie in $\mathscr{G}$. (Here $\|S\|$ denotes the total variation meansure of S (cf. [2]). This measure is defined using a riemannian metric on X. However, the family of sets of $\|S\|$-measure zero is independent of the choice of metric). The collection of $\mathscr{G}$-varieties is called the *$\mathscr{G}$-geometry* of X.

Example 1.1 (Complex geometries). Let X be a complex manifold and let $\mathscr{G} \subset G_{2k}(X)$ be the subset of *complex* k-planes (with the canonical orientation) on X. Then any $\mathscr{G}$-manifold is a complex submanifold of dimension k in X. Furthermore, by the structure theorem of King (cf. [6] or [3]) any $\mathscr{G}$-variety is, away from the support of its boundary, a complex analytic subvariety of dimension k in X.

Example 1.2 (Lagrangian geometries). Let X be a symplectic manifold, that is, a $2n$-manifold equipped with a closed 2-form ω such that ω^n is nowhere zero. We then define $\mathscr{G} \subset G_n(X)$ to be the set of n-planes such that $\omega|_\xi = 0$. These are the so-called *Lagrangian* planes. Associated to $\mathscr{G}$ is the geometry of Largangian submanifolds and Lagrangian subvarieties of X.

Example 1.3 (Foliation geometries). Let $\mathscr{F}$ be a foliation of dimension p_0 on X (cf. [7]), and for $p \leq p_0$ let $\mathscr{G}$ be the subset of p-planes tangent to $\mathscr{F}$. The $\mathscr{G}$-manifolds are then the submanifolds S of X which are everywhere tangent to the foliation. Hence, each connected component of S is a submanifold of some leaf of the foliation. In general, the $\mathscr{G}$-varieties are the p-dimensional subvarieties of X with the "leaf topology", i.e., the topology which makes X into an (uncountable) manifold of dimension p_0.

A general construction of $\mathscr{G}$-geometiries which is of interest for the study of minimal varieties is the following. Let X be a riemannian manifold and let $\Lambda_p(X)$ denote the pth exterior power of the tangent bundle of X. Then there is a natural embedding $G_p(X) \subset \Lambda_p(X)$ given by:

$$G_p(X) = \{\xi \in \Lambda_p(X) \colon \xi \text{ is a unit simple vector}\} \, .$$

Suppose now that φ is a differential p-form on X, i.e., a smooth section of $\Lambda^p(X) = \Lambda_p(X)^*$, and define the *comass* of φ to be

$$\|\varphi\|^* = \sup \{\varphi(\xi) \colon \xi \in G_p(X)\}$$

(cf. [2] and [12]). This is a norm on the space of p-forms. If $\varphi \neq 0$, we can renormalize φ so that $\|\varphi\|^* = 1$. We then consider the set

$$\mathscr{G}(\varphi) = \{\xi \in G_p(X) \colon \varphi(\xi) = 1\}$$

and its associated geometry. The submanifolds and subvarieties of this geometry will be called *φ-manifolds* and *φ-varieties* respectively. The fundamental observation is the following.

Basic Principle. *Suppose $d\varphi = 0$. Then any φ-variety with compact support is homologically mass minimizing in X.*

Proof. Let S be a φ-variety with compact support and suppose S' is any integral p-current such that $dS' = dS$ and $S - S'$ is homologous to zero. Then since $d\varphi = 0$, we have $S(\varphi) = S'(\varphi)$. However, using the notation of [2], we see that

$$S(\varphi) = \int \varphi(\vec{S}_x) d\|S\|(x) = \int d\|S\| = M(S)$$

and

$$S'(\varphi) = \int \varphi(\vec{S}'_x) d\|S'\|(x) \leqq \int d\|S'\| = M(S')$$

since $\varphi(\xi) \leqq 1$ and $\varphi(\xi) = 1$ exactly when $\xi \in \mathscr{G}(\varphi)$. This completes the proof.

For those who are unfamiliar with the language of currents, the above proof can be interpreted for an embedded, oriented C^1 submanifold S by the following dictionary: $S(\varphi) = \int_S \varphi$; $\vec{S}_x =$ the oriented unit tangent plane to S at $x\ (\in \Lambda_p(X)_x)$; $d\|S\| =$ Hausdorff p-measure restricted to S. Note that in this case, $M(S)$ is just the volume of S.

For a general current the mass can be interpreted as the "weighted volume". By the basic structure thorem of Federer [2, ch. 3, 4], any integral p-current S can be written as

$$S = \sum_{n=1}^{\infty} nS_n ,$$

where $\{S_n\}_{n=1}^{\infty}$ is a family of mutually disjoint, oriented C^1 submanifolds such that the closure of $\bigcup_n S_n$ is compact in X. The mass is given by

$$M(S) = \sum_{n=1}^{\infty} n \operatorname{vol}(S_n) .$$

Note that the argument given above actually proves the following result.

Theorem 1.4. *Suppose $d\varphi = 0$ and let S be any φ-variety with compact support in X. Then for any intergral p-current S' in X such that $dS' = dS$ and $S - S'$ is homologous to zero, one has that*

$$M(S) \leqq M(S')$$

with equality if and only if S' is also a φ-variety.

This has the following immediate consequences.

Corollary 1.5. *Suppose $d\varphi = 0$ and let S be a compactly supported φ-variety without boundary in X. Then S is a current of least mass in its homology class. Furthermore every integral current of least mass in the homology class of S must also be a φ-variety.*

Corollary 1.6. *Let φ be a closed differential p-form with $\|\varphi\|^* = 1$ in euclidean space $\mathbf{R}^n$, and let S be a compactly supported φ-variety with boundary $dS = B$. Then S is a solution to the Plateau problem for B, that is, S is a current of least mass among all currents in $\mathbf{R}^n$ having boundary B.*

Remark 1.7. In the two corollaries above the φ-variety can actually be shown to minimize mass among all homologous de Rham currents with compact support. The mass of a general de Rham current $\mathscr{S}$ is defined to be $M(\mathscr{S}) = \sup \{\mathscr{S}(\varphi) : \|\varphi\|^* = 1\}$.

Remark 1.8. Observe that any φ-manifold of class C^2 is a classical minimal submanifold of X. In fact, the basic regularity results in [10] state that if X is class C^k for $2 \leqq k \leqq \omega$, then any φ-manifold of class C^1 is actually of class C^k and hence a minimal submanifold of X.

Example 1.9. Let X be a Kähler manifold with Kähler form ω, and for a given k, set

$$\varphi = \frac{1}{k!}\omega^k .$$

The algebraic Wirtinger inequality (See Corollary 2.3 below) states that $\|\varphi\|^* = 1$ and that $\mathscr{G}(\varphi)$ is the set of complex tangent k-planes with the canonical orientation. Hence, the φ-geometry is exactly the geometry given in Example 1.1. Since, $d\varphi = 0$, the theorems above yield the well-known result [1] that the canonically oreinted complex subvarieties of a Kähler manifold are homologically mass minimizing.

In the next few sections we shall be concerned with the special case where, $X = \mathbf{R}^n$, euclidean n-space, and where φ is a parallel form, i.e., a form with "constant coefficients". Understanding this case is essential to understanding the more general ones.

In particular we will be concerned with the case of $\mathbf{R}^{2n} \cong \mathbf{C}^n$, i.e., of euclidean space furnished with a complex structure. We shall define forms (and therefore geometries) in terms of this complex structure, which are invariant under the group. The constructions will carry over directly to any Ricci-flat Kähler manifold. The existence of such manifolds is guaranteed by the recent results of Yau [13].

Before continuing, let us fix some notation. Note that the flat metric on $\mathbf{R}^n$ gives a canonical splitting

$$G_p(\boldsymbol{R}^n) = \boldsymbol{R}^n \times G(p, n) ,$$

where $G(p, n)$ denotes the Grassmannian of oriented p-planes in $\boldsymbol{R}^n$. Elements of $G(p, n)$ will always be represented as unit simple p-vectors in $\boldsymbol{R}$.

Given a parallel p-form φ of comass 1 in $\boldsymbol{R}^n$, we again get a canonical splitting

$$\mathscr{G}_p(\varphi) = \boldsymbol{R}^n \times G(\varphi) ,$$

where

$$G(\varphi) = \{\xi \in G(p, n): \varphi(\xi) = 1\} .$$

It is in general an interesting and non-trivial problem to determine the structure of the set $G(\varphi)$. For many forms φ the key to understanding $G(\varphi)$ is the following elementary proposition.

§ 2. A normal form for p-planes over U_n. We consider a complex inner product space as a pair $(\boldsymbol{R}^{2n}, J)$, where $J: \boldsymbol{R}^{2n} \to \boldsymbol{R}^{2n}$ is an orthogonal transformation with $J^2 = -1$. By a *unitary basis* of this space we mean an orthonormal basis of $\boldsymbol{R}^{2n}$ of the form $(e_1, Je_1, \cdots, e_n, Je_n)$.

Proposition 2.1. *Let $\xi \in G(p, 2n)$ be a unit simple vector with $p \leqq n$, and let $J: \boldsymbol{R}^n \to \boldsymbol{R}^n$ be any orthogonal complex structure on $\boldsymbol{R}^{2n}$. Then there exists a unitary basis $(e_1, Je_1, \cdots, e_n, Je_n)$ for $(\boldsymbol{R}^{2n}, J)$ and angles $0 \leqq \theta_1 \leqq \theta_2 \leqq \cdots \leqq \theta_{[p/2]} \leqq \pi/2$ such that*

$$\begin{aligned} \pm\xi = e_1 \wedge (Je_1 \cos\theta_1 + e_2 \sin\theta_1) \wedge e_3 \wedge (Je_3 \cos\theta_2 + e_4 \sin\theta_2) \wedge \\ \cdots \wedge \begin{cases} e_{p-1} \wedge (Je_{p-1} \cos\theta_{p/2} + e_p \sin\theta_{p/2}) & \text{if } p \text{ is even} \\ e_p & \text{if } p \text{ is odd} . \end{cases} \end{aligned} \tag{2.1}$$

Proof. Consider the Kähler form $\omega(X, Y) \equiv \langle JX, Y\rangle$ resiricted to the oriented p-plane P_ξ corresponding to ξ. Then there exists an orthonormal basis $\varepsilon_1, \cdots, \varepsilon_p$ of P_ξ such thet

$$\omega|_{P_\xi} = \sum_{j=1}^{[p/2]} \lambda_j \varepsilon^*_{2j-1} \wedge \varepsilon^*_{2j} ,$$

where $\lambda_j = \langle J\varepsilon_{2j-1}, \varepsilon_{2j}\rangle$ and $1 \geqq \lambda_1 \geqq \lambda_2 \geqq \cdots \geqq \lambda_{[p/2]} \geqq 0$. (Note that if p is even, one cannot assume this basis is properly oriented).

We consider the case where $1 > \lambda_1$ and p is even, and we define

$$\begin{aligned} e_{2j-1} &= \varepsilon_{2j-1} \\ e_{2j} &= (\varepsilon_{2j} - \langle J\varepsilon_{2j-1}, \varepsilon_{2j}\rangle J\varepsilon_{2j-1})/\|\varepsilon_{2j} - \langle J\varepsilon_{2j-1}, \varepsilon_{2j}\rangle J\varepsilon_{2j-1}\| \end{aligned}$$

for $j = 1, \cdots, p/2$. Since $\omega(\varepsilon_i, \varepsilon_j) = \langle J\varepsilon_i, \varepsilon_j\rangle = 0$ unless $\{i, j\} = \{2k-1,$

$2k\}$ for some k, it follows that the vectors $e_1, Je_1, \cdots, e_p, Je_p$ are orthonormal. We can therefore choose vectors $e_{p+1}, \cdots, e_n$ so that $(e_1, Je_1, \cdots, e_n, Je_n)$ is a unitary basis.

For each $j = 1, \cdots, p/2$ there exists $\theta_j \in [0, \pi/2]$ such that $\cos\theta_j = \lambda_j = \langle J\varepsilon_{2j-1}, \varepsilon_{2j}\rangle$. Then $\sin\theta_j = \sqrt{1-\lambda_j^2} = \|\varepsilon_{2j} - \langle J\varepsilon_{2j-1}, \varepsilon_{2j}\rangle J\varepsilon_{2j-1}\|$ and so

$$\begin{aligned} \varepsilon_{2j-1} &= e_{2j-1} \\ \varepsilon_{2j} &= Je_{2j-1}\cos\theta_j + e_{2j}\sin\theta_j \end{aligned}$$

for $j = 1, \cdots, p/2$. Since $\xi = \pm\varepsilon_1 \wedge \cdots \wedge \varepsilon_p$, the proof for this case is complete. The argument for the remaining cases is straightforward and is left to the reader.

Remark 2.2 (Refined canonical form). The ambiguity in sign in (2.1) can be eliminated by allowing $\cos\theta_{p/2}$ to be negative when p is even. That is, when p is even, we require that $0 \leqq \theta_1 \leqq \cdots \leqq \theta_{(p/2)-1} \leqq \pi/2$ and $0 \leqq \theta_{p/2} \leqq \pi$. In this case we have that (2.1) holds without the $\pm$sign.

Corollary 2.3 (The Wirtinger Inequality). *Let J be an orthogonal complex structure on $\boldsymbol{R}^{2n}$ with associated Kähler form $\omega(X, Y) \equiv \langle JX, Y\rangle$. Then for any unit simple vector $\xi \in G(2p, 2n)$,*

$$\frac{1}{p!}\omega^p(\xi) \leqq 1$$

and equality holds if and only if ξ corresponds to a canonically oriented complex p-plane.

Proof. Let $\theta_1, \cdots, \theta_p$ be the angles associated to the refined canonical form for ξ. Since $\omega = \sum e_j^* \wedge (Je_j)^*$ for any unitary basis, $(e_1, Je_1, \cdots, e_n, Je_n)$, we have that $(\omega^p/p!)(\xi) = \cos\theta_1 \cdots \cos\theta_p$. This proves the inequality. Equality holds if and only if $\cos\theta_1 = \cdots = \cos\theta_p = 1$, that is if and only if $\xi = e_1 \wedge Je_1 \wedge \cdots \wedge e_p \wedge Je_p$. This completes the proof.

§ 3. Special Lagrangian geometries.

In this section we shall be concerned with complex euclidean space $\boldsymbol{C}^n = \{(z_1, \cdots, z_n): z_j \in \boldsymbol{C}\}$ and the geometry determined by the parallel form

$$\varphi = \mathrm{Re}\,\{dz_1 \wedge \cdots \wedge dz_n\}\,. \tag{3.1}$$

Note that $\boldsymbol{C}^n$ is presented with a distinguished unitary basis $(e_1, Je_1, \cdots, e_n, Je_n)$, where $e_j = (0, \cdots, 0, 1, 0, \cdots, 0)$ with 1 in the jth place. It is also presented with a distinguished real subspace $\boldsymbol{R}^n \subset \boldsymbol{C}^n$ fixed by conjugation. This oriented space corresponds to the p-vector

$$\xi_0 = e_1 \wedge \cdots \wedge e_n\,.$$

Our first assertion is that $\|\varphi\|^* = 1$. In fact, we prove the following.

Proposition 3.1. *For all oriented n-planes* $\xi \in G(n, 2n)$

$$\varphi(\xi) \leqq 1 .$$

Furthermore, $\varphi(\xi) = 1$ *if and only if* $\xi = g(\xi_0)$ *for some* $g \in SU_n$.

Proof. Consider the form $\Phi = dz_1 \wedge \cdots \wedge dz_n$ and set $\psi = \operatorname{Im}(\Phi)$ so that $\Phi = \varphi + i\psi$. Note that for any $g \in U_n$, $g^*\Phi = \det(g)\Phi$.

Let $(e_1', Je_1', \cdots, e_n', Je_n')$ be a unitary basis with respect to which ξ is in refined canonical form (cf. Remark 2.2), and let g be the unitary transformation carring this basis to the distinguished one. Then we have

$$g(\xi) = e_1 \wedge (Je_1 \cos\theta_1 + e_2 \sin\theta_1) \wedge e_3 \wedge \cdots .$$

Observe that with respect to the distinguished dual basis,

$$\varphi = \operatorname{Re}\{\Phi\} = \operatorname{Re}\{(e_1^* + iJe_1^*) \wedge \cdots \wedge (e_n^* + iJe_n^*)\} .$$

Note that in the expansion of $g(\xi)$ every term involving a Je_k also involves e_k. However, no term in the expansion of φ contains any pair $e_k^* \wedge Je_k^*$. It follows that

$$\varphi(g\xi) = \sin\theta_1 \cdots \sin\theta_{[n/2]} .$$

On the other hand every term in the expansion of $\psi = \operatorname{Im}(\Phi)$ contains at least one Je_k^* but no pairs of type $e_k^* \wedge Je_k^*$. Consequently,

$$\psi(g\xi) = 0 .$$

It follows that

$$|\Phi(g\xi)| = \varphi(g\xi) = \sin\theta_1 \cdots \sin\theta_{[n/2]} . \tag{3.2}$$

However,

$$|\Phi(g\xi)| = |(g^*\Phi)(\xi)| = |(\det_C g^{-1})\Phi(\xi)| = |\Phi(\xi)| \geqq \varphi(\xi) \tag{3,3}$$

with equality if and only if $\psi(\xi) = 0$ and $\varphi(\xi) > 0$. Combining (3.2) and (3.3) we see that $\varphi(\xi) \leqq 1$ with equality if and only if

$$\begin{cases} \theta_1 = \cdots = \theta_{[n/2]} = \pi/2 \quad \text{and} \\ \det_C(g) = 1 . \end{cases} \tag{3.4}$$

This completes the proof.

Note that condition (3.4) is equivalent to the following:

$$(3.5)\qquad \begin{cases} \theta_1 = \cdots = \theta_{[n/2]} = \pi/2 \\ \psi(\xi) = 0 \\ \varphi(\xi) > 0 . \end{cases}$$

We shall now investigate the collection of oriented n-planes that satisfy this condition.

An oriented n-plane is called *Lagrangian* if the canonical coordinates $\theta_1 = \cdots = \theta_{[n/2]} = \pi/2$. This is equivalent to the condition that

$$(3.6)\qquad Jv \perp \xi \qquad \text{for all } v \in \xi .$$

It is also equivalent to the condition that

$$(3.7)\qquad \omega|_\xi = 0$$

where ω is the Kähler form (cf. Example 1.2).

It is clear from Proposition 2.1 that the unitary group U_n acts transitively on the set of oriented Lagrangian n-planes. Hence, this set, which we denote Lag, can be expressed as a homogeneous space

$$\text{Lag} \cong U_n/SO_n .$$

There is a fibration

$$(3.8)\qquad d\colon \text{Lag} \longrightarrow S^1$$

where $S^1 = \{e^{i\theta} : 0 \leq \theta < 2\pi\}$, given by the complex determinant on U_n. The fiber $d^{-1}(1)$ is called the space of *special Lagranigan n*-planes and will be denoted $S(\text{Lag})$. Proposition 3.1 can now be restated as follows.

Proposition 3.1′. *For all oriented n-planes $\xi \in G(n, 2n)$,*

$$\varphi(\xi) \leqq 1$$

with equality if and only if ξ is special Lagrangian.

It follows that:

$$(3.9)\qquad G(\varphi) = S(\text{Lag}) \cong SU_n/SO_n .$$

Remark 3.2. For each $\theta \in [0, 2\pi]$ we can of course consider the geometry given by the form $\varphi_\theta = \text{Re}\{e^{-i\theta} dz_1 \wedge \cdots \wedge dz_n\}$. In this case, the corresponding set of n-planes will be $d^{-1}(e^{i\theta})$ where d is the fibration (3.8) above.

The geometry determined by the form φ will be called the *special Lagranigian geometry*, and the varieties belonging to this geometry will be called *special Lagrangian varieties.*

We shall now focus attention on the local structure of these varieties. Suppose that M is a special Lagrangian submanifold of $\boldsymbol{C}^n$, and consider M expressed locally as the graph of a function over a tangent plane. All special Lagrangian planes are equivalent under SU_n to the plane ξ_0. Hence, up to allowable coordinate changes, we can consider M to be given as the graph of a function

$$y = f(x)$$

where $z_k = x_k + iy_k$ for $k = 1, \cdots, n$ and where f is defined in some open subset $\Omega \subseteq \boldsymbol{R}^n$. Our first observation is the following classical fact.

Lemma 3.3. *The graph of f is a Lagrangian submanifold of $\boldsymbol{C}^n$ if and only if the Jacobian matrix $((\partial f^i/\partial x_j))$ is symmetric, that is, if and only if*

$$\frac{\partial f^i}{\partial x_j} = \frac{\partial f^j}{\partial x_i} \tag{3.10}$$

for all i, j.

Proof. Consider a (real) linear map $L: \boldsymbol{R}^n \to \boldsymbol{R}^n$, and let $\Gamma = \{(x, L(x)): x \in \boldsymbol{R}^n\}$ denote its graph in $\boldsymbol{R}^n \oplus \boldsymbol{R}^n \cong \boldsymbol{C}^n$. Recall that Γ is Lagrangian if and only if $Jv \perp \Gamma$ for all $v \in \Gamma$ (cf. 3.6). With respect to the coordinates (x, y) we have that $J(x, y) = (-y, x)$. Hence, Γ is Lagrangian if and only if $\langle (x, L(x)), (-L(x'), x') \rangle = -\langle x, L(x') \rangle + \langle L(x), x' \rangle = 0$ for all $x, x' \in \boldsymbol{R}^n$, i.e., if and only if L is symmetric. Letting L correspond to the Jacobian matrix of f at a given point completes the proof.

Since our considerations are purely local we may assume that the domain of f is simply connected.

Corollary 3.4. *Let $f: \Omega \to \boldsymbol{R}^n$ be a C^1 map where Ω is a simply connected domain in $\boldsymbol{R}^n$. Then the graph of f is a Lagrangian submanifold of $\boldsymbol{C}^n$ if and only if there is a C^2 function $F: \Omega \to \boldsymbol{R}$ such that $f = \nabla F$, i.e., such that*

$$f^j = \frac{\partial F}{\partial x_j} \tag{3.11}$$

for $j = 1, \cdots, n$.

In order that the graph of f be special Lagrangian it must be Lagrangian and satisfy one further condition. Our main result is the following.

Theorem 3.5. *Let $f: \Omega \to \boldsymbol{R}^n$ be a C^1 map, where Ω is a simply connected domain in $\boldsymbol{R}^n$. Then the graph of f is a special Lagrangian submanifold of $\boldsymbol{C}^n$ if and only if there exists a C^2 function $F: \Omega \to \boldsymbol{R}$ such that $f = \nabla F$ and such that F satisfies the differential equation*

$$\sum_{k=0}^{[(n-1)/2]} (-1)^k \sigma_{2k+1}(F_{**}) = 0 . \tag{3.12}$$

Here F_{**} denotes the Hessian matrix of F and $\sigma_j(F_{**})$ denotes the j'^{th} elementary symmetric function of the eigenvalues of F_{**}. In particular, for $n = 3$ equation (3.12) becomes simply the equation

$$\Delta F = \det (F_{**}) , \tag{3.13}$$

where $\Delta = \sum \partial^2/\partial x_j^2$ is the negative Laplace operator.

Proof. Let $(e_1, Je_1, \cdots, e_n, Je_n)$ be the natural unitary basis for $\boldsymbol{C}^n$. Then $(e_1, \cdots, e_n)$ and $(Je_1, \cdots, Je_n)$ form bases for the domain and range of f with coordinates $(x_1, \cdots, x_n)$ and $(y_1, \cdots, y_n)$ respectively. We consider a linear map $L: \boldsymbol{R}^n \to \boldsymbol{R}^n$ with matrix L_{ij} with respect to these bases. We want to establish conditions necessary and sufficient for the graph of L in $\boldsymbol{C}^n \equiv \boldsymbol{R}^n \oplus \boldsymbol{R}^n$ to be special Lagrangian.

We know from the proof of Lemma 3.3 that L must be symmetric, i.e., $L_{ij} = L_{ji}$ for all i, j. This corresponds to the first condition in (3.5). For the remaining conditions in (3.5) we proceed as follows. Consider the action of SO_n on $\boldsymbol{C}^n = \boldsymbol{R}^n \oplus \boldsymbol{R}^n$ given by $g(x, y) = (g\cdot x, g\cdot y)$ for $g \in SO_n$. This action preserves the set of special Lagrangian n-planes. Hence, the graph of f is special Lagrangian if and only if only if for any $g \in SO_n$ the graph of $g^t \circ L \circ g$ is special Lagrangian. Since L is symmetric, it therefore suffices to consider the case where L is diagonal, i.e., where

$$L(e_j) = \lambda_j Je_j$$

for $j = 1, \cdots, n$.

Recall now that $\Phi = \varphi + i\psi = dz_1 \wedge \cdots \wedge dz_n = (e_1 + iJe_1) \wedge \cdots \wedge (e_n + iJe_n)$. The plane given by the graph of L, with orientation induced from ξ_0, corresponds to (positive multiples of) the n-vector

$$\begin{aligned} \xi_L &= (e_1 + L(e_1)) \wedge \cdots \wedge (e_n + L(e_n)) \\ &= (e_1 + \lambda_1 Je_1) \wedge \cdots \wedge (e_n + \lambda_n Je_n) . \end{aligned}$$

It follows that

$$\begin{aligned} \Phi(\xi_L) &= (1 + i\lambda_1) \cdots (1 + i\lambda_n) \\ &= \sum_{k=0}^{n} i^k \sigma_k(\lambda_1, \cdots, \lambda_n) \\ &= \sum_{k=0}^{[n/2]} (-1)^k \sigma_{2k}(L) + i \sum_{n=0}^{[(n-1)/2]} (-1)^k \sigma_{2k+1}(L) \qquad (3.14) \\ &= \varphi(\xi_L) + i\psi(\xi_L) , \end{aligned}$$

where $\sigma_k(\lambda_1, \cdots, \lambda_n) = \sigma_k(L)$ is the kth elementary symmetric function of the eigenvalues of L.

The special Lagrangian conditions (3.5) are now that $\psi(\xi_L) = 0$ and $\varphi(\xi_L) > 0$. Note from (3.14) that whenever $\psi(\xi_L) = 0$, we have $\varphi(\xi_L) \neq 0$. Consequently, the condition $\varphi(\xi_L) > 0$ can be insured by choosing the appropriate orientation for the graph. The special Lagrangian condition now becomes just the equation $\psi(\xi_L) = 0$. Setting $L = F_{**}$, we obtain equation (3.12), and the proof is complete.

Note. When $\psi(\xi_L) = 0$ we have either $\varphi(\xi_L) > 0$ or $\varphi(\xi_L) < 0$. If $n \geqq 3$, both situations can occur. Consequently, it is necessary in general to orient the graph of $f = \nabla F$ properly. However, if coordinates are chosen so that $\nabla f = 0$ at some point $x_0 \in \Omega$, then the orientation on the graph induced from ξ will be the appropriate one. This is the situation if, for example, we graph the manifold locally over one of its tangent planes.

Corollary 3.6. *Let F be a real valued function of class $\mathscr{C}^2$ defined in an open set $\Omega \subseteq \boldsymbol{R}^n$. If F satisfies the differential equation* (3.12), *then the graph of F is an absolutely volume minimizing submanifold of $\boldsymbol{R}^{2n}$. In particular, from the regularity results of Morrey* [10], *any $\mathscr{C}^2$ solution of* (3.12) *is real analytic.*

We shall now briefly consider solutions to these equations. We begin by looking for solutions of the special form

$$\nabla F(x) = \rho(|x|)x \tag{3.15}$$

where $\rho = \rho(r)$ is a real valued function of one real variable. Such solutions are invariant under the diagonal action of SO_n on $\boldsymbol{R}^n \times \boldsymbol{R}^n$. In fact, such solutions are all obtained by rotating a curve in the (x_1, y_1)-plane via this action. A straightforward analysis shows that the curves which generate special Lagrangian manifolds in this manner are all given implicitly by the equation

$$\operatorname{Im}\{(x_1 + iy_1)^n\} = c \tag{3.16}$$

for real constants c. (The resulting manifolds must then be properly oriented). This means that the general solution ρ to (3.15) is an algebraic function given by

$$r^n \operatorname{Im}\{(1 + i\rho)^n\} = c\,. \tag{3.17}$$

These equations give a family of complete algebraic manifolds in the special Lagrangian geometry.

We note that, in fact, the special Lagrangian geometry is quite rich; that is, there are many solutions to the equation (3.12). To see this we note that the linearization of (3.12) at the solution $F_0 = 0$, is just the equation $\Delta F = 0$. Therefore, by the Implicit Function Theorem one can solve the Dirichlet Problem for (3.12) for all smooth boundary data sufficiently close to zero.

We conclude this section by establishing an interesting relationship between special Lagrangian manifolds in $\boldsymbol{R}^6$ and classical minimal surfaces in $\boldsymbol{R}^3$. Suppose we look for solutions to equation (3.13) of the special form

$$F(x_1, x_2, x_3) = G(x_1, x_2)x_3 .$$

Then a straightforward calculation shows that equation (3.13) for F is completely equivalent to the equation

$$(3.18) \qquad (1 + G_2^2)G_{11} - 2G_1G_2G_{12} + (1 + G_1^2)G_{22} = 0$$

for G. Notice that (3.18) is just the classical minimal surface equation. This gives the following result.

Theorem 3.7. *Let $G(x_1, x_2)$ be a real-valued function defined in an open subset $\Omega \subseteqq \boldsymbol{R}^2$ with the property that the graph of G is a minimal surface in $\boldsymbol{R}^3$. Then the graph of the map $F\colon \Omega \times \boldsymbol{R} \to \boldsymbol{R}^3$, where*

$$F(x_1, x_2, x_3) = (G(x_1, x_2), G_1(x_1, x_2)x_3, G_2(x_1, x_2)x_3)$$

is an absolutely minimal 3-manifold in $\boldsymbol{R}^6$.

The observation above shows that equation (3.12) is related in an intimate way to the work of H. Lewy [9] who has made a profound analysis of the case $\Delta F = \det(F_{**}) = 0$. In his analysis classical minimal surfaces also play a central role.

§ 4. A generalization of the complex geometries. In this section we shall be concerned with the complex euclidean space, $\boldsymbol{C}^{2n} = \{(z_1, \cdots, z_{2n}): z_j \in \boldsymbol{C}\}$, and we shall study the geometry determined by the parallel form

$$(4.1) \qquad \Omega = \frac{1}{n!}\omega^n + \varphi ,$$

where $\varphi = \mathrm{Re}\{dz_1 \wedge \cdots \wedge dz_{2n}\}$ is the form studied in § 3 and $\omega = (i/2)\sum dz_j \wedge d\bar{z}_j$ is the Kähler form. We already know that each of the forms $\omega^n/n!$ and φ has comass 1. The surprising fact is that their sum Ω also has comass 1.

Theorem 4.1. *For all oriented real $2n$-planes $\xi \in G(2n, 4n)$,*

$$\Omega(\xi) \leqq 1 .$$

Furthermore, when $n > 2$, $\varphi(\xi) = 1$ *if and only if* ξ *is either a special Lagrangian plane or a canonically oriented compex plane.*

Proof. Let $(e_1, Je_1, \cdots, e_{2n}, Je_{2n})$ denote the distinguished unitary basis of C^{2n} (cf. § 3). Fix $\xi \in G(2n, 4n)$ and let $(e_1', Je_1', \cdots, e_{2n}', Je_{2n}')$ be a unitary basis with respect to which ξ is in refined canonical form (cf. § 2). Let g be the unitary transformation carrying this basis to the distinguished one, so that

$$g(\xi) = e_1 \wedge (Je_1 \cos\theta_1 + e_2 \sin\theta_1) \wedge e_3 \cdots .$$

As in the proof of Proposition 3.1, we define $\Phi = dz_1 \wedge \cdots \wedge dz_{2n} = \varphi + i\psi$ and observe that

$$\Phi(g\xi) = \varphi(g\xi) = \sin\theta_1 \cdots \sin\theta_n .$$

Using the fact that $\omega = \sum e_j^* \wedge Je_j^*$ we deduce easily that

$$\frac{\omega^n}{n!}(g\xi) = \cos\theta_1 \cdots \cos\theta_n .$$

Combining these two equations and using the fact that $g^*\omega = \omega$, we see that

$$\begin{aligned} \Omega(\xi) &\leqq \frac{\omega^n}{n!}(\xi) + |\Phi(\xi)| \\ &= \frac{\omega^n}{n!}(g\xi) + |\Phi(g\xi)| \qquad (4.2) \\ &= \cos\theta_1 \cdots \cos\theta_n + \sin\theta_1 \cdots \sin\theta_n \\ &\leqq 1 . \end{aligned}$$

For $n \geqq 3$, equality occurs in (4.2) if and only if $\psi(\xi) = 0$, $\varphi(\xi) \geqq 0$, and either $\cos\theta_1 \cdots \cos\theta_n = 1$ or $\sin\theta_1 \cdots \sin\theta_n = 1$. The theorem now follows easily.

Theorem 4.1 states that when $n \geqq 3$, the geometry determined by the Ω form is the "disjoint union" of the complex and special Lagrangian geometries. In the case that $n = 2$, however, the geometry determined by Ω is more complicated.

§ 5. The case of dimension 8.

We now focus attention on the space $C^4 \cong R^8$ and the geometry given by the 4-form

$$\Omega = \frac{\omega^2}{2} + \varphi \qquad (5.1)$$

defined above. We want to study the set $G(\Omega)$ of oriented 4-planes on which Ω takes the value 1. Observe that for any such 4-plane ξ there exists an oriented orthonormal basis $(e_1, \cdots e_4)$ of ξ such that

$$\omega|_\xi = \lambda e_1^* \wedge e_2^* + \mu e_3^* \wedge e_4^* ,$$

where $\lambda \geqq 0$. The numbers λ and μ are called the *eigenvalues of* $\omega|_\xi$ *with respect to an oriented basis.*

Theorem 5.1. *For all oriented 4-planes* $\xi \in G(4, 8)$, $\Omega(\xi) \leqq 1$. *Furthermore,* $\Omega(\xi) = 1$ *if and only if*:

(i) $\omega|_\xi$ *has equal eigenvalues with respect to an oriented basis.*

(ii) $\Phi(\xi) \geqq 0$, *where* $\Phi = dz_1 \wedge \cdots \wedge dz_4$.

Proof. From the proof of Theorem 4.1 we know that $\Omega(\xi) \leqq 1$ and that $\Omega(\xi) = 1$ if and only if $\Phi(\xi) \geqq 0$ and

$$\cos\theta_1 \cos\theta_2 + \sin\theta_1 \sin\theta_2 = \cos(\theta_1 - \theta_2) = 1$$

where θ_1, θ_2 are the coordinates of ξ in refined canonical form. We then observe that the eigenvalues of ξ with respect to an oriented basis are exactly $\lambda = \cos\theta_1$ and $\mu = \cos\theta_2$. This completes the proof.

Note that the condition $\Phi(\xi) \geqq 0$ is equivalent to the two conditions $\varphi(\xi) \geqq 0$ and $\psi(\xi) = 0$, where $\Phi = \varphi + i\psi$.

It is clear that the geometry associated to the form Ω is special in nature. An extensive analysis of this case has been carried out by the authors and will appear in [5]. What follows is a brief summary of these results.

The first somewhat surprising result concerning Ω is that its invariance group is substantially larger than SU_4. In fact, we have the following.

Theorem 5.2. *The subgroup of elements in* SO_8 *which fix* Ω *is isomorphic to* Spin_7. *This group acts transitively on the set* $G(\Omega)$, *and there is a natural homeomorphism*

$$G(\Omega) \cong \mathrm{Spin}_7/H ,$$

where $H \cong SU_2 \times SU_2 \times SU_2/\boldsymbol{Z}_2$ *and* $\boldsymbol{Z}_2$ *is generated by the element* $(-1, -1, -1)$.

It follows in particular that the Ω-geometry is preserved by the group Spin_7 acting on $\boldsymbol{R}^8$. The effective 8-dimensional representation of Spin_7 is intimately connected with the Cayley numbers $\boldsymbol{O}$. It is no surprise, therefore, that the form Ω is also connected with $\boldsymbol{O}$. Recall that the Cayley numbers are defined as pairs of quaternions with multiplication given as follows:

$$x_1 \cdot x_2 = (u_1 u_2 - \bar{v}_2 v_1, v_2 u_1 + v_1 \bar{u}_2)$$

for $x_j = (u_j, v_j) \in \boldsymbol{H} \times \boldsymbol{H} = \boldsymbol{O}$. This product is non-commutative and non-associative. However, it does have several nice properties. There is a conjugation on $\boldsymbol{O}$ given by setting $\bar{x} = (\bar{u}, -v)$ for $x = (u, v)$, and for all $x, x_1, x_2 \in \boldsymbol{O}$ we have that:

(i) $x\bar{x} = \bar{x}x = \|x\|^2$

(ii) $\overline{x_1 x_2} = \bar{x}_2 \bar{x}_1$

(iii) $\|x_1 x_2\| = \|x_1\| \|x_2\|$

(iv) $x_1\bar{x}_2 + x_2\bar{x}_1 = 2\langle x_1, x_2\rangle$

where $\langle\cdot,\cdot\rangle$ is the natural inner product on $\boldsymbol{H} \times \boldsymbol{H}$ and $\|\cdot\|$ is its associated norm. Furthermore, there is the following important theorem (cf. [11]):

(v) The subalgebra with unit generated by any two elements of $\boldsymbol{O}$ is associative.

We now consider the real trilinear map $A: \boldsymbol{O} \times \boldsymbol{O} \times \boldsymbol{O} \to \boldsymbol{O}$ given by setting

$$A(x_1, x_2, x_3) = \tfrac{1}{2}[x_1 \cdot (\bar{x}_2 \cdot x_3) - x_3 \cdot (\bar{x}_2 \cdot x_1)] \ . \tag{5.2}$$

It follows directly from (v) above that $A = 0$ whenever two of its arguments coincide. Consequently A is skew-symmetric, that is, A defines an $\boldsymbol{O}$-valued exterior 3-form on $\boldsymbol{O}$. Using further properties of the Cayley numbers, one can then show that the map $(x_1, \cdots, x_4) \to \langle x_1, A(x_2, x_3, x_4)\rangle$ is also alternating.

Theorem 5.3. *For a suitably chosen system of unitary coordinates on $\boldsymbol{O} \cong \boldsymbol{C}^4$, the form Ω is given precisely as an alternating 4-linear function by the formula*

$$\Omega(x_1, x_2, x_3, x_4) = \langle x_1, A(x_2, x_3, x_4)\rangle \ .$$

Note that the systems of unitary coordinates with respect to which the expression (5.1) gives the same 4-form are not all unitary related. Theorem 5.2 says precisely that we may conjugate one such system by any element of Spin_7 and obtain another such system. This process leads to a family $\mathscr{C}$ of complex structures on $\boldsymbol{R}^8 \cong \boldsymbol{O}$ of the form

$$\mathscr{C} \cong \mathrm{Spin}_7 | SU_4 \ .$$

Corollary 5.4. *The Ω-geometry contains all the complex subvarieties of complex dimension 2 and all the special Lagrangian varieties for each of the complex structures in the family $\mathscr{C}$ above.*

One might ask whether there exist more subvarieties in the Ω-geometry than those given by Corollary 5.4. The naswer to this question can be deduced from the following analysis.

We shall now examine the system of partial differential equations whose solutions represent Ω-varieties in $\boldsymbol{O} = \boldsymbol{H} \times \boldsymbol{H}$. We consider C^1 functions $f: \boldsymbol{H} \to \boldsymbol{H}$ and define two natural differential operators on the functions. The first, called the *Dirac operator*, is given by the formula

$$Df = \sum_i (\nabla_{e_i} f) \cdot \bar{e}_i \tag{5.3}$$

where $(e_0, \cdots, e_3)$ is any orthonormal basis of $\boldsymbol{H}$. This is a linear, elliptic, first-order operator. It is the quaternionic analogue of the operator $\partial/\partial z$ for functions $g: \boldsymbol{C} \to \boldsymbol{C}$. With respect to the quaternionic inner product, the adjoint of D is the operator $\bar{D}$ given by

$$\bar{D}f = \sum (\nabla_{e_j} f) \cdot e_j \,.$$

They satisfy the relationship

$$D\bar{D} = \bar{D}D = \Delta \cdot \mathbf{1}$$

where Δ is the scalar Laplacian and $\mathbf{1}$ is the 4×4 identity matrix.

The second operator, called the *generalized Monge-Ampère operator*, is defined as follows. Consider the alternating trilinear map $A: \boldsymbol{H} \times \boldsymbol{H} \times \boldsymbol{H} \to \boldsymbol{H}$ defined, as the one above, by the formula $A(x_1, x_2, x_3) = \frac{1}{2}[x_1(\bar{x}_2 x_3) - x_3(\bar{x}_2 x_1)]$. The map A gives an isomorphism $\Lambda^3\boldsymbol{R}^4 \rightleftarrows \boldsymbol{R}^4$. For any 3-vector $v_1 \wedge v_2 \wedge v_3$, we set $A(\nabla_{v_1 \wedge v_2 \wedge v_3} f) = A(\nabla_{v_1} f, \nabla_{v_2} f, \nabla_{v_3} f)$ and note that since A is alternating, thisdefinition is independent of the decomposition of the 3-vector. The Monge-Ampère operator is then given by setting

$$\sigma f = \sum_I A(\nabla_{e_I} f) \cdot \overline{A(e_I)} \tag{5.4}$$

where the sum is taken over an orthonormal basis of $\Lambda^3\boldsymbol{R}^4$. σ is a homogeneous, first order operator of degree three. It is, in fact, a linear function of the cofactors of the jacobian matrix of f. Note that there are striking formal similarities between the expressions (5.3) and (5.4). The main result concerning these operators is the following.

Theorem 5.5. *Let $f: U \to \boldsymbol{H}$ be a C^1 function where U is an open subset of $\boldsymbol{H}$. Then the graph of f is an Ω-manifold in $\boldsymbol{O} = \boldsymbol{H} \times \boldsymbol{H}$ if and only if f satisfies the differential equation*

$$Df = \sigma f \,. \tag{5.5}$$

From the fact that Ω-manifolds are volume minimizing we know that any C^1-solution to (5.5) is real analytic. However, it is proved in [4] (see also [8]) that the function $f_0: \boldsymbol{H} \to \operatorname{Im}(\boldsymbol{H})$ obtained by taking the cone over the Hopf map $\eta: S^3 \to S^2$ (for spheres of appropriate radius) is a solution to (5.5). This shows the existence of Lipschitz, non-C^1 solutions to (5.5). It also shows that there are more Ω-varieties than those given by Corollary 5.4.

References

[1] H. Federer, *Some theorems on integral currents*, Trans. A.M.S. **117** (1965), 43–67.
[2] H. Federer, Geometric Measure Theory, Springer-Verlag, New York, 1969.
[3] R. Harvey, *Holomorphic chains and their boundaries*, Proc. Symp. Pure Math., A.M.S., 1975.
[4] R. Harvey & H. B. Lawson, Jr., *A constellation of minimal varieties defined over the group* G_2, (to appear).
[5] R. Harvey & H. B. Lawson, Jr., *The geometries defined by certain non-linear partial differential equations,* (to appear).
[6] J. King, *The currents defined by analytic varieties*, Acta Math. **127** (1971), 185–220.
[7] H. B. Lawson, Jr., *Foliations*, Bull. Amer. Math. Soc. **80** (1974), 369–417.
[8] H. B. Lawson, Jr. & R. Osserman, *Non-existence, non-uniqueness and irregularity of solutions to the minimal surface system*, Acta Math. **139** (1977), 1–17.
[9] H. Lewy, *On the non-vanishing of the Jacobian of a homeomorphism by harmonic gradients*, Ann. of Math. **88** (1968), 578–529.
[10] C. B. Morrey, *Second order elliptic systems of partial differential equations*, pp. 101–160, "Contributions to the Theory of Partial Differential Equations", Ann. of Math. Studies N° 33, Princeton Univ. Press, Princeton, 1954.
[11] R. Schafer, Introduction to Non-associative Algebras, Academic Press, New York, 1966.
[12] H. Whitney, Geometric Integration Theory, Princeton Univ. Press, Princeton, N. J., 1957.
[13] S. T. Yau, *Calabi's conjecture and some new results in algebraic geometry*, Proc. Nat. Acad. Sci. U.S.A. **74** (1977), 1798–1799.

RICE UNIVERSITY
HOUSTON, TEXAS, 77001
U.S.A.
INSTITUT DES HAUTES ETUDES SCIENTIFIQUES
35, ROUTE DE CHARTRES
91440-BURES-SUR-YVETTE
FRANCE

Minimal Submanifolds and Geodesics
Kaigai Publications, Tokyo, 1978, 61–73

LOWER BOUNDS ON THE FIRST EIGENVALUE OF THE LAPLACIAN OF RIEMANNIAN SUBMANIFOLDS

DAVID HOFFMAN*)

§ 0. Introduction

In this paper, we make use of the generalized Sobolev inequality of [10] to construct lower bounds on λ_1, the first eigenvalue of the Laplacian of a submanifold (with boundary) in a Riemannian manifold. The bounds involve various geometric quantities: curvature, diameter, volume, injectivity radius, dimension. The well known fact that

$$\lambda_1 = \inf_{f|\partial M=0} \frac{\int |\nabla f|^2}{\int f^2}$$

is exploited as well as the estimate of Cheeger [4],

$$\lambda_1 \geqq 1/4\, h^2 ,$$

where h is an isoperimetric constant (defined in § 3). The results below require some restriction on the volume of the submanifold. That restriction depends on the dimension of the submanifold, an upper bound for the mean curvature of the submanifold, bounds on the sectional curvature (when the ambient manifold has some positive curvature) and the injectivity radius of the ambient manifold.

In section 2, lower bounds are derived for λ_1 on a minimal submanifold of a Riemannian manifold. Here, various isoperimetric constants are utilized. Estimates are derived for minimal submanifolds in $\boldsymbol{R}^n$, $\boldsymbol{H}^n$ and S^n, where no volume restrictions are necessary. Particular emphasis is placed on surfaces where somewhat sharper results are obtained. Section 3 uses the results of section 2 to prove stability theorems of the type of [2] and [16]: Let M be an m-dimensional minimal submanifold of a Riemannian manifold N of dimension n, and suppose the sectional curvature of N is bounded above by b^2. Let $\beta^2 = mb^2 + \|B\|^2$, where B is the

---

*) Partially supported by NSF Grant MCS-7606752.

second fundamental form of M. We show that if $\|\beta\|_m$ is "small enough", then M is stable. See Theorems 5 and 6 and Corollary 2.

§ 1. Lower bounds on λ_1 for submanifolds of bounded mean curvature

Let M be a Riemannian manifold of dimension m isometrically immersed in a Riemannian manifold N of dimension n. We will use the following notation.

ω_m = volume of the unit m-ball.
$\bar{R}(M)$ = supremum of the injectivity radius of N, taken over all points in M.
$V = V(M)$ = volume of M.
H = mean curvature vector field of the immersion $M \subsetneq N$.
$c_\alpha = c_\alpha(m) = 2^m\alpha^{-1}\cdot m/m-1\cdot((1-\alpha)\omega_m)^{-1/m}$, where $0<\alpha<1$.
δ = extrinsic diameter of M in $N = \sup\{d_N(x, y) \mid x, y \in M\}$.
b = a real nonnegative, or purely imaginary number.
$\bar{K}$ = sectional curvature function of N.

In Hoffman and Spruck [8], the following Sobolev inequality was proved.

Theorem 1. *Suppose $\bar{K} \leq b^2$. Let h be a nonnegative C^1 function on M, vanishing on ∂M. Then*

$$\left(\int_M h^{m/m-1} dV_M\right)^{(m-1)/m}$$

$$(1) \qquad \leq \begin{cases} c_\alpha \int_M [|\nabla h| + h|H|] dV_M\,, & b \text{ imaginary,} \\ \frac{\pi}{2} c_\alpha \int_M [|\nabla h| + h|H|] dV_M\,, & b \text{ real,} \end{cases}$$

provided $b^2(1-\alpha)^{-2/m}\omega_m^{-1}\,\mathrm{Vol}\,(\mathrm{supp}\, h)^{2/m} \leq 1$ *and*

$$\rho_0 \leq \bar{R}(M)/2\,,$$

where

$$\rho_0 = \begin{cases} b^{-1}\sin^{-1} b(1-\alpha)^{-1/m}(\omega_m^{-1}\,\mathrm{Vol}\,(\mathrm{supp}\, h))^{1/m}\,, & \text{for } b \text{ real,} \\ (1-\alpha)^{-1/m}(\omega_m^{-1}\,\mathrm{Vol}\,(\mathrm{supp}\, h))^{1/m}\,, & \text{for } b \text{ imaginary.} \end{cases}$$

We may use this estimate to bound from below the first eigenvalue of the Laplacian of M in terms of an upper bound for $|H|$.

Theorem 2. *Suppose M is a compact Riemannian manifold isometrically immersed in a Riemannian manifold N of nonpositive curvature.*

Suppose for some nonnegative constant κ and for some α, $0 < \alpha < 1$, that the following conditions are satisfied.

$$V(M) \le \omega_m\left(\frac{\bar{R}(M)}{2}\right)^m (1-\alpha)\,, \qquad |H| \le \kappa\,, \tag{2}$$
$$\text{and} \quad \kappa^m V(M) < c_\alpha^{-m}\,.$$

Then $\lambda_1 \ge \frac{1}{4}[c_\alpha^{-1}V^{-1/m} - \kappa]^2$.

Proof. Since

$$\lambda_1 = \inf_{f|\partial M=0} \frac{\int |\nabla f|^2}{\int f^2}\,,$$

the theorem will follow from the following estimate. Let f be any C^1 function satisfying $f|\partial M = 0$, and let W = volume of support of f. Then

$$\left[\int_M |\nabla f|^2\right]^{1/2} \ge \frac{1}{2}[c_\alpha^{-1}W^{-1/m} - \kappa]\left[\int_M f^2\right]^{1/2}.$$

To prove this we may use Theorem 1, for the first inequality of (2) is enough to satisfy the conditions of Theorem 1 when $b^2 \le 0$. Together with the Hölder inequality this yields

$$\begin{aligned}\int_M f^2 &\le \left[\int_M f^{2m/m-1}\right]^{(m-1)/m} W^{1/m}\\ &\le c_\alpha W^{1/m}\int_M [|\nabla f^2| + f^2|H|]\\ &\le c_\alpha W^{1/m}\int_M [2|f||\nabla f| + \kappa f^2]\\ &\le c_\alpha W^{1/m}\left[2\left[\int_M |f|^2\right]^{1/2}\left[\int_M |\nabla f|^2\right]^{1/2} + \int_M \kappa f^2\right].\end{aligned}$$

Thus

$$\frac{1 - c_\alpha \kappa W^{1/m}}{2c_\alpha W^{1/m}}\left[\int_M f^2\right]^{1/2} \le \left[\int_M |\nabla f|^2\right]^{1/2}. \qquad \blacksquare$$

A corresponding theorem holds for isometric immersions into Riemannian manifolds N with curvature bounded above by a positive constant b^2.

Theorem 3. *Let M be a compact Riemannian manifold isometrically immersed in a Riemannian manifold N whose curvature is bounded above by a positive constant b^2. Suppose for some real nonnegative constant κ and some α, $0 < \alpha < 1$,*

$$(3)\qquad \begin{aligned} &V(M) \le (1-\alpha)w_m b^{-m}\,, \qquad |H| \le \kappa\,, \\ &\text{and}\quad \kappa^m V(M) \le (1-\alpha)\pi/2(c_\alpha^{-m})\,, \end{aligned}$$

$$(3')\qquad b^{-1}\arcsin\,[b(1-\alpha)^{-1}\omega_m V(M)]^{1/m} \le \bar{R}(M)/2\,.$$

Then

$$\lambda_1 \ge \frac{1}{4}\left[\frac{2}{\pi}c_\alpha^{-1}V^{-1/m} - \kappa\right]^2 .$$

The proof is similar to that of Theorem 2. The extra condition (3′) is necessary in order to use Theorem 1 when $b^2 > 0$. The extra factor of $\pi/2$ comes from the Sobolev inequality of Theorem 1 when $b^2 > 0$. We may also substitute for (3′) the stronger condition

$$(3'')\qquad \bar{R}(M) \ge \pi b^{-1}\,.$$

Thus Theorem 3 is applicable to standard spheres ($K \equiv b^2 > 0$) and to compact manifolds with K satisfying $b^2/4 \le K \le b^2$.

§2. Lower bounds on λ_1 for minimal submanifolds

Theorem 4. *Let M be a compact m-dimensional minimal submanifold (with boundary ∂M) of a Riemannian manifold N whose curvature is uoupositive.*

(i) *If* Vol $(M) = V \le \omega_m[\bar{R}(M)/2]^m(1-\alpha)$, *then*

$$\lambda_1 \ge \tfrac{1}{4}c_\alpha^{-2}V^{-2/m}\,.$$

(ii) *If M satisfies the volume restraint of* (i) *and M lies in an embedded geodesic ball of radius r,*

$$\lambda_1 \ge (m/2r)^2\,.$$

(iii) *If* $N = \boldsymbol{R}^n$, $N = \boldsymbol{H}^n$, *or the radius of injectivity of N is everywhere infinite,*

$$\lambda_1 \ge \max\left[\frac{1}{4}c_{m/m+1}^{-2}V^{-2/m}, \left(\frac{m}{\delta}\right)^2\right],$$

where $\delta = \max\,\{d(p,q) = \text{distance from } p \text{ to } q \text{ in } N \,|\, p, q \in M\}$.

Proof. Statement (i): Follows directly from Theorem 2.

Statement (ii): Let p_0 denote the center of the geodesic ball $B_r(p_0)$ containing M, and let $X(q)$ be the radial vector field at q (centered at p_0). Since M, is a minimal submanifold,

$$\int_{\partial M} X \cdot \nu = \int_M \operatorname{div}_M X ,$$

where $\nu =$ exterior normal vector field on ∂M. The curvature of N being nonpositive, $\operatorname{div} X \geq m$. Since $M \subset B_r(p_0)$, $|X| < r$ on M. Hence

$$r \int_{\partial M} X \cdot \nu = \int_M \operatorname{div} X \geq m \int_M 1 .$$

Thus $\operatorname{Vol}_{m-1}(\partial M) / \operatorname{Vol} M \geq m/r$. Moreover, the same inequality will hold for *any* open submanifold $\tilde{M}$ of M, with boundary $\partial \tilde{M}$. Therefore, the isoperimetric constant

$$h = \inf \left\{ \frac{\operatorname{Vol}_{m-1}(S)}{\operatorname{Vol}_m \tilde{M}} \,\middle|\, \begin{array}{l} S \text{ is a smooth hypersurface in } M \text{ divid-} \\ \text{ing } M \text{ into } \tilde{M} \text{ and } M^1, \\ \partial \tilde{M} = S = \partial \tilde{M} \cap \partial M^1,\ \partial \tilde{M} \cap \partial M = \varnothing \end{array} \right\}$$

is not less than m/r. Hence we may use the estimate of Cheeger

$$\lambda_1 \geq h^2/4$$

to prove statement (ii).

Statement (iii). If $N = \boldsymbol{R}^n$ or $\boldsymbol{H}^n$, then $\bar{R}(N) = +\infty$ and M always lies in an embedded geodesic ball of radius $\delta/2$. Hence (i) is applicable, with any choice of α, and (ii) is applicable with $r = \delta/2$. The constant $c_\alpha(m)$ is minimized for $\alpha = m/m+1$. ■

Remark 1. Osserman points out that statement (i) of Theorem 4 also follows from the Cheeger estimate together with an appropriate isoperimetric inequality. Under the hypothesis of Theorem 4, the Sobolev inequality (1) of Theorem 1 is

$$\left(\int h_M^{m/m-1} \right)^{m-1/m} \leq c_\alpha \int_M |\nabla h| , \quad \text{all} \quad h \in H_0^1(M) ,$$

and this is equivalent to the isoperimetric inequality

$$\operatorname{Vol}_m (M)^{m-1/m} \leq c_\alpha \operatorname{Vol}_{m-1}(\partial M) .$$

The same inequality holds for any open submanifold $\tilde{M} \subset M$ with $\partial \tilde{M} \cap \partial M = \varnothing$ ($\operatorname{Vol}(\tilde{M}) \leq \operatorname{Vol}(M)$). Writing this inequality as

$$\frac{\operatorname{Vol}_{m-1}(\partial \tilde{M})}{\operatorname{Vol}_m(\tilde{M})} \geq c_\alpha^{-1} \operatorname{Vol}_m (M)^{-1/m}$$

it is easily seen that $h^2/4 \geq \frac{1}{4} c_\alpha^{-2} \operatorname{Vol}_m (M)^{-2/m}$, giving a lower bound for λ_1.

If M is a minimal *surface* with boundary in $\boldsymbol{R}^n$, then we may substitute the value $(2\pi)^{-1/2}$ for c in Theorems 1 and 2. This is a result of L. Simon. If M is simply connected, then the classical isoperimetric inequality of Rado implies that we may use the value $c = (4\pi)^{-1/2}$ in Theorems 1 and 2. Recently, Osserman and Schiffer [11] and Feinberg, [7] have shown that the isoperimetric inequality

$$L^2 \geq 4\pi A \ , \quad L = \text{Length } \partial M \ , \quad A = \text{Area } M \ ,$$

is valid for doubly connected minimal surfaces in $\boldsymbol{R}^n$. This implies that we may also use, the value $c = (4\pi)^{-1/2}$ for such surfaces in Theorems 1 and 2. By using these values in the above Theorem we have the following Corollary.

Corollary 1. *If M is a minimal surface in $\boldsymbol{R}^n$ with area A and diameter δ, then*

$$\lambda_1 \geq \max\,[\pi/2A^{-1}, (2/\delta)^2] \geq \max\,[\pi^2 L^{-2}, (2/\delta)^2] \ .$$

If, in addition, M is simply or doubly connected,

$$\lambda_1 \geq \max\,[\pi A^{-1}, (2/\delta)^2] \geq \max\,[4\pi^2 L^{-2}, (2/\delta)^2] \ .$$

Remark 2. For plane domains, D, the inequality of Faber-Krahn gives $\lambda_1 \geq \pi j^2 A^{-1}$, where j is the first zero of the zero*th* Bessel function. This inequality is sharp; equality is achieved if and only if D is a disk. The same estimate holds for domains on simply connected minimal surfaces in $\boldsymbol{R}^n$, for $K \leq 0$ on such a surface and we may apply the following theorem (Peetre [15], Bandle [1], Chavel and Feldman [3]). Let M be a surface whose curvature is bounded above by K_0 and suppose D is a bounded domain in M, with piecewise-smooth boundary, contained in a simply connected domain $\tilde{D} \subset M$ whose area $\tilde{A}$ satisfies

$$\tilde{A} < \begin{cases} 4\pi/K_0 & K_0 > 0 \\ \infty & K_0 \leq 0 \ . \end{cases}$$

Then $\lambda_1(D) \geq \lambda_1(B(\rho, K_0))$, where $B(\rho, K_0)$ is the geodesic disk of radius ρ in the space form of constant curvature K_0, the value of ρ being determined by requiring the area of $B(\rho, K_0)$ to equal the area of D. Equality is obtained if and only if $D = B(\rho, K_0)$.

It would be interesting to know whether or not this stronger estimate is also true without the hypothesis of simple connectivity.

We note that for minimal submanifolds in spaces whose curvature is bounded above by a positive constant, it is possible to prove a theorem similar to Theorem 4.

Theorem 4′. *Let M be a compact Riemannian manifold with boundary minimally immersed in a Riemannian manifold whose curvature is bounded above by a positive constant b^2. Suppose, for some real constant α, $0 < \alpha < 1$,*

$$V(M) \le (1 - \alpha)\omega_m b^{-m} \tag{4}$$

and

$$b^{-1} \arcsin [b(1 - \alpha)^{-1}\omega_m^{-1}V(M)]^{1/m} \le \bar{R}(M)/2 \quad (\text{or } \pi b^{-1} \le \bar{R}(M)/2) . \tag{4'}$$

Then

$$\lambda_1 \ge [(\pi c_\alpha)^{-1}V^{-1/m}]^2 \ge [(\pi c_\alpha)^{-1}(1 - \alpha)^{-1/m}\omega_m^{-1/m}b]^2 .$$

Proof. Similar to the proof of Theorem 4.

§ 3. Stability of Minimal Submanifolds

Consider a compact manifold M, with boundary ∂M, minimally immersed in a Riemannian manifold N. M is said to be *infinitesimally stable* if, for all normal variations M_t which fix the boundary, the function $V(t)$ = Volume M_t satisfies $V''(0) > 0$. Let X denote the variation vector field of the variation, and let $e_1, \cdots, e_j, \cdots, e_m, e_{m+1}, \cdots, e_\alpha, \cdots, e_n$ be an orthonormal framing of TN on M, fitted to TM: $e_1, \cdots, e_j, \cdots, e_m$, $1 \le j \le m$, frames TM and $e_{m+1}, \cdots, e_\alpha, \cdots, e_n$, $m + 1 \le \alpha \le n$, frames NM, the normal bundle of M in N. It is well known (see Chern [6] pp. 46–47, or Lawson [9] page 48) that

$$V''(0) = -\int_M \langle \Delta X, X\rangle + |X|^2 \sum_{i,\alpha} \bar{R}_{i\alpha i\alpha} + \|B(X)\|^2 ,$$

where ΔX denotes the Laplacian of X in NM, $B(X)$ is the second fundamental form in the direction X,

$$\|B(X)\|^2 = \sum_{\substack{i=1\\j=1}}^{m} \langle \bar{\nabla}_{e_i} e_j, X\rangle^2 ,$$

and $\bar{R}_{i\alpha i\alpha}$ is the sectional curvature in N of the plane spanned by e_i and e_α. Writing $X = un$ where $n = X/|X|$ and $u = |X|$,

$$\begin{aligned} A''(0) &= -\int_M u\Delta u + u^2\Delta n\cdot n + u^2 \sum_{i,\alpha} \bar{R}_{i\alpha i\alpha} + u^2 \|B(n)\|^2 \\ &= \int_M |\nabla u|^2 + u^2 \sum_{i=1}^{m} |\nabla e_i n|^2 - u^2\Big[\sum_{i,\alpha} \bar{R}_{i\alpha i\alpha} + \|B(n)\|^2\Big] \end{aligned}$$

$$\geq \int_M |\Delta u|^2 - u^2 \left[\sum_{i,\alpha} R_{i\alpha i\alpha} + \|B(n)\|^2 \right]$$

$$\geq \int_M |\nabla u|^2 - u^2 \left[\sum_{i,\alpha} R_{i\alpha i\alpha} + \|B\|^2 \right] .$$

Here Δu is Laplacian of M applied to the function u. The first equalities follow from the identities (see, e.g. Spruck [13])

$$\Delta(un) = (\Delta u)n + u\Delta n$$

$$n \cdot \Delta n = \sum_{i=1}^{m} |\nabla_{e_i} n|^2 ,$$

which are easily verified. The two inequalities are straightforward.

$$\|B\|^2 = \sum_{i,\alpha} \langle \bar{\nabla}_{e_i} e_j, e_\alpha \rangle^2$$

Suppose $K \leq b^2$ on N. Then

$$A''(0) \geq \int_M |\nabla u|^2 - u^2[mb^2 + \|B\|^2] .$$

Thus stability is implied by the condition

$$(5) \qquad \int_M \beta^2 u^2 < \int_M |\nabla u|^2 , \quad \text{all} \quad u \in H_0'(M) ,$$

where $\beta^2(x) = [mb^2 + \|B_x\|^2]$, $x \in M$.

We may now use the eigenvalue estimates of the previous section.

Theorem 5. *Let M be an m-dimensional minimal submanifold with boundary in a Riemannian manifold N whose curvature is bounded above by $b^2 > 0$. Suppose conditions* (4) *and* (4′) *of* § 3 *are satisfied. Namely*

$$V(M) \leq (1 - \alpha)\omega_m b^{-m} \qquad \textit{and}$$

$$b^{-1} \arcsin [b(1 - \alpha)^{-1}\omega_m^{-1} V(M)^{1/m} \leq \bar{R}(M)/2$$

$$(\textit{or } \pi b^{-1} \leq \bar{R}(M)/2) .$$

Then

(i) *If $m > 2$, the condition*

$$\|\beta\|_m < \pi^{-1} \frac{m-2}{m-1} c_\alpha^{-1}$$

implies stability.

(ii) *If $m = 2$, then*

(6) $$\|\beta\|_2 < (\pi c_\alpha)^{-1}\lambda_1^{1/2}$$

implies stability. Thus

$$\|\beta\|_2 < \pi^{-2}c_\alpha^{-2}V^{-1}$$

or

$$\begin{aligned}\|\beta\|_2 &< \pi^{-2}c_\alpha^{-2}(1-\alpha)^{-1/2}(\pi)^{-1/2}b \\ &= (16\pi)^{-3/2}(1-\alpha)^{3/2}\alpha^2 b\end{aligned}$$

implies stability.

Proof. (i) $m > 2$. Using Hölder's inequality and Theorem 1,

$$\begin{aligned}\left[\int_M \beta^2 u^2\right]^{1/2} &\leq \left[\int_M \beta^m\right]^{1/m}\left[\int_M u^{2m/m-2}\right]^{m-2/2m} \\ &\leq \pi/2\cdot c_\alpha \frac{2(m-1)}{m-2}\|\beta\|_m\left[\int_M |\nabla u|^2\right]^{1/2}.\end{aligned}$$

Hence M is stable if

$$\|\beta\|_m < \pi^{-1}\frac{m-2}{m-1}c_\alpha^{-1}.$$

(ii) $m = 2$. We have $\beta^2 = [2b^2 - \|B\|^2]$ and

$$\begin{aligned}\int_M \beta^2 u^2 &\leq \left[\int_M \beta^2\right]^{1/2}\cdot\left[\int_M u^4\right]^{1/2} \\ &\leq \|\beta\|_2 \pi/2\cdot c_\alpha \int_M |\nabla u^2| \\ &\leq \|\beta\|_2 \pi/2\cdot c_\alpha 2\int_M |u||\nabla u| \\ &\leq \|\beta\|_2 \pi\cdot c_\alpha\left[\int_M u^2\right]^{1/2}\left[\int_M |\nabla u|^2\right]^{1/2} \\ &\leq \|\beta\|_2 \pi\cdot c_\alpha \lambda^{-1/2}\int_M |\nabla u|^2.\end{aligned}$$

Hence M is stable if

$$\|\beta\|^2 < (\pi c_\alpha)^{-1}\lambda_1^{1/2}.$$

The remainder of statement (ii) follows from Corollary 2 of the last section and the fact that for $m = 2$,

$$c_\alpha = 8\alpha^{-1}(1-\alpha)^{-1/2}\pi^{-1/2}. \quad \blacksquare$$

Corollary 2. *Let M be an m-dimensional minimal submanifold of the standard n-sphere S^n.*

(i) *Suppose $m > 2$. If $\mathrm{Vol}(M) \le (1-\alpha)\omega_m$ and*

$$\|\beta\|_m < 2/\pi \frac{m-2}{m-1} c_\alpha^{-1} ,$$

where $\beta^2 = m + \|B\|^2$, then M is stable.

(ii) *Suppose $m = 2$. If $A = \mathrm{area}\, M \le (1-\alpha)\pi$ and $\|\beta\|_2$ satisfies inequality (6) or $\int_M (4-2K) < \pi c_\alpha^{-1}\lambda_1^{1/2}$, then M is stable. For any minimal surface in S^n, the condition $\lambda_1 + 2K - 4 > 0$ also implies stability. In particular, if D is a region on a totally geodesic S^2 in S^n whose area is less than 2π, then D is stable.*

Proof. Statement (i) follows directly from the previous theorem. The first assertion of (ii) follows from the same theorem together with the fact that for minimal surfaces in S^n, $\beta^2 = 2 + \|B\|^2 = 2 + 2(1-K) = 4 - 2K$. To prove the second assertion consider any $u \in H_0^1(M)$. Then

$$\int_M |\nabla u|^2 \ge \lambda_1 \int_M u^2 > \int_M (4-2K)u^2 = \int_M \beta^2 u^2$$

which is precisely the stability equation (5). The third assertion of (ii) now follows from the theorem quoted in Remark 2 above. ■

Remark. It is easily seen that if D is a domain with area $A > 2\pi$ on a totally geodesic 2-sphere in S^n, then D is unstable.

Considering minimal submanifolds in spaces of nonpositive curvature, we see that by the remark above $\|\beta\|^2 \le \|B\|^2$. Therefore stability is implied by the condition

$$\int_M \|B\|^2 u^2 < \int_M |\nabla u|^2 , \qquad u \in H_0^1(M) .$$

For surfaces $\|B\|^2 = 2(\bar{K} - K) \le -2K$, and hence stability is implied by

$$(7) \qquad 2\int |K|\, u^2 = \int_M -2Ku^2 < \int_M |\nabla u|^2 , \qquad u \in H_0^1(M) .$$

Barbosa and do Carmo [1] have shown that for minimal surfaces in $\boldsymbol{R}^3$, $\int_M |K| < 2\pi$ implies stability. The example of the catenoid shows that this bound is sharp. The integral $\int_M |K|$ is the area of the Gaussian image of M. Spruck, [13] considered the question of stability for minimal surfaces in $\boldsymbol{R}^n$. By using the generalized Gauss map into the complex hyperquadric, he was able to show that there exists a constant ε (depending on n) such that

$$\int_M |K| < \varepsilon$$

implies stability. Analogously, for minimal submanifolds (dimension m) of $\boldsymbol{R}^n$, there exists a constant

$$\eta = \frac{m-2}{2(m-1)} c_\alpha^{-1}$$

such that $\|B\|_m < \eta$ implies stability. We may use the method of Theorem 5 and Corollary 2 to prove similar results.

Theorem 6. *Suppose M is an m-dimensional minimal submanifold of N, a Riemannian manifold of nonpositive. Let α be some constant $0 < \alpha < 1$*

(i) *If $m > 2$ and $V(M) \leq (1-\alpha)\omega_m(\bar{R}(M)/2)^m$, then the condition*

$$\|B\|_m < \frac{m-2}{m-1} c_\alpha^{-1}$$

implies stability. If $N = \boldsymbol{H}^n$, $N = \boldsymbol{R}^n$, or if the radius of injectivity of N is everywhere infinite, then the volume restriction on M is unnecessary. (For $N = \boldsymbol{R}^n$, this was proved by Spruck, loc. cit.)

(ii) *Let $m = 2$. Suppose* Area $M \leq (1-\alpha)\pi(\bar{R}(M))^2$, *and*

$$\int |K| < (\pi^{-1/2}/16)\alpha(1-\alpha)^{1/2} \cdot \lambda^{1/2} .$$

Then M is stable. The estimates of Theorem 4 *give lower bounds on λ_1 and hence provide an explicit stability constant. In particular, if M lies in a geodesic ball of radius r, the condition*

$$\int_M |K| < (\pi^{-1/2}/16)\alpha(1-\alpha)^{1/2} \cdot r^{-1} = 1/16[\pi r^2]^{-1/2}\alpha(1-\alpha)^{1/2}$$

implies stability. If $N = \boldsymbol{R}^n$, $N = \boldsymbol{H}^n$ or the radius of injectivity of N is everywhere infinite, we may omit the restriction on the area of M and choose $\alpha = 2/3$.

(iii) *For minimal surfaces in $\boldsymbol{R}^n$*

$$\int_M |K| < 1/24\sqrt{3} \max [(\sqrt{\pi}/\sqrt{2})A^{-1/2}, 2/\delta]$$

implies stability. If M is simply or doubly connected, the constant may be increased by a factor of $\sqrt{2}$.

Proof. Analogous to Corollary 3 and Theorem 5. To estimate λ_1, use Theorem 4 and Corollary 1.

References

[1] C. Bandle, Konstruction isoperimetrischer Ungleichung der mathematischen Physik aus solchen Geometrie, Comm. Math. Helv., **46** (1971) 182–213.

[2] J. L. Barbosa and M. do Carmo: On the size of a stable minimal surface in $\boldsymbol{R}^n$, in Proc. Symp. in Pure Math. 27, Part 2 A. M. S., Providence, R. I. (1975).

[3] I. Chavel and E. Feldman, Isoperimetric inequalities on curved surfaces, (preprint).

[4] M. Berger.: Eigenvalues of the Laplacian, Proc. Symp. Pure Math., 16 A. M. S. Providence, R. I. (1970) 121–125.

[5] M. Berger, P. Gauduchon, E. Mazet, Le Spectre d'une Variete Riemanniene, Lecture Notes in Math., 194, Springer Verlag, 1971.

[6] J. Cheeger, A lower bound for the smallest eigenvalue of the Laplacian, in *Problems in Analysis, a symposium in honor of S. Bochner*, Princeton University Press, Princeton, N. J. (1970).

[7] S.-Y. Cheng, Eigenvalue comparison theorems and its geometric applications, Math. Z. **143** (1975) 289–297.

[8] S. S. Chern, Minimal Submanifolds in a Riemannian Manifold. Lecture notes, Tech. Report 19, Dept. of Mathematics, University of Kansas, Lawrence Kansas, 1968.

[9] J. Feinberg, The isoperimetric inequality for doubly connected minimal surfaces in $\boldsymbol{R}^n$, (to appear).

[10] D. Hoffman and J. Spruck, Sobolev and isoperimetric inequalities for Riemannian submanifolds. CPAM, Vol. XXVII, (1974) 715–727.

[11] H. B. Lawson, Lectures on Minimal Submanifolds, Volume I, I. M. P. A. Rio de Janeiro, 1970.

[12] J. Michael and L. Simon, Sobolev and mean value inequalities on generalized submanifolds of R^n, CPAM (1973) 361–369.

[13] R. Osserman and M. Schiffer: Doubly-connected minimal surfaces. Arch. Rational Mech. Anal. **58** (1975) 285–307.

[14] T. Otsuki: A remark on the Sobolev inequality for Riemannian submanifolds. Proc. of Japan Academy 51 Supp. **1** (1975) 785–789.

[15] J. Peetre, A generalization of Courant's nodal domain theorem, Math. Scand. **5** (1957) 15–20.

[16] J. Spruck: Remarks on the stability of minimal submanifolds of R^n (to appear).

UNIVERSITY OF MASSACHUSETTS
AMHERST, MA 01003
U.S.A.

Minimal Submanifolds and Geodesics
Kaigai Publications, Tokyo, 1978, 73–76

THE WEIERSTRASS FORMULA FOR SURFACES OF PRESCRIBED MEAN CURVATURE

KATSUEI KENMOTSU

In this paper we announce some results on the Gauss maps of immersed surfaces in the Euclidean 3-space $\boldsymbol{R}^3$. Complete proofs will appear elsewhere. For comparison with our later results on any surfaces we list some of the well-known facts about minimal surfaces in $\boldsymbol{R}^3$.

(i) The Gauss map Ψ of a minimal surface is a holomorphic map of the surface into the unit sphere S^2 considered as the Riemann sphere.

(ii) The Weierstrass formula: Every simply-connected (generalized) minimal surface in $\boldsymbol{R}^3$ can be represented in the form

$$x_k(\zeta) = \mathrm{Re}\left\{\int_0^\zeta \phi_k(z)dz\right\} + c_k\,, \qquad k = 1, 2, 3\,,$$

where $\phi_1 = (1/2)f(1 - g^2)$, $\phi_2 = (i/2)f(1 + g^2)$, $\phi_3 = fg$, $f(z)$ is an analytic function on a domain D and $g(z)$ is a meromorphic function on D. D is an unit disk or $\boldsymbol{R}^2$. And at each point where $g(z)$ has a pole of order m, $f(z)$ has a zero of order at least $2m$ ([1]).

We remark that in this formula $g(z)$ is the Gauss map of the surface. Thus, by the Weierstrass formula, for a given holomorphic map $g(z)$ of D into S^2, we can construct a minimal immersion of D into $\boldsymbol{R}^3$ such that the Gauss map is $g(z)$. These two results have played a major role in the theory of minimal surfaces in $\boldsymbol{R}^3$.

Now in order to study a class of more general surfaces, in particular, surfaces with constant mean curvature, we generalize the above mentioned results to an arbitrary surface in $\boldsymbol{R}^3$.

Throughout this paper we use the following notation: Let M be an oriented 2-dimensional Riemannian manifold with a Riemannian metric $ds^2 = \lambda^2|dz|^2$ in an isothermal coordinate and $\boldsymbol{R}^3$ the Euclidean 3-space. $x: M \to \boldsymbol{R}^3$ will denote an isometric immersion and $\Psi: M \to S^2$ the Gauss map of x, where the unit sphere S^2 is considered as the Riemann sphere. Let $n(z) := (n_1(z), n_2(z), n_3(z)) = (1/\lambda^2)(\partial x/\partial\xi_1) \times (\partial x/\partial\xi_2)$, $z = \xi_1 + i\xi_2$, be the unit normal vector field on M.

Since we consider Ψ as a complex mapping of M into the Riemann sphere, we can set

$$\Psi(z) := \begin{cases} \dfrac{n_1(z) + in_2(z)}{1 - n_3(z)} (\equiv \Psi_1(z)) , \\ \qquad\qquad \text{if } n(z) \in U_1 : = S^2 - \{\text{north pole}\} ; \\ \dfrac{n_1(z) - in_2(z)}{1 + n_3(z)} (\equiv \Psi_2(z)) , \\ \qquad\qquad \text{if } n(z) \in U_2 : = S^2 - \{\text{south pole}\} , \end{cases}$$

where we consider $n(z)$ as the image of the parallel translation of $n(z)$ to the origin of $\boldsymbol{R}^3$. Then we have, at first,

Theorem 1. *The Gauss map* Ψ *of any immersed surface satisfies the Beltrami eqation*:

$$H\frac{\partial \Psi}{\partial z} = \phi\frac{\partial \Psi}{\partial \bar{z}} ,$$

where H *is the mean curvature and* ϕ *is defined by* $\phi = \frac{1}{2}(h_{11} - h_{22}) - ih_{12}$, h_{ij}*'s being the coefficients of the 2nd fundamental form for the orthonormal frame* $\{(1/\lambda)(\partial x/\partial \xi_1), (1/\lambda)(\partial x/\partial \xi_2), n\}$.

We remark that the zero point of ϕ is an umbilical point and the Beltrami equation is closely related to a quasiconformal mapping. For instance, if x is non-umbilical and there is a positive constant $k(<1)$ such that $|H/\phi| \leq k$, then Ψ is a quasiconformal mapping of M into S^2.

We can calculate the norm of these complex vectors $\partial\Psi/\partial z$ and $\partial\Psi/\partial\bar{z}$:

$$\left|\frac{\partial \Psi}{\partial z}\right|^2 = \frac{\lambda^2}{4}(1 + \Psi\bar{\Psi})^2|\phi|^2 , \qquad \left|\frac{\partial \Psi}{\partial \bar{z}}\right|^2 = \frac{\lambda^2}{4}(1 + \Psi\bar{\Psi})^2 H^2 .$$

By virtue of these facts we can say that a zero point of the mean curvature is a point of $\Psi_{\bar{z}} = 0$ and an umbilical point of x is a point of $\Psi_z = 0$. Therefore we can characterize a zero point of the mean curature and an umbilical point by the first derivatives of the Gauss map.

Theorem 1 can be proved by the following lemma:

Lemma. *Under the above notation, we have*

$$\frac{\partial \Psi_1}{\partial \bar{z}} = -\frac{H}{2}(1 + \Psi_1\bar{\Psi}_1)^2\left(\frac{\partial x_1}{\partial \bar{z}} + i\frac{\partial x_2}{\partial \bar{z}}\right) ,$$

$$\frac{\partial \Psi_1}{\partial z} = -\frac{\phi}{2}(1 + \Psi_1\bar{\Psi}_1)^2\left(\frac{\partial x_1}{\partial \bar{z}} + i\frac{\partial x_2}{\partial \bar{z}}\right) .$$

The above lemma follows from the definition of Ψ_1 by a direct and elementary calculation. But this is also useful to obtain a generalize Weierstrass formula. Because the lemma says that the vector $\partial x_1/\partial\bar{z} + i\partial x_2/\partial\bar{z}$ can be represented by the Gauss map Ψ_1. We have

Theorem 2 (Weierstrass formula). *For any isometric immersion* $x: M \to \boldsymbol{R}^3$ *we have*

$$H\frac{\partial x_1}{\partial \bar z} = f(\overline{\Psi}_1^2 - 1)\,,$$

$$H\frac{\partial x_2}{\partial \bar z} = if(\overline{\Psi}_1^2 + 1)\,,$$

$$H\frac{\partial x_3}{\partial \bar z} = 2f\overline{\Psi}_1\,,$$

where we set

$$f = \frac{1}{(1 + \Psi_1\overline{\Psi}_1)^2}\frac{\partial \Psi_1}{\partial \bar z}\,.$$

(*For* Ψ_2 *we have similar formulas*).

Remark 1. In the case of minimal immersions, we can show that the above formula reduces to the famous Weierstrass formula.

As the completely integrability condition of this system, we get

Theorem 3. *The Gauss map* Ψ *of an arbitrary immersed surface must satisfy*

$$H\left(\frac{\partial^2 \Psi}{\partial z \partial \bar z} - \frac{2\overline{\Psi}}{1 + \Psi\overline{\Psi}}\frac{\partial \Psi}{\partial z}\frac{\partial \Psi}{\partial \bar z}\right) = \frac{\partial H}{\partial z}\frac{\partial \Psi}{\partial \bar z}\,.$$

Remark 2. If we assume $H =$ constant ($\neq 0$), then this formula tells us that Ψ is a harmonic mapping of M into S^2. That is, *the harmonicity is the integrability condition of the system with* $H =$ constant ($\neq 0$).

When isometric immersions are of constant mean curvature, as a corollary of the above obtained results, we have our main theorem.

Theorem 4. *Let* $x: M \to \boldsymbol{R}^3$ *be an isometric immersion with constant mean curvature* $H \neq 0$. *Then we have*

(i) $ds^2 = \dfrac{4}{H^2}\dfrac{|\Psi_{\bar z}|^2}{(1 + \Psi\overline{\Psi})^2}|dz|^2$;

(ii) $\phi = H\dfrac{\Psi_z}{\Psi_{\bar z}}$;

(iii) $\Psi_{z\bar z} - \dfrac{2\overline{\Psi}}{1 + \Psi\overline{\Psi}}\Psi_z\Psi_{\bar z} = 0$.

(*Namely, the* 1*st and* 2*nd fundamental forms of x are determined by the Gauss map and the non-zero constant H.*) *Conversely, for any non-zero constant H and any harmonic map* Ψ *of D into* S^2, *by the generalized Weierstrass formula, we can construct a branched immersion with constant mean curvature H of D into* $\boldsymbol{R}^3$ *such that the Gauss map is* Ψ.

Reference

[1] R. Osserman, *A survey of minimal surfaces,* Van Nostrand Reinhold Company, 1969.

College of General Education,
Tohoku University,
Sendai, 980
Japan

Minimal Submanifolds and Geodesics
Kaigai Publications, Tokyo. 1978, 77–84

STABLE AND UNSTABLE PERIODIC MOTIONS

W. KLINGENBERG

We consider the geodesic flow on the tangent bundle of a riemannian manifold. The orbits of the flow describe the motion of a point if there are no exterior forces. Although this is a rather special mechanical system it represents many of the typical features of a general mechanical system. This is true in particular for the periodic orbits, as was pointed out already by Poincaré in his classical investigation of the geodesics on a convex surface.

For simplicity we will assume that the underlying riemannian manifold M is compact. This is particularly useful for the following reason: A periodic orbit in the geodesic flow on the tangent bundle TM of M determines a closed geodesic on M and vice versa. Now, the closed geodesics can be viewed as critical points of the energy functional E in the space ΛM of all closed H^1-curves on M. Closed H^1-curves are mappings $c: I = S = [0, 1]/\{0, 1\} \to M$ which are absolutely continuous and have a square integrable covariant derivative. ΛM carries in a canonical way the structure of a complete Hilbert manifold with a riemannian metric.

The negative gradient flow determined by the differentiable function

$$E: \Lambda M \to \boldsymbol{R}; c \mapsto \tfrac{1}{2} \int_S \langle \dot{c}(t), \dot{c}(t) \rangle dt$$

is defined for all positive times and satisfies the condition (C) of Palais and Smale. Thus, one can apply the Morse-Lusternik-Schnirelmann theory to prove the existence of critical points for E. And, as we stated already, the critical points of E, which are not constant mappings, are precisely the closed geodesics on M.

It is important to observe that with a closed geodesic $c: S \to M$ also its m-fold covering $c^m: S \to M; t \mapsto c(mt)$, is a closed geodesic, m an integer ≥ 1. c^m, $m > 1$, is a point different from c in ΛM. This corresponds to the fact that a periodic orbit of period ω determines a periodic orbit of period $m\omega$ by going around m times.

For each closed geodesic c we have an underlying prime closed geodesic c_0 which is not any more the proper covering of a geodesic. c_0 corresponds to a periodic orbit of minimal or prime period. The closed geodesics therefore occur in "towers" $\{c_0^m; m = 1, 2, \cdots\}$ consisting of the coverings of a prime closed geodesic c_0. The energy integral grows with m^2: $E(c_0^m) = m^2E(c_0)$.

The great difficulty in proving the existence of different prime closed geodesics or different towers lies in the fact that the Morse-Lusternik-Schnirelmann theory of the gradient flow on ΛM does not automatically give information on the multiplicity of critical points=closed goedesics. One has to bring into play additional special features of $\{\Lambda M, -\text{grad}\, E\}$ to overcome this difficulty:

First of all, we observe the existence of a canonical S-action and a Z_2-action on ΛM: The S-action is the change of the parameter: $c(t)$ is carried into $c(t+r)$ under the element $e^{2\pi ir} \in S$. And the Z_2-action is generated by the orientation reversing map $c(t) \mapsto c(1-t)$. Both of these actions are compatible with the gradient flow on ΛM.

Although the S-action is not differentiable but only of class H^1, there nevertheless exists a slice for this action. Thus one can speak of a Morse-Lusternik-Schnirelmann theory on the space $\tilde{\Pi}M$ of S-orbits. Taking in addition the quotient with respect to the Z_2-action one gets the space ΠM of non-parameterized H^1-curves on M. For example, if M is the standard sphere S^n, ΠM will contain the space ΓS^n of circles, i.e., the intersections of $S^n \subset R^{n+1}$ with 2-dimensional planes.

The problem of finding many, possibly infinitely many prime closed geodesics on a compact riemannian manifold can now be successfully attacked: Essentially different methods have to be used depending on whether the fundamental group $\pi_1 M$ of M is infinite or finite.

If $\pi_1 M$ is infinite, i.e., if the universal covering $\tilde{M}$ of M is non-compact, there are still some open problems: The main difficulty is presented by manifolds where all higher homotopy groups $\pi_k M$, $k > 1$, vanish. In this case, only various special hypothesis allow a satisfactory answer. For instance, if there exists a riemannian metric on the underlying differentiable manifold with negative curvature K (or, possibly more general, a riemannian metric with geodesic flow of Anosov type), then there are infinitely many prime closed geodesics. Actually, their number is growing exponentially when one orders them according to their length. On the other hand, if M has some non-vanishing homotopy group $\pi_k M$, $k > 1$, it should be possible to prove the existence of infinitely many closed geodesics using the fact that for every riemannian metric on S^k there are infinitely many closed geodesics, see below.

In the case $\pi_1 M$ finite, i.e., $\tilde{M}$ compact, one can always prove the existence of infinitely many prime closed geodesics. The proof uses a number of various deep results, among others the Gromoll-Meyer theorem, the Sullivan theory of the minimal model with a particular feature for the space ΛM, due to joint work of Sullivan and Vigué and, last not least, the structure of the Morse-complex $\mathcal{M}M$ of ΛM in the case that all critical orbits on ΛM are non-degenerate.

Under this latter hypothesis (which is generically satisfied) one has well-defined unstable manifolds for each critical orbit, their dimension being equal to the index of a point on this orbit, plus 1. The union of these unstable manifolds defines the Morse-complex $\mathcal{M}M$. The grad E-flow carries $\mathcal{M}M$ into itself. Actually, its restriction to $\mathcal{M}M$ is defined for all $s \in \boldsymbol{R}$ and each orbit $\phi_s c$ has a α- and a ω-limit point which is of course a critical point. The S- and $\boldsymbol{Z}_2$-action on ΛM restricts to $\mathcal{M}M$. Thus, $\mathcal{M}M$ contains the complete relevant structure of the space ΛM with its gradient flow and its group actions in a pure and condensed form.

Using certain $\boldsymbol{Z}_2$-cycles in ΛM and their homotopies, representing $\boldsymbol{Z}_2$-cycles in ΠM, one proves the existence of a sequence $\{c(r), c'(r)\}$, $r = 0, \cdots$ of pairs of critical points in ΛM such that the multiplicity $m'(r)$ of $c'(r)$ divides the multiplicity $m(r)$ of $c(r)$—the multiplicity of a closed geodesic being the number of times that this geodesic covers the underlying prime closed geodesic. Moreover,

$$E(c(r)) < E(c'(r)) < E(c(r+1)) .$$

Once this fundamental Divisibility Lemma is established, one easily derives a contradiction from the assumption that there are only finitely many "towers" in ΛM. The Gromoll-Meyer theorem then allows to extend this result to the case that not all closed geodesics on M are non-degenerate.

Thus much for a report on the existence of prime closed geodesics. We now go back to the interpretation of a closed geodesic as a period orbit in the geodesic flow. To a periodic orbit $\phi_t X_0$, $0 \leq t \leq \omega$, in a Hamiltonian system we associate the so-called Poincaré map:

$$\mathscr{P} : (\Sigma_0, X_0) \to (\Sigma_\omega, X_0) .$$

$\mathscr{P}$ is defined as follows: First, restrict the flow to a non-degenerate hypersurface of constant kinetic energy, e.g. in the geodesic flow case to the unit tangent bundle $T_1 M$. Let Σ be a small flow-transversal hypersurface through X_0 and associate to every X in a small neighborhood Σ_0 of X_0 on Σ the first point, for $t > 0$, of intersection of the orbit $\phi_t X$ with Σ. For sufficiently small Σ_0 the thus defined map $\mathscr{P}$ will carry Σ_0 diffeomorphically into a neighborhood Σ_ω of X_0 on Σ, Σ_ω in general being different from Σ_0.

The importance of $\mathscr{P}$ lies in the fact that its periodic points X are in $1:1$ correspondence with periodic orbits $\phi_t X$ of the geodesic flow which remain near $\phi_t X_0$.

$\mathscr{P}$ preserves the induced symplectic structure on Σ. In particular, the linear approximation $P = D\mathscr{P}(X_0)$ of $\mathscr{P}$ in X_0 is symplectic. Hence, its eigenvalues occur in quadrupels $\{\lambda, \bar{\lambda}, \lambda^{-1}, \bar{\lambda}^{-1}\}$ which may or may not all be pairwise different.

The periodic orbit $\phi_t X_0$, $0 \leq t \leq \omega$, is called hyperbolic if none of the eigenvalues of the associated P is on the unit circle. In this case there will be two n-dimensional submanifolds $W_s^n = W_s^n(X_0)$ and $W_u^n = W_u^n(X_0)$, $n = \dim M - 1$, in Σ through X_0, the stable and the unstable manifold, which are transformed into itself by $\mathscr{P}$: On W_s^n, $\mathscr{P}$ is operating as contraction, whereas on W_u^n, $\mathscr{P}$ operates as expansion. A hyperbolic periodic orbit therefore is also called unstable: No orbit $\phi_t X$, $X \neq X_0$, will stay for all t in a neighborhood of the periodic orbit $\phi_t X_0$.

On the other hand, if some of the eigenvalues of P are on the unit circle —their number must be even, say $2q$, when counted with their multiplicity— there exist on Σ a local submanifold $W_{\mathrm{ce}}^{2q} = W_{\mathrm{ce}}^{2q}(X_0)$ the so-called center manifold, invariant under $\mathscr{P}$ such that the linear approximation P_{ce} of $\mathscr{P} \mid W_{\mathrm{ce}}^{2q}$ at X_0 has all eigenvalues on the unit circle. If $p = n - q > 0$, there also exist a stable and unstable manifold W_s^p and W_u^p on Σ through X_0, invariant under $\mathscr{P}$ on which $\mathscr{P}$ operates by contraction and expansion, respectively.

So, while for a non-hyperbolic periodic orbit $\phi_t X_0$ there still may be many nearby orbits $\phi_t X$ which do not remain near $\phi_t X_0$ for all t, those orbits $\phi_t X$ which start at the center manifold W_{ce}^{2p} might have this property. At least, this is true for the linear approximative P_{ce} of the map $\mathscr{P}_{\mathrm{ce}}$. Therefore, a non-hyperbolic (or: partially elliptic) period orbit is called infinitesimaly (partially) stable.

In the case of $\dim M = 2$, non-hyperbolic means elliptic, i.e., all eigenvalues of P (since there are only two of them) are on the unit circle.

The problem of whether an infinitesimally stable orbit $\phi_t X$, $X \in W_{\mathrm{ce}}^{2q}$, is stable in the sense that it remains near $\phi_t X_0$ for all t has been studied extensively since Poincaré by Whittacker, Birkhoff, Siegel and many others. Only the more recent work of Kolmogorov, Arnold and Moser has lead to a rather complete satisfactory answer. First of all —and this was known long time ago— it does not suffice to make an hypothesis only on the linear approximation P_{ce} of $\mathscr{P}_{\mathrm{ce}}$: excluding finitely many relations for the eigenvalues of P_{ce} (in particular one has to exclude that the eigenvalues are third of fourth roots of unity) one establishes the so-called Birkhoff normal form for $\mathscr{P}_{\mathrm{ce}}$. This is a normal form for the 3rd order approximation of $\mathscr{P}_{\mathrm{ce}}$. If $z_k, \bar{z}_k$, $1 \leq k \leq q$, are complex-conjugate coordinates on W_{ce}^{2q} this normal form reads

$$(*) \qquad z_k^* = z_k \exp\left(2\pi i a_k + 2\pi i \sum_l b_{kl} z_l \bar{z}_l\right) + O(|z|^4)$$

with a_k, b_{kl} real. If now $\det(b_{kl}) \neq 0$ —a condition which does not depend on the choice of the normal form— $\mathscr{P}_{ce}$ is called of twist type.

(*) should be viewed as a perturbation of an elementary twist map $\mathscr{P}_{ce}^{elem}$. i.e., a mapping (*) for which the remainder term is zero. We introduce action-angle variables $\tau_k > 0$ and $\theta_k \bmod 2\pi$ by

$$\tau_k = z_k \bar{z}_k / 2 , \qquad \theta_k = \arg z_k .$$

Then $\mathscr{P}_{ce}^{elem}$ reads:

$$\tau_k^* = \tau_k ; \theta_k^* = \theta_k + 2\pi\left(a_k + \sum_l 2b_{kl}\tau_l\right)$$

Thus, $\mathscr{P}_{ce}^{elem}$ leaves the q-tori $\tau = (\tau_k) = \text{const}$ invariant and on each of these tori it operates by a translation mod 2π. For the corresponding elementary geodesic flow this means that a neighborhood of the periodic orbit $\phi_t X_0$ on $\phi_t W_{ce}^{2q}$ decomposes into invariant $(q + 1)$-tori on each of which the flow is quasi periodic.

Three cases are to be distinguished: Call

$$\omega(\tau) = \left(a_k + \sum_l 2b_{kl}\tau_l\right) \bmod \boldsymbol{Z}^q \in \boldsymbol{R}^q / \boldsymbol{Z}^2$$

the period of the torus $\tau = \text{const}$.

(i) $\omega(\tau)$ is free over the rationals. In this case the flow on the torus is ergodic.

(ii) There are r, $1 < r < q - 1$, independent linear relations for $\omega(\tau)$. Then the $(q + 1)$-torus $\tau = \text{const}$ factors into a product of a $(q + 1 - r)$-torus and a r-torus such that the flow on each of the r-parameter family of $(q + 1 - r)$-tori is ergodic.

(iii) $\omega(\tau)$ is the multiple of a q-tupel of rational numbers. In this case, all orbits are periodic.

The fact that the $\mathscr{P}_{ce}^{elem}$ is of twist type garantuees that the ergodic tori are dense of measure equal to the measure of all tori.

If now $\mathscr{P}_{ce}$ is a perturbation of the elementary Poincaré map $\mathscr{P}_{ce}^{elem}$, the Kolmogorov-Arnold-Moser theorem states that in a small neighborhood of X_0 a Cantor set type family of ergodic tori survives, its measure being only a little less than the measure of the full neighborhood, chosen sufficiently small.

On the other hand, the Birkhoff-Lewis-Moser fixed point theorem asserts that the set of complementary tori, where the elementary flow is not ergodic, under the perturbation gives rise to infinitely many periodic orbits. Actually, the closure of the periodic orbits contains all the quasiperiodic orbits on the surviving invariant $(q + 1)$-tori and hence has a measure almost equal to the measure of the full neighborhood of $\phi_t X_0$ in $\phi_t \Sigma_{ce}$.

This picture is complicated further by the fact that among the periodic orbits $\phi_t X$ determined by the periodic points X of $\mathscr{P}_{ce}$, there are both, non-hyperbolic and hyperbolic ones. Much remains to be done to clarify further the situation. Of particular interest is the case of a 4-dimensional Hamiltonian system, where the Poincaré map $\mathscr{P}$ operates on a 2-dimensional surface Σ, the invariant ergodic tori actually bound 3-dimensional solid tori around the periodic orbit $\phi_t X_0$ from which an orbit cannot escape. In this case, we therefore have for a periodic orbit $\phi_0 X_0$ of twist type stability in the most complete sense imaginable: All orbits remain in the neighborhood of the periodic orbit for all time.

All this shows the great interest of the question as to whether a closed goedesic (i.e., its associated periodic orbit) is hyperbolic or (partially) elliptic.

In the case that the curvature K of the riemannian manifold M is negative or —more generally— if the geodesic flow on TM is of Anosov type, all closed geodesics are hyperbolic. Moreover, the periodic orbits are dense. Thus, an orbit sufficiently near a periodic orbit will not stay in the neighborhood of the periodic orbit. But in general it will come back to it time and again. More generally, a geodesic flow of Anosov type is ergodic in the sense that almost all its orbits form a dense set in the tangent bundle.

It would be wrong, however, to expect that in general a manifold M with positive curvature K has only non-hyperbolic or even elliptic periodic orbits. As we saw already, in every neighborhood of a partially elliptic orbit of twist type there are both, partially elliptic and hyperbolic orbits. Actually, the existence of even a single non-hyperbolic closed geodesic on a compact manifold M with finite fundamental group $\pi_1 M$ is still an open problem —recall that the fundamental group of a manifold with geodesic flow of Anosov type is infinite, it even has exponential growth.

A number of partial results lead to the conjecture that on a compact riemannian manifold M of elliptic there always exists a partially elliptic closed geodesic. Here we call a compact manifold M elliptic if its universal covering $\tilde{M}$ is compact, i.e., if $\pi_1 M$ if finite.

For instance, if M has positive curvature and $\pi_1 M$ is non-trivial, then a non-nullhomotopic closed geodesic on M of minimal length in its homotopy class is partially elliptic.

Thorbergsson has shown that on a riemannian manifold of the form $M = (S^n, g)$ (i.e., the n-sphere with a riemannian metric g) for which the curvature K satisfies $0{,}64 \leq K \leq 1$ there exists a partially elliptic closed geodesic having length in the interval $[2\pi, 2\pi/(0,8)[$. In the case $n = 2$, the hypothesis $0, 25 \leq K \leq 1$ satisfies to conclude that the shortest closed geodesic on M is elliptic (it has length in the interval $[2\pi, 4\pi[$).

Poincaré had shown with the help of bifurcation theory that on an analytic convex surface $M = (S^2, g)$ there always exists an elliptic closed geodesic, provided that the metric g is sufficiently near the canonical metric on S^2.

It seems possible to extend Poincaré's argumentation so as to cover the case of an arbitrary differentiable metric g on S^2. More generally, this extension allows to prove the existence of a partially elliptic closed geodesic for every metric g on a manifold M of elliptic type provided there exists on the universal covering $\tilde{M}$ of M a riemannian metric $\tilde{g}_0$ satisfying the following property (E).

First observe that the closed geodesics on $\tilde{M} = (\tilde{M}, \tilde{g})$ can be viewed as the critical points of the energy integral $\tilde{E}$ on the Hilbert manifold $\Lambda\tilde{M}$ of closed H^1-curves on $\tilde{M}$, lying outside the set $\Lambda^0\tilde{M}$ of constant (= trivial) closed curves. Then for $\tilde{M}_0 = (\tilde{M}, \tilde{g}_0)$ the critical points on $\Lambda\tilde{M}_0$ up to a prescribed $\tilde{E}$-value κ^* shall all be non-degenerate. This allows to define the Morse complex $\mathcal{M}^{\kappa^*}\tilde{M}_0$ in $\Lambda\tilde{M}_0$.

Since $\tilde{M}_0$ is compact and $\pi_1\tilde{M}_0 = 0$ there exists a first k, $2 \leq k + 1 \leq \dim \tilde{M}_0$, such that $\pi_{k+1}\tilde{M}_0 \neq 0$. Let $f: S^{k+1} \to \tilde{M}_0$ a homologically non-trivial map. It determines a non-trivial cycle

$$\Lambda f: (D^k, \partial D^k) \to (\Lambda\tilde{M}_0, \Lambda^0\tilde{M}_0)$$

which we can assume to be realized in the Morse complex $\mathcal{M}^{\kappa^*}M_0$. Property (E) now states that the critical points of index k at which this cycle remains hanging all are partially elliptic. (A weaker property would also do, but it is more complicate to formulate).

For simplicity we restrict ourselves to the case $M = (S^2, g)$. Here $M_0 = (S^2, g_{\text{ell}})$, an ellipsoid with three pairwise different axes, all of length approximately equal to one, is a metric with property (E): Indeed, the ellipse of minimal length on M_0 is an elliptic closed geodesic at which the generating cycle of $H_1(\Lambda S^2, \Lambda^0 S^2)$ remains hanging.

Assume now that there were a metric g on S^2 such that on $M = (S^2, g)$ all geodesics are hyperbolic. There exists a homotopy g_t, $0 \leq t \leq 1$, from $g_0 = g$ to $g_1 = g_{\text{ell}}$ with finitely many bifurcations, all of generic type (this might be an additional hypothesis) and such that

$$\frac{1}{a}|g_0| \leq |g_t| \leq a\,|g_0|$$

for all t with an a priori bound $a \geq 1$. The decisive fact now is that there can be no bifurcations at the eigenvalue -1 of the (linear) Poincaré map of geodesics of a priori bounded E-value for a such a homotopy g_t, $0 \leq t \leq 1$. Thus, as was observed already by Poincaré, the difference between the elliptic and the hyperbolic closed geodesics lying on the cycle Λf in the Morse complex of (M, g_t), t not a bifurcation point, is constant. Since it is equal to 1 for $t = 1$, and negative for $t = 0$, we get a contradiction

The existence of a partially elliptic closed geodesic on a manifold M of elliptic type would imply that generically for such a manifold the closure of the periodic orbits of the geodesic flow on TM has positive $(2q + 2)$-dimensional Hausdorff measure, $q > 0$. It is well known that for general Hamiltonian systems generically the periodic orbits are dense—this is a consequence of Pugh's Closing Lemma. Whether this is true also for the very restricted class of Hamiltonian systems formed by the geodesic on riemannian manifolds of elliptic type we don't feel able to judge. Quite possibly our above result for surfaces of the type of S^2 can be interpreted as an indication that it is true at least for surfaces.

Note. For proofs, further results, references and other background material we refer to the forthcoming "Lectures on Closed Geodesics". Grundlehren der Mathematischen Wissenschaften, vol. 230. Springer-Verlag, Berlin-Heidelberg-New York, 1978.

Mathematisches Institut
der Universität Bonn
Wegelerstrasse 10
D 5300 Bonn
Germany

Minimal Submanifolds and Geodesics
Kaigai Publications, Tokyo, 1978, 85–92

PROJECTIVE STRUCTURES OF HYPERBOLIC TYPE

SHOSHICHI KOBAYASHI*)

1. Introduction

In my recent paper [5], I associated a projectively invariant pseudo-distance d_M to every projectively flat manifold M. In this paper, I shall extend the construction to a manifold M with a (normal) projective connection and consider the problem of finding simple geometric conditions on M for d_M to be a bona fide distance. The main result of the paper is the following theorem.

Let M be a (complete) Riemannian manifold with Riemannian metric ds^2_M and the Ricci tensor Ric_M. *If* Ric_M *is negative and bounded away from zero, i.e.,* $\mathrm{Ric}_M \leqq -c^2 ds^2_M$ *with* $c > 0$, *then the pseudodistance d_M associated with the projective structure defined by ds^2_M is a (complete) distance.*

In my talk at the U. S.-Japan Seminar, I stated this as a conjecture. The conjecture became more plausible after Professor Yano had taught me a result in his old paper (see (5.1) of the present paper). After the Seminar, I have given similar talks in Hokkaido, Osaka and Nagoya. But by the time I returned to Berkeley, I had obtained a complete proof of the theorem above. This paper is therefore an improved version of the actual lecture delivered at the Seminar.

2. Projective parameters for geodesics

Let $\Gamma = (\Gamma_j{}^i{}_k)$ and $\bar{\Gamma} = (\bar{\Gamma}_j{}^i{}_k)$ be torsionfree affine connections on M. They are said to be *projectively related* or *projectively equivalent* if there exists a 1-form $\psi = \Sigma \psi_i dx^i$ such that

$$(2.1) \qquad \bar{\Gamma}_j{}^i{}_k - \Gamma_j{}^i{}_k = \delta^i_j \psi_k + \delta^i_k \psi_j \,.$$

Two such connections define the same system of geodesics. More precisely, a geodesic $x(t) = (x^i(t))$ for Γ satisfies the differential equation

$$(2.2) \qquad \frac{d^2x^i}{dt^2} + \Sigma \Gamma_j{}^i{}_k \frac{dx^j}{dt} \frac{dx^k}{dt} = \sigma \frac{dx^i}{dt} \,,$$

*) Guggenheim Fellow, partially supported by NSF Grant MCS 76-01692.

where σ is a function of t. It is clear that a solution of (2.2) satisfies

$$\frac{d^2x^i}{dt^2} + \Sigma \bar{\Gamma}_j{}^i{}_k \frac{dx^j}{dt}\frac{dx^k}{dt} = \bar{\sigma}\frac{dx^i}{dt} \tag{2.3}$$

with $\bar{\sigma} = \sigma + 2\Sigma\psi_k(dx^k/dt)$ and hence is a geodesic for $\bar{\Gamma}$.

By choosing the parameter t suitably, we can always make $\sigma = 0$ in (2.2). Such a parameter t, called an *affine parameter* of the geodesic, is unique up to an affine change $t \to at + b$. If t is an affine parameter with respect to the connection Γ, it needs not be an affine parameter with respect to $\bar{\Gamma}$. In other words, a parameter t which makes $\sigma = 0$ may not satisfy $\bar{\sigma} = 0$. Thus, the concept of affine parameter is not a projective invariant.

We shall now explain the concept of projective parameter considered by J. H. C. Whitehead [6], Berwald [1], Haantjes [4], E. Cartan [2] and others. We shall follow the route taken by Whitehead, Berwlad and Haantjes. Although it is not as geometric as that of Cartan, it is shorter and does not require the theory of projective connections. The equivalence of the two approaches has been verified by Yano [7].

Let $x(t) = (x^i(t))$ be a geodesic with affine parameter t of a torsionfree affine connection Γ. Then a *projective parameter* p of this geodesic is defined as a solution of the following differential equation:

$$\{p, t\} = \frac{2}{n-1}\Sigma R_{jk}\frac{dx^j}{dt}\frac{dx^k}{dt}, \tag{2.4}$$

where $\{p, t\}$ is the Schwarzian derivative:

$$\{p, t\} = \frac{p'''}{p'} - \frac{3}{2}\left(\frac{p''}{p'}\right)^2 \tag{2.5}$$

with primes denoting derivatives with respect to t. A projective parameter p is unique up to a linear fractional transformation:

$$p \to (ap + b)/(cp + d) . \tag{2.6}$$

If $\bar{\Gamma}$ is another torsionfree affine connection projectively related to Γ, and if $\bar{x}(\bar{t})$ is a geodesic with affine parameter $\bar{t}$ of $\bar{\Gamma}$ representing the same geodesic as $x(t)$ except for its parametrization, then a projective parameter $\bar{p}$ defined by $(\bar{x}(\bar{t}), \bar{\Gamma})$ is related to p by $\bar{p} = (ap + b)/(cp + d)$. Hence, the concept of projective parameter is a projective invariant.

3. Intrinsic pseudo-distance d_M

Let M be a manifold with a torsionfree affine connection Γ. We recall the definition of the projectively invariant pseudo-distance d_M associated with Γ, (see [5]).

Let I be the open interval $\{u; -1 < u < 1\}$ with "Poincaré metric"

$$ds_I^2 = \frac{4du^2}{(1-u^2)^2}. \tag{3.1}$$

The Poincaré distance between two points $u_0, u_1 \in I$ is given by

$$\rho(u_0, u_1) = \left|\log \frac{1+u_1}{1-u_1}\,\frac{1-u_0}{1+u_0}\right| = |\log(1, -1; u_0, u_1)|, \tag{3.2}$$

where $(1, -1; u_0, u_1)$ denotes the cross-ratio between u_0, u_1 with respect to $1, -1$.

A mapping $f: I \to M$ is said to be *projective* if f describes a gedoesic in M and if the natural parameter u for I is a projective parameter for this geodesic.

We are now in a position to define d_M. Given two points $x, y \in M$, we choose a chain α of geodesic segments consisting of

(a) a sequence of points $x = x_0, x_1, \cdots, x_k = y$ in M,

(b) pairs of points $a_1, b_1, \cdots, a_k, b_k$ in I,

(c) projective maps $f_1, \cdots, f_k$ from I into M such that

$$f_i(a_i) = x_{i-1}, \quad f_i(b_i) = x_i \qquad \text{for } i = 1, \cdots, k.$$

The length $L(\alpha)$ of the chain α is defined to be

$$L(\alpha) = \Sigma_i \rho(a_i, b_i). \tag{3.3}$$

The pseudo-distance $d_M(x, y)$ is given by

$$d_M(x, y) = \inf L(\alpha), \tag{3.4}$$

where the infimum is taken over all chains α from x to y.

Clearly, d_M depends only on the projective equivalence class of Γ, not on Γ itself. The following is also immediate from the very construction of d_M.

(3.5) **Proposition.** (1) *If $f: I \to M$ is projective, then*

$$\rho(a, b) \geqq d_M(f(a), f(b)) \qquad \textit{for } a, b \in I;$$

(2) *If δ_M is any pseudo-distance on M with the property that*

$$\rho(a, b) \geqq \delta_M(f(a), f(b)) \qquad \textit{for } a, b \in I$$

and for all projective maps $f: I \to M$, *then*

$$\delta_M(x, y) \leqq d_M(x, y) \qquad \textit{for } x, y \in M \ .$$

We say that the projective structure, i.e., the projective equivalence class defined by Γ is of *hyperbolic type* if d_M is a distance. It is said to be of *complete* hyperbolic type if d_M is a complete distance.

4. An analogue of Schwarz' lemma

In this section we establish the following lemma.

(4.1) **Lemma.** *Let* M *be a Riemannian manifold with metric* $ds_M^2 = \Sigma g_{jk}dx^j dx^k$ *and Ricci tensor* R_{jk}. *Assume*

$$(R_{jk}) \leqq -c^2(g_{jk}) \qquad \textit{(as matrices)} \ .$$

Then every projective map $f: I \to M$ *satisfies*

$$f^*ds_M^2 \leqq \frac{n-1}{4c^2} ds_I^2 \ .$$

Proof. We consider the function

$$h = (f^*ds_M)/ds_I \tag{4.2}$$

in $I = \{-1 < u < 1\}$ and try to find its upper bound. For the geodesic $f(I)$ in M we use both the projective parameter u and the usual arc-length parameter s. Thus, s may be considered as a function of u, and vice-versa. Then, from (3.1) we obtain

$$h = \frac{1-u^2}{2} \frac{ds}{du} \ . \tag{4.3}$$

Assuming for the moment that h attains its maximum in the interior of the interval I, we calculate $(d/du)(\log h)$ and $(d^2/du^2)(\log h)$ at that point. Denoting ds/du, d^2s/du^2, d^3s/du^2 by s', s'', s''', we can write

$$\frac{d}{du}(\log h) = \frac{-2u}{1-u^2} + \frac{s''}{s'} \ , \tag{4.4}$$

$$\frac{d^2}{du^2}(\log h) = \frac{-2(1+u^2)}{(1-u^2)^2} + \{s, u\} + \frac{1}{2}\left(\frac{s''}{s'}\right)^2 \ . \tag{4.5}$$

At the maximum point of h, (4.4) vanishes and, hence, $s''/s' = 2u/(1-u^2)$.

Substitute this into (4.5). On the other hand, (4.5) is non-positive at the maximum point of h. Hence,

$$\frac{d^2}{du^2}(\log h) = \frac{-2}{(1-u^2)^2} + \{s, u\} \leqq 0 \tag{4.6}$$

at the maximum point of h.

Now we make use of the following general formula for Schwarzian derivatives of composite functions:

$$\{s, u\} = (\{s, t\} - \{u, t\})\left(\frac{dt}{du}\right)^2 . \tag{4.7}$$

Setting $s = t$, we obtain

$$\{s, u\} = -\{u, s\}\left(\frac{ds}{du}\right)^2 . \tag{4.8}$$

Substituting (4.8) into (4.6), we obtain

$$-\frac{2}{(1-u^2)^2} - \{u, s\}\left(\frac{ds}{du}\right)^2 \leqq 0 , \quad \text{(at the maximum point of } h) . \tag{4.9}$$

Hence,

$$(1-u^2)^2\left(\frac{ds}{du}\right)^2 \leqq -\frac{2}{\{u, s\}} , \quad \text{(at the maximum point of } h) . \tag{4.10}$$

According to (4.3), the left hand side of (4.10) is equal to $4h^2$. On the other hand, the right hand side of (4.10) is equal to

$$-(n-1)\Big/\left(\Sigma R_{jk}\frac{dx^j}{ds}\frac{dx^k}{ds}\right)$$

by (2.4). By our assumption on the Ricci tensor, this is bounded by $(n-1)/c^2$. Hence,

$$4h^2 \leqq (n-1)/c^2 . \tag{4.11}$$

This completes the proof when h attains its maximum in I.

In the general case, we take a positive number $r < 1$ and the interval $I_r = \{-r < u < r\}$ with Poincaré metric $ds^2_{I_r} = 4r^2du^2/(r^2 - u^2)^2$. Then the function

(4.12) $$h_r = (f^* ds_M)/ds_{I_r}$$

vanishes on the boundary of the interval I_r and hence takes its maximum in the interior of I_r. We apply the argument above to h_r and obtain

(4.13) $$4h_r^2 \leqq (n-1)/c^2 .$$

To obtain the lemma, we have only to let $r \to 1$. Q.E.D.

(4.14) **Corollary.** *Let M and ds_M^2 be as in the lemma above. Let M be the distance induced by ds_M^2. Then every projective map $f: I \to M$ satisfiis*

$$\rho(a, b) \geq \frac{2c}{\sqrt{n-1}} \delta_M(f(a), f(b)) \qquad for\ a, b \in I .$$

From (4.14) and (2) of (3.5) we obtain the main theorem:

(4.15) **Theorem.** *Let M be a (complete) Riemannian manifold with metric $ds_M^2 = \Sigma g_{jk} dx^j dx^k$ and Ricci tensor R_{jk} such that*

$$(R_{jk}) \leqq -c^2(g_{jk}) .$$

Then the projective structure defined by ds_M^2 is of (complete) hyperbolic type, i.e., d_M is a (complete) distance.

We mention one immediate application.

(4.16) **Corollary.** *Let M be a compact Riemannian manifold with negative definite Ricci tensor. Then the group of projective transformations of M is finite.*

Proof. Previously it has been known that the group is discrete, see Couty [3]. Since the group leaves the distance d_M invariant, it must be compact. Hence, it is finite. Q.E.D.

5. Projective parameters for Ricci parallel affine connections

This last section is essentially due to Yano [8]. Let M be a manifold with a torsionfree affine connection such that the symmetric part of the Ricci tensor $S_{jk} = \frac{1}{2}(R_{jk} + R_{kj})$ is parallel. Then the right hand side of the differential equation (2.4) is constant along the geodesic $x(t)$. It is therefore possible to solve the equation explicity and express a projective parameter p as an elementary function of t.

(5.1) **Proposition.** (i) *If $\{p, t\} = \frac{1}{2}k^2$ with $k > 0$, then*

$$p = b \tan \frac{kt + a}{2} + c \quad with \quad b \neq 0 ;$$

(ii) *If $\{p, t\} = -\frac{1}{2}k^2$ with $k > 0$, then*

$$p = \frac{b}{1 - ae^{kt}} + c \quad \textit{with} \quad a, b \neq 0 \ ;$$

(iii) *If* $\{p, t\} = 0$, *then*

$$p = \frac{b}{t + a} + c \ .$$

Proof. Set $f = p''/p'$ so that $\{p, t\} = f' - \frac{1}{2}f^2$. Then it is easy to find f by integration. Q.E.D.

In both (i) and (iii), as t moves in a certain interval, p moves from $-\infty$ to $+\infty$. This means that, in these cases, there is a (nonconstant) projective map from the entire line $\boldsymbol{R}$ into the geodesic $x(t)$ provided that M is geodesically complete. Hence,

(5.2) **Theorem.** *Let* M *be a manifold with a torsionfree complete affine connection such that the symmetric part of the Ricci tensor* $S_{jk} = \frac{1}{2}(R_{jk} + R_{kj})$ *is parallel. If* (S_{jk}) *is positive semi-definite, then the projectively invariant pseudo-distance* d_M *is trivial, i.e., identically zero.*

If (S_{jk}) is negative definite, we consider the Riemannian metric $\Sigma g_{jk}dx^j dx^k$ with $g_{jk} = -S_{jk}$. The given connection is the Levi-Civita connection for this metric. By (4.15), d_M is a complete distance in this case. This is consistent with the fact that in case (ii) of (5.1) the parameter p moves only in a proper subinterval

This raises the following question on the differential equation

$$(5.3) \qquad \{p, t\} = g(t) \ , \qquad -\infty < t < \infty \ .$$

Question. If $g(t) \geqq 0$, does p move from $-\infty$ to $+\infty$ at t moves in some interval?

In the following paper with T. Sasaki, an affirmative answer will be given and Theorem (5.2) will be extended to the case where M is a manifold with a torsionfree complete affine connection with positive semi-definite Ricci tensor.

Bibliography

[1] L. Berwald, *On the projective geometry of paths,* Ann. of Math. **37** (1936), 879–898.

[2] E. Cartan, Leçcons sur la théorie des espaces à connexion projective, Gauthier-Villars, Paris (1937).

[3] R. Couty, *Sur les transformations des variétés riemanniennes et kaehleriennes,* Ann. Inst. Fourier, Grenoble **9** (1959), 147–248; see also: Transformations infinitésimales projectives, C. R. Acad. Sci. Paris **247** (1958), 804–806.

[4] J. Haantjes, *On the projective geometry of paths,* Proc. Edinburgh Math. Soc. **5** (1937), 103–115.

[5] S. Kobayashi, *Intrinsic distances associated with flat affine or projective structures*, J. Fac. Sci. Univ. of Tokyo IA **24** (1977), 129–135.
[6] J. H. C. Whitehead, *The representation of projective spaces*, Ann. of Math. **32** (1931), 327–360.
[7] K. Yano, *Sur les équations des géodésiques dans une variété à connexion projective*, C. R. Acad. Sci. Paris 205 (1937), 829–831.
[8] ——, *Sur la théorie des espaces à connexion conforme*, J. Fac. Sci. Imp. Univ. of Tokyo, Sec. I **4** (1939), 1–59.

UNIVERSITY OF CALIFORNIA,
BERKELEY, CA 94720
U.S.A.

Minimal Submanifolds and Geodesics
Kaigai Publications, Tokyo, 1978, 93–99

PROJECTIVE STRUCTURES WITH TRIVIAL INTRINSIC PSEUDO-DISTANCE

SHOSHICHI KOBAYASHI* & TAKESHI SASAKI

1. Introduction

In the preceding paper [3] it was shown that if M is a (complete) Riemannian manifold with metric ds_M^2 and Ricci tensor Ric_M such that $\mathrm{Ric}_M \leqq -c^2 ds_M^2$ (with $c > 0$), then the projectively invariant pseudo-distance d_M is a true (complete) distance. This paper is concerned with the pseudo-distance of an affinely connected manifold M with positive semi-definite Ricci tensor. The Ricci tensor $\mathrm{Ric}_M = (R_{ij})$ of an affine connection need not be symmetric. So we say that the Ricci tensor is positive semi-definite if its symmetric part $S_{ij} = \frac{1}{2}(R_{ij} + R_{ji})$ is positive semi-definite. Then our main result can be stated as follows.

Theorem. *Let M be a manifold with a torsionfree complete affine connection. If the Ricci tensor is positive semi-definite, then the projectively invariant pseudo-distance d_M is trival, i.e., $d_M = 0$.*

The theorem above answers completely the question raised at the end of the preceding paper [3] in which the theorem was verified when (S_{ij}) is parallel and positive semi-definite. In the proof given below, the case where $S_{ij} = 0$ will be excluded since it falls into the special case considered in [3].

For unexplained definitions and notation such as d_M in this paper, we refer the reader to [3].

2. Proof of Theorem

The following is clear from the definition of the projectively invariant pseudo-distance d_M.

Lemma 1. *Let M be a manifold with a torsionfree affine connection. Let $x_0, x_1 \in M$. If there exists a geodesic $x(u)$ with projective parameter u, $-\infty < u < \infty$, such that $x_0 = x(u_0)$ and $x_1 = x(u_1)$ for some $u_0, u_1 \in$ **R**, then*

$$d_M(x_0, x_1) = 0 .$$

*) Guggenheim Fellow, partially supported by NSF Grant MCS 76-01692.

Remark. The essential point in the assumption of Lemma 1 is that u is not an affine parameter but is a projective parameter and that $x(u)$ is defined for all real values of u, $-\infty < u < \infty$.

Lemma 2. *Let M be as in Lemma 1, and $x(t)$, $-\infty < t < \infty$, a geodesic with affine parameter t. Assume that there exists a (finite or infinite) sequence of open intervals $I_i = (a_i, b_i)$, $i = 0, \pm 1, \pm 2, \cdots$, such that*

(i) $a_{i+1} \leqq b_i$, $\lim_{i \to -\infty} a_i = -\infty$, $\lim_{i \to \infty} b_i = \infty$ *(so that* $\bigcup_i \bar{I}_i = (-\infty, \infty)$)

(ii) *in each interval $I_i = (a_i, b_i)$, a projective parameter u moves from $-\infty$ to ∞ as t moves from a_i to b_i.*

Then, for any pair of points x_0, x_1 on thisg eodesic, we have

$$d_M(x_0, x_1) = 0 .$$

Proof. We should remark that we have not excluded the possibility of $a_h = -\infty$ for some h and $b_k = \infty$ for some k.

It follows from Lemma 1 that the distance between any two points in the same interval I_i is zero. Two consecutive intervals I_i and I_{i+1} have either a point in common (if $a_{i+1} < b_i$) or a boundary point in common (if $a_{i+1} = b_i$). In either case, given $\varepsilon > 0$ there exist points $x_i \in I_i$ and $x_{i+1} \in I_{i+1}$ such that $d_M(x_i, x_{i+1}) < \varepsilon$. Hence, the distance between any two points on this geodesic is zero. Q.E.D.

Suppose that we are given a geodesic $x(t) = (x^i(t))$ with affine parameter t. Then a projective parameter u is a solution of the differential equation (see [3]):

$$\{u, t\} = 2Q(t) , \tag{1}$$

where $\{u, t\}$ denotes the Schwarzian derivative and

$$Q(t) = \frac{1}{n-1} \sum R_{jk} \frac{dx^j}{dt} \frac{dx^k}{dt} . \tag{2}$$

Lemma 3. *If y_1, y_2 are linearly independent solutions of the differential equation*

$$y''(t) + Q(t)y(t) = 0 , \tag{3}$$

then the general solution of (1) *is given by*

$$u(t) = (\alpha y_1 + \beta y_2)/(\gamma y_1 + \delta y_2) \quad \textit{with} \quad \alpha\delta - \beta\gamma \neq 0 . \tag{4}$$

Proof. This is classical (Hille [2; p. 648]). We remark first that a solution of (1) is unique up to a linear fractional transformation

$$u \to (\alpha u + \beta)/(\gamma u + \delta)\,. \tag{5}$$

This follows from the following identity on the Schwarzian derivative of the composite function:

$$\{u, v\} = \left(\frac{dt}{du}\right)^2 [\{v, t\} - \{u, t\}]\,. \tag{6}$$

This identity shows that if both u and v are solutions of (1), then $\{v, u\} = 0$, and hence v is obtained from u by a linear fractional transformation of the form (5).

Because of the remark just made, it suffices to show that $u = y_1/y_2$ is a solution of (1). Since the coefficient of $y'(t)$ in (3) is zero, the Wronskian $W(y_1, y_2)$ is constant. We may assume that $W(y_1, y_2) = 1$. Then

$$u' = 1/y_2^2 \tag{7}$$

so that

$$u''/u' = -2y_2'/y_2 \tag{8}$$

and

$$\left(\frac{u''}{u'}\right) = -2\left(\frac{y_2''}{y_2}\right) + 2\left(\frac{y_2'}{y_2}\right)^2 = 2Q + \frac{1}{2}\left(\frac{u''}{u'}\right)^2. \tag{9}$$

completing the proof of Lemma 3. Q.E.D.

Lemma 4. *If $Q(t) \geqq 0$ for $-\infty < t < \infty$, then every solution $y(t)$ of the differential equation* (3) *has at least one zero unless $Q(t) = 0$ and $y(t) = c \neq 0$.*

Proof. Assume that $y(t)$ is positive everywhere. Then

$$y''(t) = -Q(t)y(t) \leqq 0 \tag{10}$$

and $y(t)$ is concave downward everywhere. If $y(t)$ is not constant, $y'(a) \neq 0$ for some a. Since the graph of $y(t)$ lies below its tangent line at $t = a$, $y(t)$ has at least one zero. This is a contradiction. If $y(t) < 0$ everywhere, consider $-y(t)$. Q.E.D.

Lemma 5. *Let $y_1(t), y_2(t)$ be two linearly independent solutions of the differential equation* (3). *Then between any two consecutive zeros of $y_1(t)$, there is exactly one zero of $y_2(t)$. In other words, the zeros of $y_1(t)$ separate and are separated by the zeros of $y_2(t)$.*

This is Sturm's separation theorem (see, for example [1; p. 374]).

To complete the proof of the theorem, we have only to produce a sequence of intervals I_i satisfying the assumption of Lemma 2.

Lemma 6. *Let $y_1(t), y_2(t)$ be two linearly independent solutions of* (3). *If $t = a, b$ are two consecutive zeros of $y_2(t)$, then $u = y_1(t)/y_2(t)$ or $u = -y_1(t)/y_2(t)$ is a projective parameter in the interval (a, b) which moves from $-\infty$ to ∞ as t moves from a to b.*

Proof. This follows from Lemma 3 and Lemma 5. Q.E.D.

We can now complete the proof of the theorem when the differential equation (3) is *oscillatory* at $t = \pm\infty$. That is, if the zeros

$$\cdots < a_{-2} < a_{-1} < a_0 < a_1 < a_2 < \cdots$$

of $y_2(t)$ have the property that $\lim_{h\to-\infty} a_h = -\infty$ and $\lim_{k\to\infty} a_k = \infty$, then the sequence of intervals $I_i = (a_i, a_{i+1})$ satisfies the condition of Lemma 2.

We shall now consider the case when the differential equation (3) is nonoscillatory at $t = \infty$, i.e., the case when $y_2(t)$ does not vanish for sufficiently large t. (Because of Lemma 5, this condition is independent of the choice of a particular solution $y_2(t)$).

Lemma7. *If the differential equation* (3) *is nonoscillatory at $t = \infty$, then there is a solution $y_2(t)$ which is uniquely determined up to a constant factor by the condition that*

$$\lim_{t\to\infty} y_2(t)/y_1(t) = 0$$

for any solution $y_1(t)$ linearly independent of $y_2(t)$.

Proof. The following argument is taken from [1 ; p. 356]. Let $y(t)$ and $z(t)$ be linearly independent solutions of (3) so that $W(y, z) = yz' - y'z = c \neq 0$, or equivalently, $(z/y)' = c/y^2 \neq 0$ (wherever $y(t) \neq 0$). Hence z/y is monotone after t exceeds the largest zero of y and so $C = \lim_{t\to\infty} z(t)/y(t)$ exists if $C = \pm\infty$ is allowed. We shall show that y and z can be chosen so that $C = 0$.

If $C = \pm\infty$, interchange y and z. If C is a finite number, replace z by $z - Cy$.

Now we set $y_2 = z$ so that $\lim_{t\to\infty} y_2(t)/y(t) = 0$. If $y_1(t)$ is any solution of (3) linearly independent of $y_2(t)$, then $y_1 = \alpha y + \beta z$ with $\alpha \neq 0$. It follows that $\lim_{t\to\infty} y_2(t)/y_1(t) = 0$.

To prove the uniqueness, let $y_2^*(t)$ be another solution satisfying the requirement. If $y_2^*(t)$ is linearly independent of $y_2(t)$, then

$$\lim_{t\to\infty} y_2(t)/y_2^*(t) = 0 \quad \text{and} \quad \lim_{t\to\infty} y_2^*(t)/y_2(t) = 0 .$$

This is a contradiction. Q.E.D.

A solution $y_2(t)$ in Lemma 7 is called a *principal* solution.

Lemma 8. *Assume that the differential equation* (3) *is nonoscillatory* *at $t = \infty$ and that $Q(t) \geqq 0$. Let $y_2(t)$ be a principal solution as in Lemma 7. If a is the largest zero of $y_2(t)$ and if $y_1(t)$ is a solution linearly independent of $y_2(t)$, then $y_1(t)$ vanishes at some $t > a$.*

Proof. Assume the contrary. Considering $-y_1$ if necessary, we may assume that $y_1(t) > 0$ for $t \geqq a$. To derive a contradiction, we make use of

Second comparison theorem of sturm. *Consider two differential equations:*

(i) $y''(t) + Q_1(t)y(t) = 0$, (ii) $y''(t) + Q_2(t)y(t) = 0$

with $Q_1(t) \geqq Q_2(t)$. Let $y_1(t)$ and $y_2(t)$ be solutions of (i) *and* (ii) *respectively such that*

$$\frac{y_1'(a)}{y_1(a)} \leqq \frac{y_2'(a)}{y_2(a)} . \tag{11}$$

If $y_1(t)$ and $y_2(t)$ have no zeros in the interval $a < t < \infty$, then

$$\frac{y_1'(t)}{y_1(t)} \leqq \frac{y_2'(t)}{y_2(t)} \qquad \text{for } t > a . \tag{12}$$

This statement is weaker than the one found in standard text books but suffices for our purpose. If $y_2(a) = 0$, the term $y_2'(a)/y_2(a)$ is considered to be ∞.

Now we apply the comparison theorem to the case where $Q(t) = Q_1(t) = Q_2(t)$ and $y_1(t), y_2(t)$ are as in the statement of Lemma 8. Let $b > a$. Multiplying $y_1(t)$ by a suitable (positive) constant, we may assume that

$$y_1(b) = y_2(b) . \tag{13}$$

If we set

$$Y_i(t) = \int_b^t \frac{y_i'(r)}{y_i(r)} dr = \log y_i(t) - \log y_i(a) , \qquad i = 1, 2 , \tag{14}$$

then (12) and (13) imply

$$y_1(t) = y_1(b)e^{Y_1(t)} \leqq y_2(b)e^{Y_2(t)} = y_2(t) \qquad \text{for } t \geqq b . \tag{15}$$

This contradicts the property of $y_2(t)$ expressed in Lemma 7. Q.E.D.

We are now in a position to complete the proof of the theorem when the differential equation (3) is nonoscillatory at $t = \infty$ and or at $t = -\infty$.

If (3) is nonoscillatory at $t = \infty$ but oscillatory at $t = -\infty$, we take a principal solution $y_2(t)$ and another solution $y_1(t)$ linearly independent of $y_2(t)$. Let $\cdots < a_{-2} < a_{-1} < a_0 < a_1 < \cdots < a_k$ be the zeros of $y_2(t)$.

Then the sequence of intervals $I_i = (a_i, a_{i+1})$, $i = \cdots -2, -1, 0, 1, \cdots, k$ (with $a_{k+1} = \infty$), equipped with a projective parameter $u = y_1/y_2$ or $u = -y_2/y_2$ satisfy the requirements of Lemma 2. (Lemma 8 implies that u is a projective parameter in the last interval $I_k = (a_k, \infty)$.)

If (3) is nonoscillatory at $t = -\infty$ but oscillatory at $t = \infty$, we replace Lemmas 7 and 8 by analogous lemmas for $t = -\infty$. (They can be proved directly in a similar fashion or can be derived from Lemmas 7 and 8 by the transformation $t \to -t$.) Then we obtain a sequence of intervals I_i, $i = h, h+1, \cdots$, satisfying the requirements of Lemma 2.

Assume that (3) is nonoscillatory at $t = \pm\infty$. Let $y_2(t)$ be a principal solution for $t = \infty$. If it is not a principal solution for $t = -\infty$, let $y_1(t)$ be a principal solution for $t = -\infty$. Then $y_1(t)$ and $y_2(t)$ are linearly independent. As in Fig. 2, we obtain a sequence of intervals I_i, $i = 0, \cdots, k$, with a projective parameter $u = y_1/y_2$, $-y_1/y_2$, y_2/y_1 or $-y_2/y_1$ satisfying the requirements of Lemma 2. In this case, there are some overlaps among these intervals.

If $y_2(t)$ happens to be a principal solution for both $t = \infty$ and $t = -\infty$, then let $y_1(t)$ be any solution linearly independent of $y_2(t)$. As in Fig. 3,

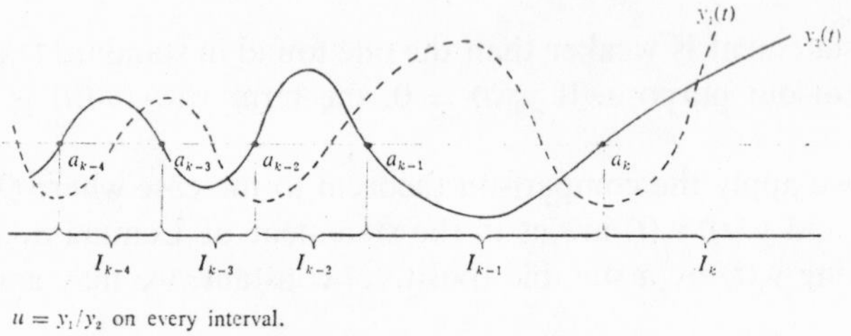

$u = y_1/y_2$ on every interval.

Fig. 1. (Nonoscillatory at $t = \infty$)
y_2: principal solution

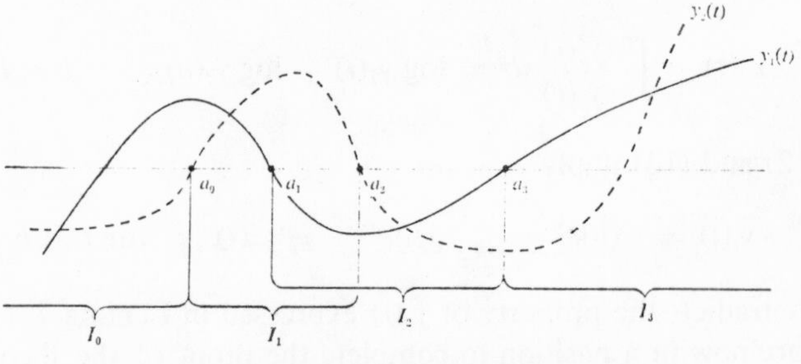

$u = -y_2/y_1$ in I_0, I_1; $u = y_1/y_2$ in I_2, I_3.

Fig. 2. (Nonoscillatory at $t = \pm\infty$)
y_1: principal solution for $t = -\infty$,
y_2: principal solution for $t = \infty$.

we obtain a sequence of intervals I_i, $i = 0, 1, \cdots, k$, with a projective parameter $u = y_1/y_2$ or $-y_1/y_2$ satisfying the requirements of Lemma 2. In this case, there are no overlaps of intervals. This completes the proof of theorem.

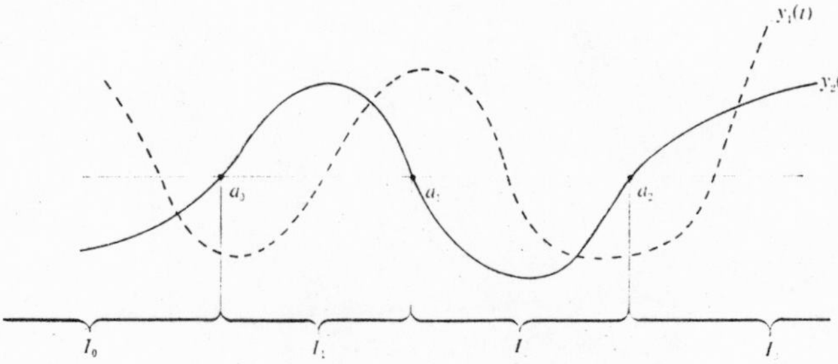

$u = y_1/y_2$ in every interval.

Fig. 3. (Nonoscillatory at $t = \pm\infty$)
y_2: principal solution for $t = \pm\infty$.

Bibliography

[1] P. Hartman, Ordinary Differential Equations, 1973, Hartman, Baltimore.
[2] E. Hille, Lectures on Ordinary Differential Equations, 1969, Addison-Wesley.
[3] S. Kaboyashi, *Projective structures of hyperbolic type,* this proceedings.

UNIVERSITY OF CALIFORNIA
BERKELEY, CA 94720
U.S.A.
AND
NAGOYA UNIVERSITY
NAGOYA, 464
JAPAN

Minimal Submanifolds and Geodesics
Kaigai Publications, Tokyo, 1978, 101–102

THE CLASSICAL PLATEAU PROBLEM AND THE TOPOLOGY OF 3-MANIFOLDS

WILLIAM H. MEEKS III[1] & SHING-TUNG YAU[2]

It was an old problem whether there is a disk with least area bounding a given closed curve in $\boldsymbol{R}^3$. This was solved about forty years ago by Douglas and Rado in a somewhat generalized sense. Only in 1969 Osserman proved that the solution of Douglas has no branch points in the interior. Later Gulliver even proved that there is no false branch point. Hence, the Douglas solution is an immersion of the disk.

It was a general question whether the solution is actually embedded when γ is a Jordan curve on the boundary of a convex region. In this note, we announce a solution of this problem in the more general case when γ lies on the convex boundary of a three-dimensional manifold. This provides an interesting proof of Dehn's lemma and the sphere theorem in 3-manifold theory. Some other new results in 3-manifold theory can also be proved with our approach.

For simplicity, all curves are C^3 and all manifolds are smooth. If γ is a Jordan curve in a Riemannian 3-manifold M^3 then we will call a conformal mapping $f: D^2 \to M^3$ a Douglas-Morrey solution to Plateau's problem if f has least energy with respect to all piecewise smooth mappings of the disk D^2 into M^3 such that $f|\partial D^2$ is a monotonic parameterization of γ.

Theorem 1. *If M^3 is a compact Riemannian 3-manifold with convex boundary and γ is a Jordan curve on the boundary which contracts to a point in M^3, then there exists a Douglas-Morrey solution to Plateau's problem for γ.*

Remark. The compactness of M^3 can be replaced by a Morrey type condition that M^3 is homogeneous regular. The above theorem is a consequence of Morrey's solution of Plateau's problem.

Theorem 2. *If M^3 is a Riemannian 3-manifold with convex boundary, γ is a Jordan curve on the boundary, and $f: D^2 \to M^3$ is a Douglas-Morrey solution to Plateau's problem then f is an embedding.*

It should be noted that the fact that the above f is an immersion was proved by R. Ossermann [3] for curves in $\boldsymbol{R}^3$ and by R. Gulliver [1] for

[1] This research was supported in part by NSF Grant No. MCS76-07147.
[2] This author was supported by the Sloan Fellowship.

curves in general 3-manifolds. Since every 3-manifold admits a nice Riemannian metric with convex boundary, the above theorems yield:

Corollary (Dehn's Lemma). *Let M^3 be a 3-manifold with boundary and let γ be any Jordan curve on the boundary. If γ is homotopically trivial, then γ bounds an embedded disk.*

Since Morrey also deals with the case of mapping from a general plane domain into M^3, the statements of the above theorems can be generalized.

Theorem 3. *Let M^3 be a compact Riemannian 3-manifold with convex boundary which may be empty. Then*

1) *There exists a $\pi_1(M^3)$-generating set for $\pi_2(M^3)$ consisting of immersions $f_i : S^2 \to M^3$ which are of least area in their homotopy class, which are globally 1–1 or 2–1, and which are pairwise disjoint.*

2) *There is an immersion $f : S^2 \to M^3$ which represents a nontrivial homotopy class of least area over all nontrivial elements in $\pi_2(M^3)$. Furthermore, the image of any such mapping is an embedded S^2 or $\boldsymbol{P}^2$. Pairwise, the images of these minimal spheres are disjoint.*

To prove the above theorem we need to use the results of *K*. Uhlenbeck and J. Sachs in [4].

R. Gulliver and J. Spruck [2] were the first ones who gave a positive result in the direction of Theorem 2. They assumed that the curve γ lies on a convex surface in $\boldsymbol{R}^3$ and the absolute total curvature of γ is not greater than 4π. We have learned that F. Almgren and L. Simon proved Theorem 2 in a special case. They proved there exists one embedded minimal disk bounding γ (with least area among embedded disks) when γ lies on the boundary of a convex body in $\boldsymbol{R}^3$. We also learned that F. Tomi and T. Tromba proved a similar result to F. Almgren and L. Simon.

References

[1] R. Gulliver, *Regularity of minimizing surfaces of prescribed mean curvature*, Ann. of Math. **97** (1973), 275–305.

[2] R. Gulliver and J. Spruck, *On embedded minimal surfaces*, Ann. of Math. **103** (1976), 331–347.

[3] R. Osserman, *A proof of regularity everywhere of the classical solution to Plateau's problem*, Ann. of Math. **91** (1970), 550–569.

[4] J. Sachs and K. Uhlenbeck, *Minimal immersions of spheres in Riemannian manifolds*, preprint.

University of California
Los Angeles, CA 90024
U.S.A.
University of California
Berkeley, CA 94720
U.S.A.

Minimal Submanifolds and Geodesics
Kaigai Publications, Tokyo, 1978, 103–115

REMARKS ON THE STABILITY OF A MINIMAL SURFACE IN S^n

HIROSHI MORI

1. Introduction

In their work [1] Barbosa and do Carmo studied the size of a stable minimal surface in the 3-dimensional Euclidean space $\boldsymbol{R}^3$. In earlier work [11] the author studied the stability of a minimal surface in the 3-dimensional unit sphere S^3.

In this note we study the stability of a minimal surface in the n-dimensional unit sphere S^n which generalizes our earlier work under some additional assumptions.

The author wishes to thank Professors T. Otsuki and H. Kitahara for their kind advice and constant encouragement.

2. Statement of result

Let M be a 2-dimensional connected orientable C^∞-Riemannian manifold. In this paper, a domain $D \subset M$ is an open connected subset with compact closure $\bar{D}$ and such that the boundary ∂D is a finite union of simple closed piecewise C^∞-curves in M. Let $f: M \to S^n$ be an isometric minimal C^∞-immersion of M into the n-dimensional unit sphere S^n. We denote by K the Gaussian curvature of M, by $\mathscr{L}^2$ the canonical measure on M associated with the Riemannian metric of M and by $B_r(x)$ the open geodesic ball in M with center x and radius r. For any open set U of M we define area U as $\mathscr{L}^2(U)$.

The purpose of this paper is to estimate the "size" of a stable minimal immersion and the result is as follows (see [10] or [14] for the definition of a stable minimal immersion).

Theorem. *Let $f: M \to S^n$ be a minimal immersion of M into S^n, D a domain in M and u an eigenfunction of the Laplacian Δ of M corresponding to the first eigenvalue $\lambda_1(D)$ of D. Assume that $\delta \leqq K \leqq 1$ for some constant δ, $0 < \delta \leqq 1$, and that all the critical points of u in $\bar{D}$ are non-degenerate. Then there exists a constant ρ in $(0, \pi/2)$ depending on δ and D such that if* area $D <$ area $B_\rho(m)$ *for some point m in M, D is stable.*

Remark 1. The dependence of the above constant ρ on D is in the sense of non-degeneracy of the eigenfunction u in $\bar{D}$.

Remark 2. We note (see [6]) that there are many minimal immersions which satisfy the conditions of the theorem, but are not totally geodesic. If M is a 2-dimensional sphere in $\boldsymbol{R}^3$ with constant Gaussian curvature δ, $0 < \delta \leqq 1$, then a sharp result is obtained.

3. Preliminaries

Let M be a 2-dimensional connected orientable C^∞-Riemannian manifold. Let δ be a constant with $0 < \delta \leqq 1$, and assume that the Gaussian curvature K of M satisfies $\delta \leqq K \leqq 1$. We denote by M_x the tangent plane of M at a point x.

We now prove some lemmas which will be used later.

Lemma 1. *Let M be as above and G be a domain in M whose boundary ∂G is a finite union of simple closed C^∞-curves in M, and assume that* area G = area $B_\sigma(m)$ *for some σ in $(0, \pi/2)$ and some point m in M. Let $h_0 > 0$ be a sufficiently small constant. Then there exists a constant $\mu_0 > 0$ depending on δ, h_0, σ and the length of ∂G such that* area $T_h(G) \geqq$ area $B_{\sigma+h\mu_0}(m)$ *for h, $0 \leqq h \leqq h_0$, where $T_h(G) = \bigcup \{B_h(x)\,;\, x \in G\}$.*

Proof. We first note (see [9]) that M is compact. We consider only the case where the boundary ∂G is a simple closed C^∞-curve in M, the other case is similar. Let $c\colon [0, l] \to M$ be a simple closed C^∞-curve parametrized by arc length which satisfies that $c(t) \in \partial G$ for all t in $[0, l]$ and l = the length of ∂G. Since M is orientable and ∂G is C^∞ we can define ξ as the outward unit normal C^∞ vector field to G on ∂G. For $\varepsilon > 0$ we define the C^∞ map ψ from the rectangle $[0, l] \times [0, \varepsilon]$ into M as $\psi(t, s) = \exp_{c(t)} s\xi(t)$, where $\exp_n$ is the exponential map of M at a point n and $\xi(t) := \xi(c(t))$ for all t in $[0, l]$. We shall see that ψ is a one to one mapping on $[0, l] \times [0, \varepsilon]$ for sufficiently small $\varepsilon > 0$. If we set $Y(t, s) = (\partial\psi/\partial t)(t, s)$, then we notice (see [3]) that for each fixed t in $[0, l]$ the C^∞ map $s \mapsto Y(t, s)$ is a (∂G)-Jacobi field along the geodesic $\psi_t(\cdot)\colon s \mapsto \psi(t, s)$. We denote by $\kappa(t)$ the curvature of the C^∞-curve c, that is, $\nabla_{\dot{c}(t)}\dot{c}(t) = \kappa(t)\xi(t)$ for each t in $[0, l]$, where ∇ is the Riemannian connection of M. Let τ_0^s be the paralell translation along the geodesic $\psi_t(\cdot)$ from $\psi_t(s)$ to $\psi_t(0) = c(t)$. We denote by the exterior multiplication in $M_{\psi(t,s)}$, considered as a real vector space (see [7]) and $|\ \ |$ the norm induced from the Riemannian metric of M and set

$$\phi(t, s) = \frac{1}{s}\left\{\tau_0^s\left(\frac{\partial\psi}{\partial t}(t, s) \wedge \frac{\partial\psi}{\partial s}(t, s)\right) - \frac{\partial\psi}{\partial t}(t, 0) \wedge \frac{\partial\psi}{\partial s}(t, 0)\right\}.$$

Then we can show that $\lim_{s\to+0} \phi(t, s) = -\kappa(t)\dot{c}(t) \wedge \xi(t)$, and set $\phi(t, 0) = \lim_{s\to+0} \phi(t, s)$. Since

$$\left\{\frac{\partial\psi}{\partial t}(t, 0) = \dot{c}(t), \frac{\partial\psi}{\partial s}(t, 0) = \xi(t)\right\}$$

is an orthonormal basis of the tangent plane $M_{c(t)}$, and

$$\frac{\partial\psi}{\partial t}(t, s) , \quad \frac{\partial\psi}{\partial s}(t, s)$$

and $\kappa(t)$ are C^∞ maps defined on compact sets, the number

$$\eta := \max \{|\phi(t, s)| \varepsilon ; (t, s) \in [0, l] \times [0, \varepsilon]\}$$

satisfies that η is an increasing function in ε and that $\eta \to 0$ as $\varepsilon \to 0$. So we see that for sufficiently small $h_0 > 0$ we have

$$\left|\frac{\partial\psi}{\partial t}(t,s) \wedge \frac{\partial\psi}{\partial s}(t,s)\right| \geqq 1 - \eta > 0$$

for all (t, s) in $[0, l] \times [0, h_0]$, and hence we get

$$\begin{aligned} \text{area } T_h(G) &= \text{area } G + \int_0^l \int_0^h \left|\frac{\partial\psi}{\partial t}(t,s) \wedge \frac{\partial\psi}{\partial s}(t,s)\right| dsdt \\ &\geqq \text{area } G + h(1 - \eta) \text{ length } (\partial G) , \qquad 0 \leqq h \leqq h_0 . \end{aligned} \tag{1}$$

On the other hand, from the conditions of M we notice (see [4]) that injectivity radius $i(M)$ of M satisfies $i(M) \geqq \pi$. Let $\{e_1, e_2\}$ be a positively orthonormal basis of M_m and $\Theta(r, \theta)$ be the Jacobian of the C^∞ map $\exp_m$ of $(0, \pi) \times [0, 2\pi]$ into M defined as $\exp_m \{r(\cos\theta e_1 + \sin\theta e_2)\}$, $0 < r < \pi$, $0 \leqq \theta \leqq 2\pi$. Then from the condition of the Gaussian curvature we can show (see [3]) that

$$\begin{aligned} &\sin r \leqq \Theta(r, \theta) \leqq (\sin\sqrt{\delta}\, r)/\sqrt{\delta}\,) \leqq r , \\ &0 < r < \pi, \ 0 \leqq \theta \leqq 2\pi . \end{aligned} \tag{2}$$

From this inequality we get

$$\begin{aligned} \text{area } B_{\sigma+\mu h}(m) &= \text{area } B_\sigma(m) + \int_\sigma^{\sigma+\mu h} \int_0^{2\pi} \Theta(r, \theta) d\theta dr \\ &\leqq \text{area } B_\sigma(m) + \frac{2\pi}{\sqrt{\delta}} \mu h \sin \frac{\sqrt{\delta}\,(2\sigma + \mu h)}{2} , \end{aligned} \tag{3}$$

for $\mu > 0$ and $0 \leqq h \leqq h_0$.

If we define μ_0 as

$$\mu_0 = \min\left\{1, \sqrt{\delta}(1-\eta)\ \text{length}\,(\partial G)/2\pi \sin \frac{\sqrt{\delta}(2\sigma + h_0)}{2}\right\}$$

for sufficiently small $h_0 > 0$, then from (1) and (3) we see that our assertion is true and μ_0 is decreasing in h_0.

Remark. From the above discussion we see that the assertion is true for any domain G in M whose boundary ∂G is not C^∞-curves, but piecewise C^∞-curves.

Lemma 2. *Let M be as in* Lemma 1 *and $h_0 > 0$ and $\sigma_0 > 0$ be sufficiently small numbers. Then there exists $\mu_0 > 0$ depending on δ, h_0 and σ_0 such that the following holds: for any domain G whose boundary ∂G is a finite union of simple closed C^∞-curves in M and which satisfies* area G $=$ area $B_\sigma(m)$, $0 < \sigma < \sigma_0$, $m \in G$, *we have* area $T_h(G) \geqq$ area $B_{\sigma + h\mu_0}(m)$, $0 \leqq h \leqq h_0$.

Proof. For small $\zeta > 0$, there exists a small constant $\sigma_0 > 0$ ($5\sigma_0 < \pi/2$) such that

$$\sin (1 + \zeta)t \geqq t \tag{4}$$

and

$$1 - \cos (1 + \zeta)t \geqq (1 + \zeta)(1 - \cos t) \tag{5}$$

for t, $0 \leqq t \leqq 5\sigma_0$.

Let G be a domain which satisfies the conditions of this lemma. We consider first the case where $G \subset B_{3\sigma_0}(m)$. Since $i(M) \geqq \pi > 5\sigma_0$ we can set $G^* = (\exp_m | B_\pi(m))^{-1}(G)$. Let $\bar{m}$ be a point of the unit sphere S^2 of dimension 2 and we identify the tangent plane M_m with the tangent plane $S^2_{\bar{m}}$ as 2-dimensional real vector space with inner product. Then from (2) we have

$$\text{area}\ T_h(G) \geqq \text{area}\ T_h\,(\exp_{\bar{m}}(G^*))\,, \qquad 0 \leqq h \leqq \sigma_0\,, \tag{6}$$

where $\exp_{\bar{m}}$ is the exponential map of S^2 at the point $\bar{m}$. We denote by $\tilde{B}_{\bar{\sigma}}(\bar{m})$ the open geodesic ball in S^2 with center $\bar{m}$ and radius $\bar{\sigma}$ and area $\tilde{B}_{\bar{\sigma}}(\bar{m}) =$ area $\exp_{\bar{m}}(G^*)$. Then we get (see [13]) that

$$\text{area}\ T_h\,(\exp_{\bar{m}}(G^*)) \geqq \text{area}\ \tilde{B}_{\bar{\sigma}+h}(\bar{m})\,, \qquad 0 \leqq h \leqq \sigma_0\,. \tag{7}$$

From the assumption on area G and (2), (4), (5) we get the following inequalities:

$$\text{area}\ B_h(m) \leqq \text{area}\ \tilde{B}_{(1+\zeta)h}(\bar{m})\,, \qquad 0 \leqq h \leqq \sigma_0\,, \tag{8}$$

and

(9) $$\text{area } B_\sigma(m) \leqq \text{area } \tilde{B}_{(1+\zeta)\bar\sigma}(\bar m) .$$

Combining (6), (7), (8) and (9) we have

(10) $$\begin{aligned}&\text{area } T_h(G) \geqq \text{area } \tilde{B}_{\bar\sigma+h}(\bar m) = 2\pi(1 - \cos(\bar\sigma + h)) ,\\ &0 \leqq h \leqq \sigma_0 ,\end{aligned}$$

and

(11) $$(1 + \zeta)^{-1}\sigma \leqq \bar\sigma \leqq (1 + \zeta)\sigma .$$

On the other hand, from the assumption on Gaussian curvature of M we have

(12) $$\text{area } B_{\sigma+\mu h}(m) \leqq \frac{2\pi}{\delta}(1 - \cos\sqrt{\delta}(\sigma + \mu h)) , \qquad 0 \leqq h \leqq \sigma_0 .$$

If we define μ_0 as $\mu_0 = \{\mathrm{Cos}^{-1}[1 - \delta(1 - \cos(\bar\sigma + h_0))] - \sqrt{\delta}\,\sigma\}/\sqrt{\delta}\,h_0$ for h_0, $0 < h_0 < \sigma_0$, then we see that μ_0 is decreasing in $h_0(>0)$, and from (10), (11) and (12) we get area $T_h(G) \geqq$ area $B_{\sigma+h\mu_0}(m)$ for h, $0 \leqq h \leqq h_0$.

Next, we consider the case where $G \not\subset B_{3\sigma_0}(m)$, $m \in G$. For sufficiently small $h_0 > 0$ we can show that $\mathscr{H}^1\{t \in [\sigma_0, 3\sigma_0]$; length $\{(T_h(G) - G) \cap B_t(m)\} \geqq 2h\} \geqq \sigma_0$, where $\mathscr{H}^1$ is 1-dimensional Hausdorff measure, $0 \leqq h \leqq h_0$. From this inequality we get area $T_h(G) \geqq$ area $G + 2h\sigma_0$, $0 \leqq h \leqq h_0$. If we define μ_0 as $\mu_0 = \min\{1, \{\mathrm{Cos}^{-1}[1 - (\delta(1 - \cos\sigma) + (\delta/\pi)h_0\sigma_0)] - \sqrt{\delta}\,\sigma\}/\sqrt{\delta}\,h_0\}$, then from (12) we see that our assertion is true. This completes the proof.

Hereafter, we assume furthermore that M is real analytic manifold, and let D be a domain in M. We denote by Δ the Laplacian of M and by $\lambda_1(D) > 0$ the first eigenvalue in D for Δ. Let $u \in C^\infty(\bar D)$ be an eigenfunction of Δ corresponding to $\lambda_1(D)$. It is well known (see [1]) that u is analytic in D with $u \neq 0$ in D and $u \equiv 0$ on ∂D.

Now, we can assume that $u > 0$ in D and define u as identically zero outside D. Let m be a point in M such that $u(m) = \max u$. For m given above we define the following quantities:

$$\begin{aligned}\alpha(x) &= \text{geodesic distance in } M \text{ from } m \text{ to } x ,\\ d &= \max\{\alpha(x) ; x \in M\} ,\\ \beta(t) &= \text{area } B_t(m) = \text{area } \overline{B_t(m)} , \qquad \text{for } 0 \leqq t \leqq d .\end{aligned}$$

We note that β: $[0, d] \to [0, \text{area } M]$ is a homeomorphism. Using Lemmas 1 and 2 we get the following

Lemma 3. *Let M, D and u be as above. Assume that* area $D =$

area $B_\rho(m)$, $0 < \rho \leqq \pi/2$, *and that L is the Lipschitz constant of u. Let* $h_0 > 0$ *and* $\sigma_0 > 0$ *be sufficiently small constants. Then there exists a number* $\mu_0 > 0$ *depending on* δ, h_0, σ_0 *and D, and a continuous function* u^* *on M with the following properties.*

1*) u^* *is a Lipschitz function with Lipschitz constant* L/μ_0.

2*) $u_\#(\mathscr{L}^2)|\mathscr{B}(\boldsymbol{R}) = u^*_\#(\mathscr{L}^2)|\mathscr{B}(\boldsymbol{R})$, *i.e., the induced measures over the Borel family of* ***R*** *coincide* (cf. [7]).

3*) *There exists a non-increasing continuous function g which maps the interval* $[0, d]$ *into* ***R*** *and satisfies* $u^*(x) = g(\alpha(x))$ *for all* $x \in M$.

Proof. We follow essentially the argument in [13]. For $\sigma \in \boldsymbol{R}$ we define the following two functions:

$$s(\sigma) = \beta^{-1}(\mathscr{L}^2\{u^{-1}([\sigma, \infty))\}) ,$$
$$r(\sigma) = \beta^{-1}(\mathscr{L}^2\{u^{-1}((\sigma, \infty))\}) .$$

Then both functions are monotone decreasing, and from the complete additivity of the measure $\mathscr{L}^2$ on M we get

$$\lim_{\tau \to \sigma - 0} s(\tau) = s(\sigma) ,$$
$$\lim_{\tau \to \sigma + 0} s(\tau) = r(\sigma) , \qquad \text{for each } \sigma \in \boldsymbol{R} .$$

We note that for each t in $[0, d]$ the set $\{\sigma \in \boldsymbol{R} ; s(\sigma) \geqq t\}$ contains always the minimum 0 of u. So we can define the function $g(t)$ as

$$g(t) = \sup \{\sigma \in [0, \max u] ; s(\sigma) \geqq t\} , \qquad \text{for } t \in [0, d] .$$

We see that g is a monotone decreasing function, and for $\tau := g(t)$ we get

$$s(\tau) \geqq t \geqq r(\tau) , \tag{13}$$

and the following holds

$$g^{-1}([\tau, \infty)) = [0, s(\tau)] , \tag{14}$$

and

$$g^{-1}(\tau) = [r(\tau), s(\tau)] , \qquad \text{for } \tau \text{ in } \boldsymbol{R} . \tag{15}$$

Now we define the function $u^* : M \to \boldsymbol{R}$ as

$$u^*(x) = g(\alpha(x)) , \qquad \text{for } x \in M .$$

We see that 2*) is true if we show that

$$\mathscr{L}^2\{u^{-1}([\sigma, \infty))\} = \mathscr{L}^2\{(u^*)^{-1}([\sigma, \infty))\} \qquad \text{for eac } \sigma \in \boldsymbol{R} .$$

At first we get

$$\begin{aligned}\max u^* &= g(0) = u^*(m) \\ &= \sup\{\sigma \in [0, \max u]\,;\, s(\sigma) \geqq 0\} \\ &= \sup\{\sigma \in [0, \max u]\,;\, \mathscr{L}^2\{u^{-1}([\sigma, \infty))\} \geqq 0\} \\ &= \max u ,\end{aligned}$$

and

$$\min u^* = 0 = \min u$$

by virtue of area (supp u) = area D = area $B_\rho(m)$, $0 < \rho \leqq \pi/2$. Next, for each $\sigma \in [0, \max u]$ we get

$$\begin{aligned}\mathscr{L}^2\{(u^*)^{-1}([\sigma, \infty))\} &= \mathscr{L}^2\{x \in M\,;\, \sigma \leqq g(\alpha(x))\} \\ &= \mathscr{L}^2\{x \in M\,;\, \alpha(x) \in g^{-1}([\sigma, \infty))\} \\ &= \mathscr{L}^2\{x \in M\,;\, \alpha(x) \leqq s(\sigma)\} \\ &= \beta(s(\sigma)) = \mathscr{L}^2\{u^{-1}([\sigma, \infty))\} .\end{aligned}$$

Finally, we will prove 1*). Using the theorem of Sard and the conditions of u and M we see that $u^{-1}(\sigma)$ is a compact, 1-dimensional C^∞-submanifold of M for almost all σ in $(0, \max u)$. And from the analyticity of u we see that for a sufficiently small constant $\sigma_0 > 0$ there exists ρ_0 in $(0, \max u)$ such that

$$\mathscr{L}^2\{u^{-1}((\rho_0, \max u])\} = \text{area } B_{\sigma_0}(m) . \tag{16}$$

At first, we will show that for sufficiently small $h_0 > 0$ there exists $\mu_1 > 0$ depending on δ, h_0, σ_0 and u such that

$$\begin{aligned}\text{area } T_h\{u^{-1}([\sigma, \infty))\} &\geqq \text{area } T_{h\mu_1}\{(u^*)^{-1}([\sigma, \infty))\} \\ &\equiv \text{area } B_{s(\sigma)+h\mu_1}(m)\end{aligned} \tag{17}$$

for $0 \leqq h \leqq h_0$ and $0 < \sigma \leqq \rho_0$.

Since the all critical points of u in $\bar{D}$ are nondegenerate, they are finite in number, which are named as $p_1, p_2, \cdots, p_k$. For $\varepsilon > 0$ we define D_ε as $D_\varepsilon = \bar{D} - \bigcup_{j=1}^k B_\varepsilon(p_j)$. We can easily show that for sufficiently small $\varepsilon > 0$ we have

$$u^{-1}([\sigma, \infty)) \cap D_\varepsilon \neq \varnothing , \qquad 0 < \sigma \leqq \rho_0 .$$

And we see that for each σ in $(0, \rho_0]$, $u^{-1}(\sigma) \cap D_\varepsilon$ consists of some simple closed C^∞-curves in M and some 1-dimensional compact C^∞-submanifolds of M with boundaries. So there exists a number $N > 0$ depending on $\varepsilon, \delta, \sigma_0$ and u such that

$$\max\{\text{curvature of } u^{-1}(\sigma) \cap D_\varepsilon\} =: \kappa(\varepsilon, \sigma) \leqq N$$

for all σ in $(0, \rho_0]$. Hence from the discussion of Lemma 1 we see that for sufficiently small $h_0 > 0$ there exists η in $(0, 1)$ such that

$$\begin{aligned}\text{area } T_h\{u^{-1}([\sigma, \infty)) \cap D_\varepsilon\} &- \text{area }\{u^{-1}([\sigma, \infty)) \cap D_\varepsilon\} \\ &\geqq h(1 - \eta) \text{ length } \partial\{u^{-1}([\sigma, \infty)) \cap D_\varepsilon\}\,,\end{aligned} \tag{18}$$

for $0 < \sigma \leqq \rho_0$ and $0 \leqq h \leqq h_0$.

If we define $\bar\mu$ as $(2\pi/\delta)(1 - \cos\sqrt{\delta}(\sigma + \bar\mu h_0)) = \text{area }\{u^{-1}([\sigma, \infty)) \cap D_\varepsilon\} + h_0(1 - \eta) \text{ length } \partial\{u^{-1}([\sigma, \infty)) \cap D_\varepsilon\}$, then we can show that $\bar\mu$ is decreasing in ε and h_0 for other arguments fixed. Thus we see that for $\mu_1 = \mu_1(\delta, h_0, \sigma_0, u) = \lim_{\varepsilon \to +0} \bar\mu(\delta, \varepsilon, h_0, \sigma_0, u)$, (17) is true.

Next, from Lemma 2 and (16) we see that for sufficiently small $h_0 > 0$ there exists $\mu_2 > 0$ depending on δ, h_0 and σ_0 such that

$$\begin{aligned}\text{area } T_h\{u^{-1}([\sigma, \infty))\} &\geqq \text{area } T_{h\mu_2}\{(u^*)^{-1}([\sigma, \infty))\} \\ &\equiv \text{area } B_{s(\sigma)+h\mu_2}(m)\end{aligned} \tag{19}$$

for $0 \leqq h \leqq h_0$ and for almost all σ in $(\rho_0, \max u)$.

Combining (18) and (19) we see that for sufficiently small $h_0 > 0$ and $\sigma_0 > 0$ we see that there exists $\mu_0 > 0$ depending on δ, h_0, σ_0 and u such that

$$\text{area } T_h\{u^{-1}([\sigma, \infty))\} \geqq \text{area } B_{s(\sigma)+h\mu_0}(m) \tag{20}$$

for $0 \leqq h \leqq h_0$ and for almost all σ in $(0, \max u)$, and μ_0 is decreasing in h_0.

We now define dist (x, y), $x, y \in M$, to be geodesic distance in M from x to y. Since M is, by assumption, complete as metric space and u^* is continuous, 1*) is proved if we show that $|u^*(x) - u^*(y)| \leqq L \operatorname{dist}(x, y)/\mu_0$ for any x, y which satisfy that both $u(x)$ and $u(y)$ are regular values and dist (x, y) is sufficiently small.

For any two regular values σ_1, σ_2 of u with $0 < \sigma_2 < \sigma_1 < \max u$, $h := (\sigma_1 - \sigma_2)/L \leqq h_0$, we define the set A as $A = u^{-1}([\sigma_1, \infty))$. For each $x \in T_h(A)$ there exists $z \in A$ such that dist $(x, z) < h$, and hence $|u(x) - u(z)| \leqq L \operatorname{dist}(x, z) < Lh$ or $u(x) > u(z) - Lh \geqq \sigma_1 - Lh = \sigma_2$. Therefore we have $x \in u^{-1}((\sigma_2, \infty))$ and hence $T_h(A) \subset u^{-1}((\sigma_2, \infty))$. From the definition of β, (2) and (20) we obtain

$$\begin{aligned}\beta(r(\sigma_2)) &= \mathscr{L}^2\{u^{-1}((\sigma_2, \infty))\} \\ &\geqq \mathscr{L}^2(T_h(A)) \\ &= \beta(s(\sigma_1) + h\mu_0) .\end{aligned}$$

Consequently, we have $r(\sigma_2) \geqq s(\sigma_1) + h\mu_0$. Taking t_1, t_2 with $0 \leqq t_1 \leqq t_2 \leqq d$, $g(t_1) = \sigma_1$, $g(t_2) = \sigma_2$ we have $t_2 - t_1 \geqq r(\sigma_2) - s(\sigma_1) \geqq h\mu_0 = \mu_0(\sigma_1 - \sigma_2)/L$ by virtue of (13). From this inequality we see that the assertion of 1*) is true. This completes the proof.

Remark. By the condition of $r(\tau)$ and $s(\tau)$ we have

$$r(0) = \beta^{-1}(\mathscr{L}^2(\operatorname{supp} u)) = \beta^{-1}(\mathscr{L}^2(B_\rho(m)) = \rho \quad \text{and} \quad s(0) = d .$$

Therefore from (15) we see that $u^*(x) = 0$ for $\alpha(x) \geqq \rho$, i.e., $\operatorname{supp} u^* \subset \overline{B_\rho(m)}$.

Lemma 4. *Let M, D, u and u^* be as in* Lemma 3. *Then there exists $\mu > 0$ depending only on δ and ρ such that*

$$\operatorname{length}\{u^{-1}(\sigma)\} \geqq \mu \operatorname{length}\{(u^*)^{-1}(\sigma)\}$$

for almost all σ in $(0, \max u)$.

Proof. Using Lemma 2 in [13] we can show this lemma, but we give an another proof here. From the theorem of Sard it is sufficient to show that for any domains Ω_j, $j = 1, \cdots, n$ in M whose boundaries $\partial\Omega_j$ are C^∞ and which satisfy $\sum_{j=1}^n \operatorname{area} \Omega_j = \operatorname{area} B_\sigma(n)$, $0 < \sigma \leqq \rho$, $n \in \Omega_j$ for some $j \in \{1, \cdots, n\}$, we have $\sum_{j=1}^n \operatorname{length} \partial\Omega_j \geqq \mu \operatorname{length} \partial B_\sigma(n)$.

First of all, we show that if $\sum_{j=1}^n \operatorname{area} B_{r_j}(n_j) \geqq \operatorname{area} B_r(n)$, $0 < r_j \leqq r \leqq \rho$, $j = 1, \cdots, n$, then we have

$$\sum_{j=1}^n \operatorname{length} \partial B_{r_j}(n_j) \geqq c_2^2 \operatorname{length} \partial B_r(n) , \tag{21}$$

where $c_2 := (\sqrt{\delta} \sin \rho)/\sin \sqrt{\delta}\,\rho$.

Since the function $f(t) = (\sqrt{\delta} \sin t)/\sin \sqrt{\delta}\, t$, if $0 < t < \pi$, and $= 1$, if $t = 0$, is decreasing in t, we have for given ρ

$$c_2 \leqq f(t) \leqq 1 , \qquad 0 \leqq t \leqq \rho . \tag{22}$$

From (2) and the assumptions of area $B_r(n)$ and area $B_{r_j}(n_j)$, $j = 1, \cdots, n$, we get

$$\begin{aligned}&\sum_{j=1}^n 2\pi \frac{1 - \cos \sqrt{\delta}\, r_j}{\delta} \geqq 2\pi(1 - \cos r) , \\ &\sum_{j=1}^n ((\sin \sqrt{\delta}\, r_j/2)/\sqrt{\delta}\,)^2 \geqq (\sin r/2)^2 , \\ &\sum_{j=1}^n (\sin \sqrt{\delta}\, r_j/2)/\sqrt{\delta} \geqq \sin r/2 , \\ &\sum_{j=1}^n (\sin \sqrt{\delta}\, r_j)/\sqrt{\delta} \geqq \sin r .\end{aligned}$$

From this inequality and (22) we get

$$c_2^{-1} \sum_{j=1}^{n} \sin r_j \geqq \sum_{j=1}^{n} (\sin \sqrt{\delta}\, r_j)/\sqrt{\delta} \geqq \sin r \geqq c_2 (\sin \sqrt{\delta}\, r)/\sqrt{\delta}\ ,$$
$$\sum_{j=1}^{n} 2\pi \sin r_j \geqq c_2^2 2\pi\, (\sin \sqrt{\delta}\, r)/\sqrt{\delta}\ .$$

From this inequality and (2) we have

$$\begin{aligned} \sum_{j=1}^{n} \text{length}\, \partial B_{r_j}(n_j) &\geqq 2\pi \sum_{j=1}^{n} \sin r_j \\ &\geqq c_2^2 2\pi (\sin \sqrt{\delta}\, r)/\sqrt{\delta} \\ &\geqq c_2^2\, \text{length}\, \partial B_r(n)\ . \end{aligned}$$

Thus we see that the assertion of (21) is true.

From (21) we see that the assertion of this lemma is true if we can show that there exists a constant $\mu_1 > 0$ depending only on δ and ρ such that for any domain Ω in M whose boundary $\partial\Omega$ is C^∞ and which satisfies area $\Omega =$ area $B_\sigma(n)$, $0 < \sigma \leqq \rho$, $n \in \Omega$, we have

$$\text{length}\, \partial\Omega \geqq \mu_1\, \text{length}\, \partial B_\sigma(n)\ .$$

We first consider the case where $\Omega \subset B_{b\rho}(n)$, where $b \geqq 1$ is a constant depending only on δ and ρ determined afterwards. Let $\bar{n}$ be a point in the 2-dimensional unit sphere S^2, and we denote by $\exp_{\bar{n}}$ the exponential map of S^2 at the point $\bar{n}$. Since the injectivity radius $i(M)$ of M satisfies $i(M) \geqq \pi$ we can define Ω^* as $\Omega^* = (\exp_n | B_\pi(n))^{-1}(\Omega)$ and Ω_1 as $\Omega_1 = \exp_{\bar{n}} \Omega^*$. Let $\bar{\sigma} > 0$ be a unique number such that area $\Omega_1 =$ area $\tilde{B}_{\bar{\sigma}}(\bar{n})$, where $\tilde{B}_{\bar{\sigma}}(\bar{n})$ is the open geodesic ball in S^2. From the fact that the function $f(t)$ is decreasing in t, $0 \leqq t \leqq \pi/2$ and $a(1 - \cos t) \leqq 1 - \cos at$ for $0 \leqq t \leqq \pi/2$, constant a, $0 < a < 1$ and (2) we have

$$\bar{\sigma} \geqq c_1 \sigma\ , \qquad c_1 = (\sqrt{\delta}\, \sin b\rho)/\sin \sqrt{\delta}\, b\rho\ .$$

From this inequality and the isoperimetric inequality on the unit 2-sphere S^2 we have

$$\begin{aligned} \text{length}\, \partial\Omega &\geqq \text{length}\, \partial\Omega_1 \\ &\geqq 2\pi \sin c_1\sigma\ , \end{aligned} \tag{23}$$

From (2) we get

$$\text{length}\, \partial B_\sigma(n) \leqq (2\pi \sin \sqrt{\delta}\, \sigma)/\sqrt{\delta}\ . \tag{24}$$

Combining (22), (23) and (24) we get

$$\text{length}\, \partial\Omega \geqq c'\, \text{length}\, \partial B_\sigma(n)\ , \qquad c' = c_{\cdot} \cdot c\ . \tag{25}$$

Next, we consider the case where $\Omega \not\subset B_{b\rho}(n)$, $n \in \Omega$. In this case we get

$$\text{length}\, \partial\Omega > 2b\rho\ . \tag{26}$$

Combining (24) and (26) we get

$$\text{length}\, \partial\Omega > c'' \,\text{length}\, \partial B_\rho(n)\ , \qquad c'' = (b\rho\sqrt{\delta})/\pi \sin\sqrt{\delta}\,\rho\ . \tag{27}$$

On the other hand, the function $\phi(b) = (b\rho \sin\sqrt{\delta}\, b\rho)/\sqrt{\delta}\, \sin b\rho \sin \rho$, $1 \leqq b < \pi/\rho$, is strictly decreasing in b such that $\phi(1) < \pi$ and $\phi(b) \to \infty$ as $b \to \pi/\rho - 0$. Thus there exists a unique constant $b > 1$ depending only on δ and ρ such that $\phi(b) = \pi$, i.e., $c' = c''$. Therefore we see that for the constant b determined above and $\mu_1 = c'$ our assertion is true by virtue of (25) and (27), and $\mu = \mu_1 c_2^2$ is decreasing in ρ. This completes the proof.

Using above lemmas and Lemma 4 in [13] we get the following

Proposition. *Let M, D, u, u^* and μ be as in* Lemma 4. *Then we have*

$$\int_M u^2 d\mathscr{L}^2 = \int_M (u^*)^2 d\mathscr{L}^2\ ,$$

$$\int_M |\nabla u|^2\, d\mathscr{L}^2 \geqq \mu \int_M |\nabla u^*|^2\, d\mathscr{L}^2\ ,$$

where ∇u (resp. ∇u^*) *is the gradient vector field of u* (resp. u^*).

Corollary. *Let M and D be as in the* Proposition. *We denote by $\lambda_1(D)$ and $\lambda_1(B_\rho(m))$ the first eigenvalue of D and $B_\rho(m)$ with respect to Δ respectively. Then we have*

$$\lambda_1(D) \geqq \mu\lambda_1(B_\rho(m))\ .$$

Proof. At first we recall the characterization of $\lambda_1(D)$ (resp. $\lambda_1(B_\rho(m))$. Let $\mathscr{F}(D)$ (resp. $\mathscr{F}(B_\rho(m))$) be the class of functions which are Lipschitzian, nonnegative and not identically zero on M and which vanish outside D (resp. outside $B_\rho(m)$). Then we have the following:

$$\lambda_1(B_\rho(m)) = \inf_{v \in \mathscr{F}(B_\rho(m))} \int_M |\nabla v|^2\, d\mathscr{L}^2 \Big/ \int_M v^2 d\mathscr{L}^2\ , \tag{28}$$

and $\lambda_1(D)$ is characterized similary.

Using the remark of Lemma 3 we see that $u^* \in \mathscr{F}(B_\rho(m))$. Since u is an eigenfunction of Δ in D we get

$$\lambda_1(D) = \int_M |\nabla u|^2\, d\mathscr{L}^2 \Big/ \int_M u^2 d\mathscr{L}^2\ . \tag{29}$$

Combining (28) and (29) and using the proposition we see that our assertion is true.

From Theorem 3.6 in [5] and this corollary we get the following. We denote by $\lambda_1(\rho)$ the first eigenvalue of an open geodesic ball in S^2 with radius ρ with respect to the Laplacian of S^2.

Lemma 5. *Let M, D, u and μ be as in* Lemma 4. *Then we have*

$$\lambda_1(\rho) \leqq \lambda_1(B_\rho(m)) \leqq \lambda_1(D)/\mu .$$

4. Proof of the Theorem

Let M, D and f be as in the theorem. We denete by $\|B\|$ the norm of the second fundmental form B of the immersion f. Let ξ be a unit normal C^∞ vector field on M with respect to f and v be a C^∞ function on M vanishing outside D. Since S^n is of constant sectional curvature 1, Lemma 1 in [10] can be reduced in the following form.

Lemma 6. *Let M, D and f be as above. Then for a normal variation with variation vector field $v\xi$, the second variation of area is given by*

$$\left.\frac{d^2A}{dt^2}\right|_{t=0} \geqq \int_M \{v\Delta v - (\|B\|^2 + 2)v^2\} d\mathscr{L}^2$$

$$\leqq \int_M \{\lambda_1(D) - (\|B\|^2 + 2)\} v^2 d\mathscr{L}^2 .$$

Proof. From the Theorem of Stokes we get

$$\int_M v\Delta v \, d\mathscr{L}^2 = \int_M |\nabla v|^2 \, d\mathscr{L}^2$$

by virtue of $v \equiv 0$ on ∂D. And we can easily show that

$$\int_M |\nabla u|^2 \, d\mathscr{L}^2 = \int_M |\nabla(|v|)|^2 \, d\mathscr{L}^2 \quad \text{and} \quad v \in \mathscr{F}(D) .$$

Therefore, using the characterization of $\lambda_1(D)$ we see that our asseetion is true.

Remark. We note that the sign of the Laplacian Δ is different from the one in [10].

We now prove the theorem. We first note that M is a compact, real analytic Riemannian manifold. From the equation of Gauss and the fact that M is minimally immersed we get $K = 1 - \frac{1}{2}\|B\|^2$ and hence the condition on K implies $\|B\|^2 \leqq 2(1 - \delta)$. On the other hand, it is known (see [8]) that $\lambda_1(\rho)$ is a strictly decreasing, continuous function in ρ, $0 < \rho < \pi$ such that $\lambda_1(\rho) \to \infty$ as $\rho \to 0$ and $\lambda_1(\pi/2) = 2$. Therefore, for a given δ, $0 < \delta \leqq 1$, there exists a unique number ρ in $(0, \pi/2]$ such that $\lambda_1(\rho)\mu(\rho, \delta) = 2 + 2(1 - \delta)$. Since the constant $\lambda_1(\sigma)\mu(\sigma, \rho)$ is stricly decreasing in σ, $0 < \sigma \leqq \pi/2$, we see that our theorem is true.

References

[1] J. L. Barbosa and M. P. do Carmo, *On the size of a stable minimal surface in* R^3, Amer. J. Math., **98** (1976), 515–528.

[2] M. Berger, P. Gauduchon and E. Mazet, Le spectre d'une variété Riemannienne, Lecture Notes in Math., Berlin-Heidelberg-New York, Springer, 1971.

[3] R. L. Bishop and R. J. Crittenden, Geometry of Manifolds, Academic Press, New York, 1964.

[4] J. Cheeger and D. G. Ebin, Comparison Theorem in Riemannian Geometry, North-Holland Publishing Company, 1975.

[5] S. Y. Cheng, *Eigenfunctions and eigenvalues of Laplacian,* Proc. Symp. in Pure Math., **27** (1975), 185–193.

[6] S. S. Chern, M. P. do Carmo and S. Kobayashi, *Minimal submanifolds of a sphere with second fundamental form of constant length,* Functional analysis and related field (1970), 59–75.

[7] H. Federer, Geometric measure theory, Berlin-Heidelberg-New York, Springer, 1969.

[8] S. Friedland and W. K. Hayman, *Eigenvalue inequalities for the Dirichlet problem on spheres and the growth of subharmonic functions,* Comment. Math. Helv. **51** (1976), 133–161.

[9] S. Kobayashi and K. Nomizu, Foundations of differential geometry, Interscience, New York, vols. 1 & 2, 1963, 1969.

[10] H. Mori, *Notes on the stability of minimal submanifolds of Riemannian manifolds,* to appear in the Yokohama Math. J.

[11] ——, *A note on the stability of minimal surfaces in the 3-dimensional unit sphere,* to appear in Indiana Univ. Math. J.

[12] T. Otsuki, *Minimal hypersurfaces in a Riemannian manifold of constant curvature,* Amer. J. Math., **92** (1970), 145–173.

[13] E. Sperner, *Zur Symmetrisierung von Funktionen auf Sphären,* Math. Z. **134** (1973), 317–327.

[14] J. Spruck, *Remarks on the stability of minimal submanifolds of* R^n, Math. Z. **144** (1975), 169–174.

FACULTY OF EDUCATION
TOYAMA UNIVERSITY
TOYAMA 930
JAPAN

Minimal Submanifolds and Geodesics
Kaigai Publications, Tokyo, 1978, 117–120

MINIMAL SURFACES IN TORI BY WEYL GROUPS. II

TADASHI NAGANO & BRIAN SMYTH

1. Introduction

In 1867, H. A. Schwarz [S] discovered a procedure for generating periodic minimal surfaces in $\boldsymbol{R}^3$ with octahedral or tetrahedral symmetry; see also Schoenflies [Sf]. The detailed structure of these surfaces —especially the question of whether any of the associates of these minimal surfaces are also periodic— was elucidated by Schwarz and Neovius [N] for only the two simplest examples; in both cases the conjugate surfaces could be found by *ad hoc* methods and happened to be periodic.

Now the analogous construction procedure developed in our earlier paper [N-S, 1] will generate periodic minimal surfaces in $\boldsymbol{R}^n$ with symmetry corresponding to any Weyl group of rank n. We outline here several results on the structure of these surfaces. As a consequence we can obtain the existence of *infinitely many* periodic associate surfaces in general; this is true in particular for all of the classical Schwarz surfaces.

2. Construction

Take a simple closed polygon P with m edges in $\boldsymbol{R}^n$ (lying in no proper subspace), the reflexion in the edges of which generate a discrete uniform subgroup G of motions of $\boldsymbol{R}^n$ with translation subgroup H. When P admits a monotone convex projection on some 2-plane P in $\boldsymbol{R}^n$, the Plateau problem for P has a solution Σ with no interior branch points, according to a theorem of Rado; this is automatically so when $m = n + 1$. Furthermore, if no vertex angle of P is obtuse and P satisfies some simple convexity conditions, then the orbit $G \cdot \Sigma$ is a singularity-free periodic minimal surface in $\boldsymbol{R}^n$. After dividing by the appropriate lattice, we obtain a compact orientable surface X and a minimal isometric immersion f of it into a flat n-torus T^n; moreover, G gives rise to a finite group W of orientation-preserving isometries acting on X and T equivariantly with respect to f.

Conversely, given a Weyl group of rank n, we have shown in [N-S, 2] how to construct a polygon P which gives rise, as above, to a compact

Supported by N.S.F. Grant MCS 77-01715.

minimal surface X in a flat n-torus T, the corresponding group W being the group of orientation-preserving elements of the given Weyl group.

When P has $n + 1$ edges we call the resulting X *primitive*; in this case, the linear part of the action of W on T^n is automatically irreducible. We confine ourselves to this case in the statement of the main results and add some remarks on the general results in the last section.

3. The main results

Considering X as a Riemann surface with the induced conformal structure, let $a: X \to A(X)$ be the natural map of X into its Jacobi variety. To within a translation, $f = h \circ a$ for some homomorphism H: $A(X) \to T^n$. Let U denote the subspace tangent to the kernel of h at the identity and let V be its maximal complex subspace. Now $\dim_R U = 2p - n$ and it can be shown that

(i) $\dim_C V = p - n$ $\quad (p = \text{genus } X)$

and

(ii) $f(X)$ is homologous to zero

when the linear part of the action of W on T^n is irreducible.

Now f lifts to a minimal immersion $\tilde{f}$: $\tilde{X} \to R^n$ on universal covers and the harmonic conjugate $\hat{f}$ of $\tilde{f}$ gives another minimal isometric immersion of $\tilde{X}$ in R^n; more generally so is $\tilde{f}_\theta = \cos\theta\tilde{f} + \sin\theta\hat{f}$, $0 \leqslant \theta < \pi$. We wish to show that $\tilde{f}_\theta$ projects to a minimal immersion f_θ of X in some flat n-torus; f_θ is then called an *associate* of f. The existence of f_θ is equivalent to saying that the subspace $e^{i\theta}U$ determines a subtorus of $A(X)$.

The important step is

Theorem 1. *If X is a primitive surface, then the subspace V is closed (i.e. determines a complex codimension n subtorus of the Jacobi variety).*

This result is proved via the Riemann-Roch theorem for orbit spaces ([A-S], Theorem 4.7, p. 566), but a classical theorem of Chevalley-Weil [C-W] can be used instead: the essential point of the proof is to show that the action of W on $f^*(H^1(T, C))$ appears just once in the representation of W on $H^1(X, C)$.

From this result we see that we have the following diagram

$$\begin{array}{ccc} X & \xrightarrow{F} & T^n_C = A(X)/V \\ & {\scriptstyle f}\searrow \;{\scriptstyle x}\; \swarrow{\scriptstyle \pi} & \\ & T^n & \end{array}$$

Theorem 2. *Each primitive surface X admits a branched covering of an elliptic curve E which factors through F.*

The main idea of the proof is to examine the group W of the surface X. Abstractly and in its action, this group corresponds to the subgroup of orientation-preserving transformations of some irreducible Weyl group of rank n, and as such is completely known. Examining its action on T_C^n we obtain a complex hypertorus H and E is the quotient T_C^n/H. Moreover, and this is used below, H is contained in a real hypertorus G which contains U.

Theorem 3. *Each primitive surface X has infinitely many associate surfaces $f_\theta : X \to T_\theta^n$.*

For infinitely many values of θ $(0 < \theta < \pi)$, the subspace $e^{i\theta}G_e$ determines a real hypertorus in $A(X)$ which we denote G_θ; observe $G_\theta \supset H$. Define $U_\theta = \bigcap_{\omega \in W} \omega G_\theta$. It is rather easy to check that this subtorus has $e^{i\theta}U$ as its tangent space at e. Thus the associate f_θ exists.

4. Remarks

Much of the foregoing holds for the general class of surfaces X arising from the procedure of Section 2; in addition, we mention some other results.

(i) Theorem 2 is true in general once V is known to be closed.

(ii) $A(X)$ is isogenous to $V^{p-n} \times T_1 \times \overset{(n)}{\cdots} \times T_1$, where T_1 is an elliptic curve, provided V is closed and W is irreducible.

(iii) As a Riemann surface, X/W is always the sphere; so the field $\mathfrak{M}(X)$ of meromorphic functions on X is a Galois extension of the rational function field.

(iv) The Euler number of X is given by the simple formula

$$\chi(X) = \sharp W\left\{1 - \frac{1}{2} \sum_{k=2,3,4,6} \frac{k-1}{k} \nu(k)\right\}$$

where $\sharp W$ is the order of W and $\nu(k)$ denotes the number of vertices in P with angle π/k ($k = 2, 3, 4$ and 6 are the only possibilities).

Bibliography

[A-S] M. F. Atiyah and I. Singer, *The index of elliptic operators. III.* Ann. of Math. **87** (1968), 546–604.

[C-W] C. Chevalley and A. Weil, *Ueber das Verhalten der Integrale 1 Gattung bei Automorphismen des Funktionenkoerpers.* Abh. Math. Sem. Hamburg **10** (1934), 358–361.

[N] E. R. Neovius, Bestimmung zweier speciellen periodischen Minimalflächen. . . . J. C. Frenckel Sohn, Helsingfor, 1883.

[N-S, 1] T. Nagano and B. Smyth, *Minimal surfaces in tori by Weyl groups. I.* Proc. Amer. Math. Soc. **61** (1976), 102–104.

[N-S, 2] T. Nagano and B. Smyth, *Periodic minimal surfaces.* Comment. Math. Helv. (to appear).

[Sf] A. Schoenflies, *Sur les surfaces minima limitées par quatre arêtes d'un quadrilatère gauche.* C. R. Acad. Sci. Paris **112** (1891), 478–480.

[S] H. A. Schwarz, Gesammelte Mathematische Abhandlungen, Band I, Berlin, 1890.

[W] A. Weil, *Ueber Matrizenringe auf Riemannschen Flaeche und den Riemann-Rochschen Satz.* Abh. Math. Sem. Hamburg **11** (1936), 110–115.

UNIVERSITY OF NOTRE DAME
NOTRE DAME, IN 46556
U.S.A.

Minimal Submanifolds and Geodesics
Kaigai Publications, Tokyo, 1978, 121–142

SOME RESULTS ON MINIMAL SURFACES WITH THE RICCI CONDITION

REIKO NAKA

§ 0. Introduction

The metric ds^2 on a minimal surface in the 3-dimensional space form $M^3(c)$ of curvature c has the property that its Gauss curvature K satisfies $K \leqq c$, and except at the points where $K = c$, the new metric $d\hat{s}^2 = \sqrt{c - K}ds^2$ is flat.

On the other hand, a metric ds^2 given over a simply connected surface $\mathscr{R}$ is realized by a continuous one-parameter family of isometric minimal immersions $\psi_\theta : \mathscr{R} \to M^3(c)$; $0 \leqq \theta < 2\pi$, if

$$(*) \qquad K < c \text{ and } d\hat{s}^2 = \sqrt{c - K}ds^2 \text{ is flat.}$$

In particular, when the metric ds^2 is originally induced by a minimal immersion of $\mathscr{R}$ into $M^3(c)$, each ψ_θ is defined all over $\mathscr{R}$ even if $K = c$ at some points of $\mathscr{R}$. In either case, moreover, the maps ψ_θ ; $0 \leqq \theta < \pi$ represent all local isometric minimal immersions of $\mathscr{R}$ into $M^3(c)$ up to congruences [6, p. 364].

The condition $(*)$ was first discovered by Ricci when $c = 0$ [9], and then generalized by H. B. Lawson. We call it the Ricci condition with respect to c. As a result, we can say:

All the minimal surfaces in $M^3(c)$ which are locally isometric to a given minimal surface M in $M^3(c)$ are obtained from the maps $\psi_\theta : \mathscr{R} \to M^3(c)$; $0 \leqq \theta < \pi$, where $\mathscr{R}$ is supposed to be the universal covering surface of M.

Then the next problem is to classify those minimal surfaces in $M^n(c)$ (= the $n(>3)$-dimensional space form of curvature c) which are locally isometric to a minimal surface in $M^3(c)$, that is, to classify those minimal surfaces in $M^n(c)$ whose metrics satisfy the Ricci condition with respect to c.

In the euclidean case, this so-called Pinl's problem was solved by H. B. Lawson [7] with the result that such surfaces either already lie in $\boldsymbol{R}^3$ or they lie fully in $\boldsymbol{R}^6$ as a special type of surfaces.

In the spherical case, let $c = 1$ and $S^n = M^n(1)$ for convenience, and call $(*)$ the spherical Ricci condition. Then by using the above mentioned family $\{\psi_\theta\}_{0\leqq\theta<\pi}$, we can construct a minimal immersion $\psi: \mathscr{R} \to S^{4m-1} \subset \boldsymbol{R}^{4m}$ by setting

$$(**) \qquad \psi = a_1\psi_{\theta_1} \oplus \cdots \oplus a_m\psi_{\theta_m}$$

where $\sum_{k=1}^{m} a_k^2 = 1$, $0 \leqq \theta_1 < \theta_2 < \cdots < \theta_m < \pi$, each ψ_{θ_k} is viewed as an $\boldsymbol{R}^4$-valued function with $|\psi_{\theta_k}|^2 = 1$, and the simbol $\oplus$ denotes the direct sum with respect to an orthogonal decomposition $\boldsymbol{R}^{4m} = \boldsymbol{R}^4 \oplus \cdots \oplus \boldsymbol{R}^4$.

Lawson's conjecture [7]. *Every minimal surface in S^n whose metric satisfies the spherical Ricci condition away from the points where $K = 1$ must be of the form $(**)$ for some integer m.*

As a matter of fact, contrary to the euclidean case, there are easy counterexamples to this conjecture. Actually, every flat minimal surface in S^n automatically satisfies the spherical Ricci condition, and there are many such surfaces not of the type $(**)$ in any odd dimensional sphere [4, 5, 8]. Under additional hypotheses, however, we shall give a partial answer to this conjecture:

Theorem I. *If $n \leqq 5$, Lawson's conjecture is true for non-flat parabolic minimal surfaces in S^n. Here by parabolic surfaces, we mean those surfaces on which there exist no non-constant upper bounded subharmonic functions.*

Theorem II. *If $n \leqq 9$, Lawson's conjecture is true for compact minimal surfaces in S^n with genus greater than one.*

By the assumption on genus, we excluded the flat case and the exceptional case of topological 2-sphere on which we can not consider the spherical Ricci condition as will be shown in Remark 3.2.

For the proof, in the euclidean case, H. B. Lawson applied Calabi's rigidity theorem to the generalized Gauss map of a minimal surface in $\boldsymbol{R}^n$, which is a holomorphic mapping into $\boldsymbol{CP}^{n-1}$ and has constant curvature when the metric of the surface satisfies the Ricci condition with respect to 0.

In the spherical case, the difficulty comes from the uselessness of the Gauss map itself. By this reason, we define a series of normal vectors $\Phi_1, \Phi_2, \cdots, \Phi_k$, so that the k-th one is orthogonal to the space H_{k-1} spanned by the position vector, tangent vectors and normal vectors $\Phi_1, \cdots, \Phi_{k-1}$. Under the spherical Ricci condition, using each Φ_k, we can define certain subharmonic functions and holomorphic differentials on the surface. Then by investigating them and using the global assumption of the surface, we obtain the above two theorems.

The auther wishes to express her hearty thanks to Professor T. Otsuki for his valuable suggestions and patient check of the contents. She would especially like to dedicate this paper to him on his 60th birthday.

§ 1. Preliminaries

In the following sections, we restrict our argument to $S^n = \{x \in \boldsymbol{R}^{n+1} : |x| = 1\}$ without loss of generality. We always mean by the spherical Ricci condition the Ricci condition with respect to 1.

To begin with, we observe that it is sufficient to prove the theorems for orientable surfaces, since the facts in them for non-orientable surfaces are inherited from their orientable coverings. Thus by a surface M in S^n, we mean a conformal C^2-immersion

$$\psi : \mathscr{R} \to S^n$$

of some Riemann surface $\mathscr{R}$, by virtue of the existence of isothermal coordinates on 2-dimensional Riemannian manifolds of class C^2 with which we are concerned. The map ψ is always considered as $\boldsymbol{R}^{n+1}$-valued with $|\psi|^2 = 1$.

Let $z = x_1 + \sqrt{-1}x_2$ be a local complex coordinate on $\mathscr{R}$ and set

$$\partial = \frac{1}{2}\left(\frac{\partial}{\partial x_1} - \sqrt{-1}\frac{\partial}{\partial x_2}\right).$$

Then the induced metric has the form

$$ds^2 = 2F\,|dz|^2$$

where $F = \langle \partial\psi, \bar{\partial}\psi \rangle$ by using the complex linearly extended inner product. Since ψ is conformal and $\langle \psi, \psi \rangle = 1$, we have

$$\langle \partial^k\psi, \partial^l\psi \rangle = 0\,, \qquad \text{for } 1 \leqq k + l \leqq 3\,. \tag{1.1}$$

The Laplace-Beltrami operator in this coordinate is $(2/F)\partial\bar{\partial}$. Hence the Gauss curvature is given by

$$K = -\frac{1}{F}\partial\bar{\partial}\log F \tag{1.2}$$

and the minimal surface equation becomes

$$\partial\bar{\partial}\psi = -F\psi\,. \tag{1.3}$$

From now on, let M be minimal, i.e., let ψ satisfy (1.3). Then ψ is a real analytic mapping [6]. Furthermore the local vector-valued function

$$\Phi := F\partial\left(\frac{1}{F}\partial\psi\right) \tag{1.4}$$

satisfies

$$(1.5)\qquad |\Phi|^2 = F^2(1-K)$$

$$(1.6)\qquad \bar{\partial}\Phi = -(1-K)F\partial\psi$$

$$(1.7)\qquad |\bar{\partial}\Phi|^2 = \frac{|\Phi|^4}{F}$$

all of which are easy to prove by using (1.1) ~ (1.3).

Remark 1.1. In fact, we know that

$$(1.8)\qquad \Phi = \tfrac{1}{2}(B_{11} - \sqrt{-1}B_{12})$$

where B_{ij} $(1 \leqq i, j \leqq 2)$ is the component of the vector-valued second fundamental form of M [6].

In another coordinate $\tilde{z} = \tilde{x}_1 + \sqrt{-1}\tilde{x}_2$, let

$$\tilde{\partial} = \frac{1}{2}\left(\frac{\partial}{\partial\tilde{x}_1} - \sqrt{-1}\frac{\partial}{\partial\tilde{x}_2}\right)$$

$\tilde{F} = \langle\tilde{\partial}\psi, \bar{\tilde{\partial}}\psi\rangle$ and $\tilde{\Phi} = \tilde{F}\tilde{\partial}((1/\tilde{F})\tilde{\partial}\psi)$. Then we have on the common coordinate neighborhood,

$$(1.9)\qquad \partial = \tilde{z}'\tilde{\partial} \quad \text{and} \quad F = |\tilde{z}'|^2\tilde{F}$$

where $\tilde{z}' = d\tilde{z}/dz$, therefore it follows

$$(1.10)\qquad \Phi = (\tilde{z}')^2\tilde{\Phi} \quad \text{or} \quad \Phi dz^2 = \tilde{\Phi}d\tilde{z}^2$$

since $\tilde{z} = \tilde{z}(z)$ is holomorphic. Now defining $f := \langle\Phi, \Phi\rangle$, we have by (1.1) and (1.6):

Lemma 1.2. *fdz^4 is a holomorphic differential form on $\mathcal{R}$.*

Proposition 1.3. *Let M be a minimal surface in S^n and let Φ and f be as above. Then M lies in some totally geodesic 3-sphere of S^n if and only if $|f|^2 = |\Phi^4|$ around every point of M.*

Proof. Since

$$|\Phi|^4 - |f|^2 = |\Phi \wedge \bar{\Phi}|^2\,,$$

we have $|f|^2 = |\Phi|^4$ if and only if the first normal space is of dimension one. This being the case, there is a real unit normal vector field η such that $\Phi = \varphi\eta$ where $\varphi^2 = f$. Using (1.6) and $\bar{\partial}\varphi = 0$, we have

$$-F(1-K)\bar{\partial}\psi = \partial\bar{\Phi} = \bar{\varphi}\partial\eta$$

so that

$$\partial\eta = -\frac{\varphi}{F}\bar{\partial}\psi$$

by (1.5). Hence the first normal space is parallel in the normal bundle with respect to the natural normal connection, and M lies in some totally geodesic 3-sphere by Erbacher's theorem [2]. Q.E.D.

Now we define a series of local vector-valued function Φ_k ($k = 0, 1, 2, \cdots$) by

$$\Phi_0 := \partial\psi$$
$$\Phi_1 := \Phi$$

$$\Phi_{k+1} := \partial\Phi_k - \frac{f_k}{|\Phi_{k-1}|^4 - |f_{k-1}|^2}(\bar{f}_{k-1}\Phi_{k-1} - |\Phi_{k-1}|^2\bar{\Phi}_{k-1})$$
$$- \frac{1}{|\Phi_k|^4 - |f_k|^2}\left\{\left(|\Phi_k|^2\langle\partial\Phi_k, \bar{\Phi}_k\rangle - \frac{\partial f_k}{2}\bar{f}_k\right)\Phi_k + \left(\frac{\partial f_k}{2}|\Phi_k|^2 - \langle\partial\Phi_k, \bar{\Phi}_k\rangle f_k\right)\bar{\Phi}_k\right\}, \tag{1.11}$$

where

$$f_l := \langle\Phi_l, \Phi_l\rangle, \qquad \text{for } l = k-1, k,$$

if $|\Phi_k|^4 \not\equiv |f_k|^2$, and

$$\Phi_{k+1} := 0 \tag{1.11$'$}$$

if $|\Phi_k|^4 \equiv |f_k|^2$. When $|\Phi_k|^4 \not\equiv |f_k|^2$ ($k \geqq 1$), Φ_{k+1} is defined only where $|\Phi_l|^4 - |f_l|^2 \neq 0$ ($l = 0, 1, \cdots, k$), but we shall see later that under the spherical Ricci condition, it is everywhere well-defined.

For the present, we assume that $\Phi_0, \cdots, \Phi_k, \cdots$ are defined around $p \in \mathscr{R}$, and let H_k be the complex linear subspace spanned by $\psi, \Phi_l, \bar{\Phi}_l$ at p for $l = 0, 1, \cdots, k$.

Lemma 1.4. *For $k = 0, 1, 2, \cdots$, we have*

$$\bar{\partial}\Phi_k \in H_k \tag{1.12}$$

$$\langle\Phi_{k+1}, X\rangle = 0, \qquad \textit{for all } X \in H_k. \tag{1.13}$$

Proof. We get $\bar{\partial}\Phi_0 \in H_0$ from (1.3). Assume (1.12) holds for $k = 0, 1, \cdots, l$. Then noting (1.11), we have

$$\bar{\partial}\Phi_{l+1} = \partial(\bar{\partial}\Phi_l) + [\Phi_{l-1}, \bar{\Phi}_{l-1}, \Phi_l, \bar{\Phi}_l]$$
$$+ [\bar{\partial}\Phi_{l-1}, \overline{\partial\Phi}_{l-1}, \bar{\partial}\Phi_l, \overline{\partial\Phi}_l] \in H_{l+1}$$

where [ ] denotes a term composed of a linear combination of vectors in [ ]. Next, $\langle \Phi_1, X\rangle = 0$ for all $X \in H_0$ follows from (1.1) and $F = \langle \partial\psi, \bar{\partial}\psi\rangle$. Assume (1.13) holds for $k = 0, 1, \cdots, l$. We may show $\langle \Phi_{l+2}, X\rangle = 0$ for $X = \psi, \Phi_0, \bar{\Phi}_0, \cdots, \Phi_{l+1}, \bar{\Phi}_{l+1}$. It follows from the assumption and (1.11) that

$$\langle \Phi_{l+2}, \psi\rangle = \langle \partial\Phi_{l+1}, \psi\rangle = -\langle \Phi_{l+1}, \Phi_0\rangle = 0 \ .$$

For $m < l$, we obtain

$$\langle \Phi_{l+2}, \Phi_m\rangle = \langle \partial\Phi_{l+1}, \Phi_m\rangle = -\langle \Phi_{l+1}, \partial\Phi_m\rangle = -\langle \Phi_{l+1}, \Phi_{m+1}\rangle = 0 \ ,$$

and

$$\langle \Phi_{l+2}, \bar{\Phi}_m\rangle = \langle \partial\Phi_{l+1}, \bar{\Phi}_m\rangle = -\langle \Phi_{l+1}, \partial\bar{\Phi}_m\rangle = 0$$

by using (1.12). For the rest we have

$$\begin{aligned}
\langle \Phi_{l+2}, \Phi_l\rangle &= \langle \partial\Phi_{l+1}, \Phi_l\rangle - \frac{f_{l+1}}{|\Phi_l|^4 - |f_l|^2}(|f_l|^2 - |\Phi_l|^4) \\
&= -f_{l+1} + f_{l+1} = 0 \ , \\
\langle \Phi_{l+2}, \bar{\Phi}_l\rangle &= \langle \partial\Phi_{l+1}, \bar{\Phi}_l\rangle - \frac{f_{l+1}}{|\Phi_l|^4 - |f_l|^2}(\bar{f}_l\,|\Phi_l|^2 - |\Phi_l|^2\,\bar{f}_l) \\
&= -\langle \Phi_{l+1}, \partial\bar{\Phi}_l\rangle = 0 \ , \\
\langle \Phi_{l+2}, \Phi_{l+1}\rangle &= \langle \partial\Phi_{l+1}, \Phi_{l+1}\rangle - \frac{1}{|\Phi_{l+1}|^4 - |f_{l+1}|^2} \\
&\quad \times \Big\{ |\Phi_{l+1}|^2 \langle \partial\Phi_{l+1}, \bar{\Phi}_{l+1}\rangle f_{l+1} - \frac{\partial f_{l+1}}{2}|f_{l+1}|^2 \\
&\qquad + \frac{\partial f_{l+1}}{2}|\Phi_{l+1}|^4 - \langle \partial\Phi_{l+1}, \bar{\Phi}_{l+1}\rangle f_{l+1}|\Phi_{l+1}|^2 \Big\} \\
&= \frac{\partial f_{l+1}}{2} - \frac{\partial f_{l+1}}{2} = 0 \ ,
\end{aligned}$$

and

$$\begin{aligned}
\langle \Phi_{l+2}, \bar{\Phi}_{l+1}\rangle &= \langle \partial\Phi_{l+1}, \bar{\Phi}_{l+1}\rangle - \frac{1}{|\Phi_{l+1}|^4 - |f_{l+1}|^2} \\
&\quad \times \Big\{ \Big(|\Phi_{l+1}|^2 \langle \partial\Phi_{l+1}, \bar{\Phi}_{l+1}\rangle - \frac{\partial f_{l+1}}{2}\bar{f}_{l+1} \Big)|\Phi_{l+1}|^2 \\
&\qquad + \Big(\frac{\partial f_{l+1}}{2}|\Phi_{l+1}|^2 - \langle \partial\Phi_{l+1}, \bar{\Phi}_{l+1}\rangle f_{l+1} \Big)\bar{f}_{l+1} \Big\} \\
&= \langle \partial\Phi_{l+1}, \bar{\Phi}_{l+1}\rangle - \langle \partial\Phi_{l+1}, \bar{\Phi}_{l+1}\rangle = 0 \ .
\end{aligned}$$

§ 2. Proof of Theorem I

In this section, we shall treat parabolic minimal surfaces under the spherical Ricci condition.

Remark 2.1. By our definition of parabolicity in Theorem I, the base Riemann surface $\mathscr{R}$ may be either open or closed.

Remark 2.2. Since we have

$$d\hat{s} = \sqrt{1-K}ds^2 = 2F\sqrt{1-K}\,|dz|^2\,,$$

the spherical Ricci condition is equivalent to $\Phi_1 \neq 0$ and $\partial\bar{\partial}\log|\Phi_1|^2 = 0$. When $\Phi_1 dz^2$ is identically zero, we know that M lies in a totally geodesic 2-sphere of S^n. Therefore throughont this paper, we assume $\Phi_1 dz^2$ is not identically zero so that the spherical Ricci condition makes sense in the non-empty open subset $M' := \{p \in M : K(p) \neq 1\}$.

Proposition 2.3. *Let M be a parabolic minimal surface in S^n whose metric satisfies the spherical Ricci condition away from the points where $K = 1$. Then*

$$|f_1|^2 = c_1|\Phi_1|^4 \tag{2.1}$$

for some constant $0 \leqq c_1 \leqq 1$ which does not depend on the choice of coordinates.

To prove this, we need the following lemma the proof of which will appear in the appendix.

Lemma 2.4. *Let D be a plane domain containing the origin with coordinate $z = x_1 + \sqrt{-1}x_2$. Let u be a real analytic non-negative function on D such that $u(0) = 0$. If u is not identically zero and satisfies $\partial\bar{\partial}\log u = 0$ away from the points where $u = 0$, then the order l' of zero of u at the origin is even, and there exists a non-zero holomorphic function ω around the origin such that*

$$u = |z|^{2l}|\omega|$$

where $l' = 2l$.

Proof of Proposition 2.3. Since $\Phi_1 \not\equiv 0$, we define

$$G_1 := \frac{|f_1|^2}{|\Phi_1|^4}$$

and show that G_1 can be considered as a subharmonic function on $\mathscr{R}$. First, G_1 is defined independently of the choice of coordinates by virtue of (1.10). Next, we must examine the points where $\Phi_1 = 0$. Choose a coordinate z around a zero point p of Φ_1 such that p corresponds to $z = 0$. Then since $|\Phi_1|^2$ satisfies the conditions in the above lemma, letting the order of zero of $|\Phi_1|^2$ at p be $2l$, we have $|\Phi_1|^2 = |z|^{2l}|\varphi_1|$ where φ_1 is the non-zero holomorphic function in a neighborhood of p. On the other hand, the holomorphic function f_1 is also zero at p. If $f_1 \equiv 0$ around p, we may take $c_1 = 0$ as $f_1 dz^4$ becomes identically zero on $\mathscr{R}$. So assume $f_1 \not\equiv 0$ and let $m(1 \leqq m < \infty)$ be the order of zero of f_1 at p. Then the Schwarz-type inequality

$$|f_1| \leqq |\Phi_1|^2 \tag{2.2}$$

shows $m \geqq 2l$. Therefore writing $f_1 = z^m \hat{f}_1$ for the non-zero holomorphic $\hat{f}_1$, we get

$$\frac{|f_1|^2}{|\Phi_1|^4} = \frac{|z|^{2(m-2l)}|\hat{f}_1|^2}{|\varphi_1|^2}$$

around p which assures us to extend G_1 analytically to p.

Now it is easy to see that G_1 is subharmonic. Since

$$\partial\bar{\partial} \log G_1 = 0$$

except the zeros of G_1 which are isolated, we get immediately

$$\partial\bar{\partial} G_1 \geqq 0$$

everywhere on $\mathscr{R}$ by the continuity. The upper boundedness of G_1 by (2.2) then implies

$$G_1 = c_1$$

for some constant $0 \leqq c_1 \leqq 1$, since M is parabolic. Q.E.D.

Proposition 2.5. *Let M be as in* Proposition 2.3, *and let $c_1 < 1$. Then we have around the points where $K \neq 1$:*

$$\Phi_2 = \partial\Phi_1 - \frac{\partial|\Phi_1|^2}{|\Phi_1|^2}\Phi_1 + \frac{f_1}{F}\bar{\partial}\psi \tag{2.3}$$

$$|\Phi_2|^2 = (1 - c_1)\frac{|\Phi_1|^4}{F} \tag{2.4}$$

$$\bar{\partial}\Phi_2 = -\frac{|\Phi_1|^2}{F}\Phi_1 + \frac{f_1}{F}\bar{\Phi}_1 \tag{2.5}$$

$$|\bar{\partial}\Phi_2|^2 = (1 - c_1)\frac{|\Phi_1|^6}{F^2}\,. \tag{2.6}$$

Remark 2.6. When $c_1 = 1$, the above formulas hold trivially. In fact, if we define Φ_2 by (2.3) instead of (1.11)′, it must be identically zero by (2.4).

Proof. From (1.6) and (1.13), it follows

$$\langle \partial\Phi_1, \bar{\Phi}_1 \rangle = \partial |\Phi_1|^2 .$$

Differentiating (2.1), we have

$$\partial f_1 \cdot \bar{f}_1 = 2c_1 |\Phi_1|^2 \partial |\Phi_1|^2 = 2|f_1|^2 \frac{\partial |\Phi_1|^2}{|\Phi_1|^2} ,$$

so that

$$|\Phi_1|^2 \langle \partial\Phi_1, \bar{\Phi}_1 \rangle - \frac{\partial f_1}{2} \bar{f}_1 = (1 - c_1) |\Phi_1|^2 \partial |\Phi_1|^2$$

and

$$\frac{\partial f_1}{2} |\Phi_1|^2 - \partial |\Phi_1|^2 \cdot f_1 = 0 .$$

Now (2.3) follows from $|\Phi_1|^4 - |f_1|^2 = (1 - c_1)|\Phi_1|^4$ in (1.11). (2.4) is obtained from

$$\begin{aligned} |\Phi_2|^2 &= |\partial\Phi_1|^2 + \frac{|\partial|\Phi_1|^2|^2}{|\Phi_1|^2} + \frac{|f_1|^2}{F} - 2\frac{|\partial|\Phi_1|^2|^2}{|\Phi_1|^2} - 2\frac{|f_1|^2}{F} \\ &= |\Phi_1|^2 \left(\partial\bar{\partial} \log |\Phi_1|^2 + \frac{|\bar{\partial}\Phi_1|^2}{|\Phi_1|^2} \right) - \frac{|f_1|^2}{F} \\ &= \frac{|\Phi_1|^4}{F} - c_1 \frac{|\Phi_1|^4}{F} \\ &= (1 - c_1) \frac{|\Phi_1|^4}{F} \end{aligned}$$

by using (1.7) and (2.1). From (2.3) together with (1.4) and (1.6), we know that $\bar{\partial}\Phi_2$ is a linear combination of $\Phi_0, \bar{\Phi}_0, \Phi_1, \bar{\Phi}_1$. But as it is easily seen that

$$\langle \bar{\partial}\Phi_2, \Phi_0 \rangle = \langle \bar{\partial}\Phi_2, \bar{\Phi}_0 \rangle = 0 ,$$

we may put

$$\bar{\partial}\Phi_2 = a\Phi_1 + b\bar{\Phi}_1 .$$

Then from

$$af_1 + b|\Phi_1|^2 = \langle \bar{\partial}\Phi_2, \Phi_1 \rangle = -\langle \Phi_2, \bar{\partial}\Phi_1 \rangle = 0$$

and

$$a|\Phi_1|^2 + b\bar{f}_1 = \langle \bar{\partial}\Phi_2, \bar{\Phi}_1 \rangle = -\langle \Phi_2, \bar{\partial}\overline{\Phi}_1 \rangle = -|\Phi_2|^2$$

we get

$$a = \frac{-|\Phi_1|^2|\Phi_2|^2}{|\Phi_1|^4 - |f_1|^2} = \frac{-|\Phi_2|^2}{(1-c_1)|\Phi_1|^2} = -\frac{|\Phi_1|^2}{F}$$

and

$$b = \frac{f_1|\Phi_2|^2}{|\Phi_1|^4 - |f_1|^2} = \frac{f_1|\Phi_2|^2}{(1-c_1)|\Phi_1|^4} = \frac{f_1}{F} .$$

(2.6) is then easy. Q.E.D.

Remark 2.7. By (2.3), Φ_2 is defined except at the points where $K = 1$. Here, however, we may define $\Phi_2 = 0$ by (2.4) preserving continuity. On the other hand, $f_2 = \langle \Phi_2, \Phi_2 \rangle$ is holomorphic by (2.5), and when it is continuously extended to the points where $K = 1$, it remains holomorphic by Radó-Behnke-Stein-Cartan's theorem [3]. Since the relation

$$\Phi_2 dz^3 = \tilde{\Phi}_2 d\tilde{z}^3 \tag{2.7}$$

between the expressions in two coordinates z and $\tilde{z}$ follows from (2.3) by using (1.9) and (1.10), we have:

Corollary 2.8. *The extended $f_2 dz^6$ is a holomorphic differential form on $\mathcal{R}$.*

Now we are ready to prove Theorem I. As the surfaces in it satisfy the conditions in Proposition 2.3, we can use all the above results.

Proof of Theorem I. When $c_1 = 1$, the result follows from Proposition 1.3. Suppose $c_1 < 1$. Then as Φ_2 is not identically zero and $n \leqq 5$, we have $\Phi_2 \wedge \bar{\Phi}_2 = 0$ and hence

$$|f_2|^2 = |\Phi_2|^4 = (1-c_1)^2 \frac{|\Phi_1|^8}{F^2} .$$

Since f_2 is holomorphic, we have

$$0 = \partial\bar{\partial} \log |f_2|^2 = \partial\bar{\partial} \log (1-c_1)^2 \frac{|\Phi_1|^8}{F^2} = 2FK .$$

which contradicts our assumption $K \not\equiv 0$. Q.E.D.

§ 3. Compact surfaces with genus $\geqq 2$

As the compactness is a special case of parabolicity by our definition, we can use the results in § 2 in the following.

Proposition 3.1. *Let M be a compact minimal surface in S^n with genus greater than one, whose metric satisfies the spherical Ricci condition away from the points where $K = 1$. Then the holomorphic differential $f_2 dz^6$ is identically zero on $\mathscr{R}$.*

Proof. Suppose $f_2 dz^6 \not\equiv 0$. Then $c_1 < 1$. Defining

$$G_2 := \frac{|f_2|^2 (1 - K)}{|\Phi_2|^4},$$

we show that G_2 can be considered as a subharmonic function on $\mathscr{R}$. First, (2.7) shows that the definition does not depend on coordinates. Next, since $\Phi_2 = 0$ if and only if $\Phi_1 = 0$ by (2.4), we may use the same coordinate z around a zero point p of Φ_2 as in the proof of Proposition 2.3, in which $|\Phi_1|^2 = |z|^{2l} |\varphi_1|$ for the non-zero holomorphic function φ_1 where $2l$ is the order of zero of $|\Phi_1|^2$ at p. This time, the Schwarz-type inequality

$$(3.1) \qquad |f_2| \leqq |\Phi_2|^2 = \frac{(1 - c_1) |\Phi_1|^4}{F}$$

shows that the order k of zero of f_2 at p satisfies $k \geqq 4l$. Hence putting $f_2 = z^k \hat{f}_2$ where $\hat{f}_2$ is non-zero holomorphic, we have

$$\frac{|f_2|^2 (1 - K)}{|\Phi_2|^4} = \frac{|f_2|^2}{(1 - c_1)^2 |\Phi_1|^6} = \frac{|z|^{2(k-3l)} |\hat{f}_2|^2}{(1 - c_1)^2 |\varphi_1|^3}$$

around p which assures us to extend G_2 analytically to p.

Now since we have

$$\partial\bar{\partial} \log G_2 = 0$$

away from the isolated zeros of G_2, it follows

$$\partial\bar{\partial} G_2 \geqq 0$$

everywhere on $\mathscr{R}$ by continuity. Hence $G_2 = c_2$ for some positive constant c_2 because M is compact and we have assumed $f_2 \not\equiv 0$. Moreover (3.1) shows

$$c_2 \leqq 1 - K$$

everywhere on M. Therefore rewriting $\partial\bar{\partial} \log |\Phi_1|^2 = 0$ as

$$\frac{1}{F}\partial\bar{\partial}\log(1-K)=2K \tag{3.2}$$

by using (1.2) and (1.5), we integrate the both sides on M to get

$$0=\int_M \frac{1}{F}\partial\bar{\partial}\log(1-K)\omega=4\pi\chi(M)$$

where $\omega=\sqrt{-1}Fdz\wedge d\bar{z}$ is the area element and $\chi(M)$ is the Euler characteristic of M. This contradicts to our assumption on genus.

Q.E.D.

Remark 3.2. Let M be a compact minimal surface in S^n whose metric satisfies the spherical Ricci condition away from the points where $K=1$. Let l_p denote the order of the zero of $|\Phi_1|^2$ (of $1-K$ at the same time) at $p\in M$ and let $N=\sum_{p\in M} l_p$. Then we have the potential theoretic formula

$$-2\pi N=\lim_{\varepsilon\to 0}\int_{M_\varepsilon}\frac{2}{F}\partial\bar{\partial}\log(1-K)\omega$$

where M_ε denotes the complement in M of an ε-neighborhood of all points where Φ_1 becomes zero. Under the spherical Ricci condition, i.e. under (3.2), we obtain

$$-2\pi N=8\pi\chi(M)\,,$$

which shows that the spherical Ricci condition does not make sense when M is a topological sphere. This is natural when we remember the fact that a minimal immersion of a topological 2-sphere into S^3 is totally geodesic, or more generally that if a topological 2-sphere fully immersed in S^n as a minimal surface, then n must be even [1].

Proposition 3.3. *Let M be as in* Proposition 3.1, *and let $c_1<1$. Then we have around the points where $K\neq 1$:*

$$\Phi_3=\partial\Phi_2-\frac{\partial|\Phi_2|^2}{|\Phi_2|^2}\Phi_2 \tag{3.3}$$

$$|\Phi_3|^2=(1-c_1)|\Phi_1|^4 \tag{3.4}$$

$$\bar{\partial}\Phi_3=-F\Phi_2 \tag{3.5}$$

$$|\bar{\partial}\Phi_3|^2=(1-c_1)F|\Phi_1|^4\,. \tag{3.6}$$

Proof. By (2.5), we have $\langle\partial\Phi_2,\bar{\Phi}_2\rangle=\partial|\Phi_2|^2$ and so (3.3) follows from $f_2\equiv 0$. Next, since $\partial\bar{\partial}|\Phi_2|^2=|\partial\Phi_2|^2-|\bar{\partial}\Phi_2|^2$, we obtain

$$\begin{aligned}|\Phi_3|^2 &= |\partial\Phi_2|^2 - \frac{|\partial|\Phi_2|^2|^2}{|\Phi_2|^2} \\ &= |\Phi_2|^2\left(\partial\bar{\partial}\log|\Phi_2|^2 + \frac{|\bar{\partial}\Phi_2|^2}{|\Phi_2|^2}\right) \\ &= (1-c_1)\frac{|\Phi_1|^4}{F}\left(FK + \frac{|\Phi_1|^2}{F}\right) \\ &= (1-c_1)|\Phi_1|^4\end{aligned}$$

by using (2.6) and (1.5). From (3.3) and (2.5), we know that $\bar{\partial}\Phi_3$ is composed of $\psi, \Phi_l, \bar{\Phi}_l$ where l is at most two. But as it is easily seen that

$$0 = \langle\bar{\partial}\Phi_3, \psi\rangle = \langle\bar{\partial}\Phi_3, \Phi_0\rangle = \langle\bar{\partial}\Phi_3, \bar{\Phi}_0\rangle = \langle\bar{\partial}\Phi_3, \Phi_1\rangle = \langle\bar{\partial}\Phi_3, \bar{\Phi}_1\rangle\,,$$

we may put $\bar{\partial}\Phi_3 = a\Phi_2 + b\bar{\Phi}_2$. Then from

$$af_2 + b|\Phi_2|^2 = \langle\bar{\partial}\Phi_3, \Phi_2\rangle = -\langle\Phi_3, \partial\Phi_2\rangle = 0$$

and

$$a|\Phi_2|^2 + b\bar{f}_2 = \langle\bar{\partial}\Phi_3, \bar{\Phi}_2\rangle = -\langle\Phi_3, \overline{\partial\Phi_2}\rangle = -|\Phi_3|^2$$

we get

$$a = -\frac{|\Phi_3|^2}{|\Phi_2|^2} = -F\,, \qquad b = 0$$

since $f_2 = 0$. (3.6) is easy. Q.E.D.

Remark 3.4. For Φ_3 and f_3, the same facts as in Remark 2.7 hold by virtue of (3.3) $\sim$ (3.5). In particular, by (1.9) and (2.7), we have

$$\Phi_3 dz^4 = \tilde{\Phi}_3 d\tilde{z}^4\,. \tag{3.7}$$

Corollary 3.5. *The extended $f_3 dz^8$ is a holomorphic differential form on $\mathscr{R}$.*

When $c_1 < 1$, let

$$G_3 := \frac{|f_3|^2}{|\Phi_3|^4}\,.$$

Then it is again shown that G_3 can be considered as a well-defined subharmonic function on $\mathscr{R}$, just as in the case of G_1 by using (3.7) and the Schwarz-type inequality around zeros of Φ_3. Hence there is some constant $0 \leqq c_3 \leqq 1$ such that

$$G_3 = c_3\,.$$

Proposition 3.6. *Let M be as in* Proposition 3.1, *and let $c_1 < 1$ and $c_3 < 1$. Then we have around the points where $K \neq 1$:*

$$\Phi_4 = \partial\Phi_3 - \frac{\partial|\Phi_3|^2}{|\Phi_3|^2}\Phi_3 + \frac{f_3}{|\Phi_2|^2}\bar{\Phi}_2 \tag{3.8}$$

$$|\Phi_4|^2 = (1 - c_1)(1 - c_3)F|\Phi_1|^4 \tag{3.9}$$

$$\bar{\partial}\Phi_4 = -F\Phi_3 + \frac{Ff_3}{(1 - c_1)|\Phi_1|^4}\bar{\Phi}_3 \tag{3.10}$$

$$|\bar{\partial}\Phi_4|^2 = (1 - c_1)(1 - c_3)F^2|\Phi_1|^4 . \tag{3.11}$$

Remark 3.7. When $c_3 = 1$, the above formulas hold trivially. Note that if we define Φ_4 by (3.8) instead of (1.11)′, it must be identically zero by (3.9).

Proof. (3.8) is a consequence of $\langle\partial\Phi_3, \bar{\Phi}_3\rangle = \partial|\Phi_3|^2$ and $\bar{\partial}G_3 = 0$ (cf. the proof of Proposition 2.5). We have

$$\begin{aligned}|\Phi_4|^2 &= |\partial\Phi_3|^2 - \frac{|\partial|\Phi_3|^2|^2}{|\Phi_3|^2} - \frac{|f_3|^2}{|\Phi_2|^2}\\ &= |\Phi_3|^2\left(\partial\bar{\partial}\log|\Phi_3|^2 + \frac{|\bar{\partial}\Phi_3|^2}{|\Phi_3|^2}\right) - \frac{c_3|\Phi_3|^4}{|\Phi_2|^2}\\ &= (1 - c_1)F|\Phi_1|^4 - c_3(1 - c_1)F|\Phi_1|^4\\ &= (1 - c_1)(1 - c_3)F|\Phi_1|^4\end{aligned}$$

by (3.6), (2.4) and (3.4). We may put $\bar{\partial}\Phi_4 = a\Phi_3 + b\bar{\Phi}_3$ as before and solve

$$\begin{aligned}af_3 + b|\Phi_3|^2 &= \langle\bar{\partial}\Phi_4, \Phi_3\rangle = 0 ,\\ a|\Phi_3|^2 + b\bar{f}_3 &= \langle\bar{\partial}\Phi_4, \bar{\Phi}_3\rangle = -|\Phi_4|^2\end{aligned}$$

to get

$$\begin{aligned}a &= \frac{-|\Phi_4|^2}{|\Phi_3|^2(1 - c_3)} = -F ,\\ b &= \frac{f_3|\Phi_4|^2}{|\Phi_3|^4(1 - c_3)} = \frac{Ff_3}{(1 - c_1)|\Phi_1|^4} .\end{aligned}$$

Q.E.D.

Remark 3.8. For Φ_4 and f_4, the same facts as in Remark 2.7 hold by virtue of (3.8) ~ (3.10). In particular,

$$\Phi_4 dz^5 = \tilde{\Phi}_4 d\tilde{z}^5 \tag{3.12}$$

follows from (1.9) and (3.7). Thus we have:

Corollary 3.9. *The extended $f_4 dz^{10}$ is a holomorphic differential form on $\mathscr{R}$.*

§ 4. Proof of Theorem II

Lemma 4.1. *Let M be as in* Proposition 3.1, *and let $c_1 < 1$. Then M lies in some totally geodesic 7-sphere of S^n if and only if $c_3 = 1$.*

Proof. Necessity is easy since

$$|\Phi_3|^4 - |f_3|^2 = |\Phi_3 \wedge \bar{\Phi}_3|^2$$

shows $c_3 = 1$ if and only if $\Phi_3 \wedge \bar{\Phi}_3 = 0$. Assume $c_3 = 1$. Then Φ_3 can be written as $\Phi_3 = \varphi_3 \xi$ for some real unit normal (local) vector field ξ where $\varphi_3^2 = f_3$. By reason of (3.5) and $\bar{\partial}\varphi_3 = 0$, we have

$$\partial\bar{\Phi}_3 = -F\bar{\Phi}_2 = \bar{\varphi}_3 \partial\xi$$

so that

$$\partial\xi = -\frac{F}{\bar{\varphi}_3}\bar{\Phi}_2$$

where $K \neq 1$. This shows that the normal subspace spanned by the first, second and third normal vectors (for definition, see Erbacher [2, p. 338]) is invariant under the parallel translation with respect to the normal connection. Since the isolated exceptional points where $K = 1$ have no influence on the conclusion, the lemma follows from the theorem of Erbacher [2, p. 339]

Proposition 4.2. *Let M be as in* Proposition 3.1. *If $n \leqq 9$, then M lies either in some totally geodesic* 3*-sphere or in some totally geodesic* 7*-sphere of S^n.*

Proof. Assume M does not lie in any totally geodesic 3-sphere, i.e. assume $c_1 < 1$. Then by (2.4) and (3.4), Φ_2 and Φ_3 are not identically zero, which means M lies in a sphere of dimension greater than six. Suppose M does not lie in any seven dimensional totally geodesic sphere. Then $c_3 < 1$ by the above lemma and hence Φ_4 is not identically zero by (3.9). Since $n \leqq 9$, $\Phi_4 \wedge \bar{\Phi}_4$ must be zero and it follows

$$|f_4|^2 = |\Phi_4|^4 = (1 - c_1)^2(1 - c_3)^2 F^2 |\Phi_1|^8 .$$

Therefore we have

$$0 = \partial\bar{\partial} \log |f_4|^2 = -2FK$$

by (1.2) and so M must be flat. This contradicts our assumption on genus. Q.E.D.

To prove Theorem II, we must investigate the case when M lies in some totally geodesic 7-sphere S^7 in the conclusion of the above proposition. In this case, note that $c_3 = 1$ and (3.8) holds for $\Phi_4 \equiv 0$ (cf. Remark 3.7).

In a coordinate neighborhood U where $\Phi_3 \neq 0$, we define

$$\psi^U := \frac{\Phi_3}{\sqrt{f_3}},$$

by choosing a square root of f_3 arbitrarily. Then as $\langle \psi^U, \psi^U \rangle = 1 = |\psi^U|^2$, ψ^U is real vector-valued and hence a map from U into S^7. Using

$$\partial \Phi_3 = \frac{\partial |\Phi_3|^2}{|\Phi_3|^2} \Phi_3 - \frac{f_3}{|\Phi_2|^2} \bar{\Phi}_2, \tag{4.1}$$

we have

$$\begin{aligned} \partial \psi^U &= \frac{1}{\sqrt{f_3}}\left(\partial \Phi_3 - \frac{\partial f_3}{2 f_3} \Phi_3\right) \\ &= -\frac{\sqrt{f_3}}{|\Phi_2|^2} \bar{\Phi}_2 \end{aligned} \tag{4.2}$$

since

$$\frac{\partial f_3}{2 f_3} = \frac{\partial |f_3|^2}{2 |f_3|^2} = \frac{\partial |\Phi_3|^4}{2 |\Phi_3|^4} = \frac{\partial |\Phi_3|^2}{|\Phi_3|^2}. \tag{4.3}$$

By (3.5), we get

$$\bar{\partial} \psi^U = -\frac{F}{\sqrt{f_3}} \Phi_2 \tag{4.4}$$

and so we obtain

$$\begin{cases} \langle \partial \psi^U, \partial \psi^U \rangle = 0 \\ \langle \partial \psi^U, \bar{\partial} \psi^U \rangle = F. \end{cases} \tag{4.5}$$

Furthermore by (4.4) and (3.3) we have

$$\begin{aligned} \partial \bar{\partial} \psi^U &= -F \frac{\partial \Phi_2}{\sqrt{f_3}} - \partial\left(\frac{F}{\sqrt{f_3}}\right) \Phi_2 \\ &= -F \psi^U - \left\{\frac{F}{\sqrt{f_3}} \frac{\partial |\Phi_2|^2}{|\Phi_2|^2} + \partial\left(\frac{F}{\sqrt{f_3}}\right)\right\} \Phi_2 \end{aligned}$$

while it follows from $|f_3| = |\Phi_3|^2 = F|\Phi_2|^2$ that

$$\begin{aligned}\frac{F}{\sqrt{f_3}}\frac{\partial|\Phi_2|^2}{|\Phi_2|^2} + \partial\left(\frac{F}{\sqrt{f_3}}\right) &= \frac{F}{\sqrt{f_3}}\partial\log\left(\frac{F}{\sqrt{f_3}}|\Phi_2|^2\right)\\ &= \frac{F}{\sqrt{f_3}}\partial\log\left(\frac{\sqrt{\bar{f}_3}}{|f_3|}\cdot F|\Phi_2|^2\right)\\ &= \frac{F}{\sqrt{f_3}}\partial\log\sqrt{\bar{f}_3} = 0\,.\end{aligned}\tag{4.6}$$

Hence we get

$$\partial\bar{\partial}\psi^U = -F\psi^U\,,\tag{4.7}$$

which with (4.5) shows that ψ^U is an isometric minimal immersion of U into S^7.

Now we define an isometric minimal immersion $\psi_\theta^U: U \to S^7$ for $0 \leqq \theta < \pi$ by

$$\psi_\theta^U := \cos\theta\psi + \sin\theta\psi^U\,.$$

Let

$$\Phi_\theta^U := \partial^2\psi_\theta^U - \frac{\partial F}{F}\partial\psi_\theta^U$$

as (1.4) for ψ_θ^U. Since we have

$$\begin{aligned}\partial^2\psi^U - \frac{\partial F}{F}\partial\psi^U &= -\frac{\sqrt{f_3}}{|\Phi_2|^2}\partial\bar{\Phi}_2 - \left\{\partial\left(\frac{\sqrt{f_3}}{|\Phi_2|^2}\right) - \frac{\partial F}{F}\frac{\sqrt{f_3}}{|\Phi_2|^2}\right\}\bar{\Phi}_2\\ &= -\frac{\bar{f}_1\sqrt{f_3}}{F|\Phi_2|^2}\Phi_1 + \frac{|\Phi_1|^2\sqrt{f_3}}{F|\Phi_2|^2}\bar{\Phi}_1\\ &= -\frac{\bar{f}_1\sqrt{f_3}}{(1-c_1)|\Phi_1|^4}\Phi_1 + \frac{\sqrt{f_3}}{(1-c_1)|\Phi_1|^2}\bar{\Phi}_1\end{aligned}$$

by (4.6), (2.5) and (2.4), it follows that

$$\Phi_\theta^U = \left(\cos\theta - \sin\theta\frac{\bar{f}_1\sqrt{f_3}}{(1-c_1)|\Phi_1|^4}\right)\Phi_1 + \sin\theta\frac{\sqrt{f_3}}{(1-c_1)|\Phi_1|^2}\bar{\Phi}_1\,.$$

Here we claim $f_1 = \alpha\sqrt{f_3}$ for some constant $\alpha \in \boldsymbol{C}$ such that $|\alpha|^2 = c_1/(1-c_1)$. This is easily shown, since f_1^2/f_3 is meromorphic on $\mathscr{R}$, and $|f_1^2/f_3| = c_1/(1-c_1)$. Finally we have by $|f_3| = |\Phi_3|^2 = (1-c_1)|\Phi_1|^4$,

$$(4.8)\qquad \Phi_\theta^U = (\cos\theta - \bar{\alpha}\sin\theta)\Phi_1 + \sin\theta\frac{\sqrt{f_3}}{(1-c_1)|\Phi_1|^2}\bar{\Phi}_1 .$$

To find out immersions ψ_θ^U of U into totally geodesic 3-spheres, we must choose θ so that $\Phi_\theta^U \wedge \bar{\Phi}_\theta^U = 0$ by Proposition 1.3, which is equivalent to

$$|\cos\theta - \bar{\alpha}\sin\theta|^2 - \left|\sin\theta\frac{\sqrt{f_3}}{(1-c_1)|\Phi_1|^2}\right|^2 = 0 ,$$

or

$$\cos^2\theta - (\alpha + \bar{\alpha})\cos\theta\sin\theta - \sin^2\theta = 0 .$$

Since $\theta = \pi/2$ does not satisfy this equation, we may consider the equation

$$\tan^2\theta + (\alpha + \bar{\alpha})\tan\theta - 1 = 0$$

which has clearly two solutions θ_1, θ_2 such that $\theta_2 = \theta_1 + \pi/2$.

For $k = 1, 2$, put $\psi_k^U := \psi_{\theta_k}^U$ and let $\boldsymbol{R}_k^4$ be the linear subspace of $\boldsymbol{R}^8$ such that the totally geodesic 3-sphere $S_k^3 := S^7 \cap \boldsymbol{R}_k^4$ contains the image $\psi_k^U(U)$, which is minimal in it. Then we claim that $\boldsymbol{R}_1^4$ and $\boldsymbol{R}_2^4$ are orthogonal to each other in $\boldsymbol{R}^8$. To see this, we may show for any $X_k \in \{\psi_k^U, \partial\psi_k^U, \bar{\partial}\psi_k^U, \Phi_{\theta_k}^U\}$; $k = 1, 2$, $\langle X_1, X_2\rangle = 0$. For instance,

$$\langle\psi_1^U, \psi_2^U\rangle = \cos\theta_1\cos\theta_2 + \sin\theta_1\sin\theta_2 = 0 ,$$

and by using Lemma 1.2, (4.2), (4.4) and (4.5), we have

$$\begin{aligned}
\langle\psi_1^U, \partial\psi_2^U\rangle &= \langle\psi_1^U, \bar{\partial}\psi_2^U\rangle = 0 ,\\
\langle\psi_1^U, \Phi_{\theta_2}^U\rangle &= -\langle\partial\psi_1^U, \partial\psi_2^U\rangle = 0 ,\\
\langle\partial\psi_1^U, \bar{\partial}\psi_2^U\rangle &= (\cos\theta_1\cos\theta_2 + \sin\theta_1\sin\theta_2)F = 0 ,
\end{aligned}$$

and

$$\langle\partial\psi_1^U, \Phi_{\theta_2}^U\rangle = \langle\bar{\partial}\psi_1^U, \Phi_{\theta_2}^U\rangle = 0 .$$

To prove $\langle\Phi_{\theta_1}^U, \Phi_{\theta_2}^U\rangle = 0$, we use

$$\tan\theta_1 + \tan\theta_2 = -(\alpha + \bar{\alpha}) \quad\text{and}\quad \tan\theta_1\tan\theta_2 = -1$$

as follows:

$$\langle \Phi^U_{\theta_1}, \Phi^U_{\theta_2} \rangle = \cos\theta_1 \cos\theta_2 \langle (1 - \bar{\alpha}\tan\theta_1)\Phi_1 + \tan\theta_1 \frac{\sqrt{f_3}}{(1-c_1)|\Phi_1|^2}\bar{\Phi}_1 ,$$

$$(1 - \bar{\alpha}\tan\theta_2)\Phi_1 + \tan\theta_2 \frac{\sqrt{f_3}}{(1-c_1)|\Phi_1|^2}\bar{\Phi}_1 \rangle$$

$$= \cos\theta_1\cos\theta_2 \Big[\{1 + \bar{\alpha}(\alpha+\bar{\alpha}) - \bar{\alpha}^2\} f_1$$

$$- \{(\alpha+\bar{\alpha}) - 2\bar{\alpha}\} \frac{\sqrt{f_3}}{1-c_1} - \frac{f_3\bar{f}_1}{(1-c_1)^2|\Phi_1|^4} \Big]$$

$$= \cos\theta_1\cos\theta_2 \left\{ (1+|\alpha|^2) f_1 - (\alpha - \bar{\alpha}) \frac{\sqrt{f_3}}{1-c_1} - \frac{\bar{\alpha}\sqrt{f_3}}{1-c_1} \right\}$$

$$= \cos\theta_1\cos\theta_2 (\alpha - \alpha + \bar{\alpha} - \bar{\alpha}) \frac{\sqrt{f_3}}{1-c_1}$$

$$= 0 .$$

Finally it turns out that $\boldsymbol{R}^8 = \boldsymbol{R}^4_1 \oplus \boldsymbol{R}^4_2$ (direct sum) and

$$\psi = \cos\theta_1 \psi^U_1 \oplus \cos\theta_2 \psi^U_2 \tag{4.10}$$

on U.

Claim. For $k = 1, 2$, let $\pi_k : \boldsymbol{R}^8 \to \boldsymbol{R}^4_k$ be the projection map, and define $\psi_k := (1/\cos\theta_k)\pi_k \cdot \psi$. Then ψ_k is an isometric minimal immersion of $\mathscr{R}$ into S^3_k such that $\psi = \cos\theta_1\psi_1 \oplus \cos\theta_2\psi_2$.

First we prove the claim on $\mathscr{R}' := \{p \in \mathscr{R} : K \neq 1 \text{ at } p\}$. Note that the points of $\mathscr{R} - \mathscr{R}'$ are isolated since

$$|f_3| = |\Phi_3|^2 = (1 - c_1)|\Phi_1|^4 = (1 - c_1)F^4(1 - K)^2$$

shows that they are the zeros of holomorphic function f_3 which is not identically zero (Remark 2.2). Let $p \in \mathscr{R}'$, and let c be a simple curve in $\mathscr{R}'$ from p to a point q of U. Then we can extend ψ^U along c to a map ψ^V where V is a neighborhood of c, and the decomposition (4.10) becomes $\psi = \cos\theta_1\psi^V_1 \oplus \cos\theta_2\psi^V_2$ on V.

Clearly we have $\psi_k = \psi^V_k$ on V and hence ψ_k is isometric and minimal in the neighborhood V of p. At a point $p \in \mathscr{R} - \mathscr{R}'$, ψ_k is analytic since $\psi = \cos\theta_1\psi_1 \oplus \cos\theta_2\psi_2$ is analytic. Then $\langle \partial\psi_k, \partial\psi_k \rangle = 0$, $\langle \partial\psi_k, \bar{\partial}\psi_k \rangle = F$ and $\partial\bar{\partial}\psi_k = -F\psi_k$ at p follow from the continuity. Thus the proof of Theorem II is completed. Q.E.D.

Remark 4.3. To prove the conjecture in more general case in the same way as above, there are two problems. First, as a necessary condition for ψ to be of the form (**), we have $G_1 = c_1$, and $G_2 = 0$, both of which we could not show without the global assumption on M. Next, when n is general, we need successive formulas similar to (2.3) ~ (2.6) for Φ_{4k+2} with $f_{4k+2} = 0$, (3.3) ~ (3.6) for Φ_{2k+1} with $|f_{2k+1}|^2/|\Phi_{2k+1}|^4 =$ const., and (3.8) ~ (3.11) for Φ_{4k} with $f_{4k} = 0$. We failed in showing $f_4 = 0$ in the above method and hence we do not know whether the conjecture is true for $n > 9$ in the case of Theorem II.

§ 5. Appendix

It is well-known that if a positive function u of class C^2 in a simply-connected plain domain D satisfies $\partial\bar{\partial} \log u = 0$, then there exists a holomorphic function ω on D such that $u = |\omega|$. We extend this fact as follows:

Lemma 2.4. *Let D be a plane domain containing the origin with coordinate $z = x_1 + \sqrt{-1}x_2$. Let u be a real analytic non-negative function on D such that $u(0) = 0$. If u is not identically zero and satisfies $\partial\bar{\partial} \log u = 0$ away from the points where $u = 0$, then the order l' of zero of u at the origin is even and there exists a non-zero holomorphic function ω around the origin such that*

$$u = |z|^{2l} |\omega|$$

where $l' = 2l$.

Proof. Since u is real analytic and is not identically zero, we can consider the order l' of zero of u at 0. Moreover as $u \geqq 0$, l' must be even. Putting $l' = 2l$, we expand u around the origin as

$$u(z) = \sum_{k=0}^{2l} a_k z^k \bar{z}^{2l-k} + O(|z|^{2l+1}) \tag{5.1}$$

where $a_k = \bar{a}_{2l-k}$ is a complex number for each $0 \leqq k \leqq 2l$, and $O(|z|^{2l+1})$ denotes the higher order terms. Rewrite $\partial\bar{\partial} \log u = 0$ as

$$u\partial\bar{\partial}u = \partial u \bar{\partial} u\ , \tag{5.2}$$

and substitute (5.1) into (5.2). Then comparing the least order terms on the both hand sides, we obtain

$$\left(\sum_{k=0}^{2l} a_k z^k \bar{z}^{2l-k}\right)\left(\sum_{k=1}^{2l-1} k(2l-k) a_k z^{k-1} \bar{z}^{2l-k-1}\right)$$
$$= \left(\sum_{k=1}^{2l} k a_k z^{k-1} \bar{z}^{2l-k}\right)\left(\sum_{k=0}^{2l-1} (2l-k) a_k z^k \bar{z}^{2l-k-1}\right),$$

from which it follows $a_k = 0$ for all $k \neq l$ through a careful investigation. Thus (5.1) becomes

$$u = a|z|^{2l} + O(|z|^{2l+1})$$

for some positive constant a. Defining

$$\frac{u}{|z|^{2l}} = a$$

we obtain a continuous positive function $u/|z|^{2l}$ in a simply-connected neighborhood V of the origin.

Now we can define a continuous function $h(z)$ in V by

$$h(z) := \log \frac{u}{|z|^{2l}} - \log a$$

which is evidently harmonic where $z \neq 0$.

Claim. $h(z)$ is harmonic everywhere in V.

For convenience, we may consider $V = \{z : |z| < 1\}$, in the closure of which $h(z)$ is continuous. Then the function $g(z)$ given by

$$g(z) := \frac{1}{2\pi} \int_0^{2\pi} \frac{1 - |z|^2}{|\exp(\sqrt{-1}t) - z|^2} h(\exp(\sqrt{-1}t))dt \qquad |z| < 1$$

$$g(z) := h(z) \qquad |z| = 1$$

is harmonic in V and continuous in $\bar{V}$. Letting f be a holomorphic function in D such that $|f(z)| \leqq 1$ in $\bar{V}$, $f(0) = 0$ and $f(z) \neq 0$ for $z \in V - \{0\}$, we define

$$H(z; \alpha) := g(z) - h(z) + \alpha \log |f(z)|$$

for $\alpha \in \boldsymbol{R}$. Then we have

(1) $H(z; \alpha)$ is harmonic in $V - \{0\}$,

(2) For any $\varepsilon > 0$,

$$\limsup_{z \to 0} H(z; \varepsilon) \leqq 0 \quad \text{and} \quad \liminf_{z \to 0} H(z; -\varepsilon) \geqq 0\,,$$

the latter follows from the fact that $g(z) - h(z)$ is bounded in $\bar{V}$ and that $\lim_{z \to 0} \log |f(z)| = -\infty$. Then by the maximum principle for harmonic functions, we obtain

$$H(z; \varepsilon) \leqq 0 \quad \text{and} \quad H(z; -\varepsilon) \geqq 0$$

in $V - \{0\}$ and hence

$$|g(z) - h(z)| \leqq -\varepsilon \log |f(z)|$$

for $z \in V - \{0\}$ and any $\varepsilon > 0$. It turns out that $g(z) = h(z)$ in $V - \{0\}$ and so everywhere in V by the continuity. Finally we know that $\log (u/|z|^{2l})$ can be considered as a harmonic function in V. Thus there exists a holomorphic function ω in V such that $|\omega| = u/|z|^{2l}$ and the proof is completed.

References

[1] J. L. M. Barbosa, *On minimal immersions of S^2 in S^{2m}*, Thesis, Univ. of of California (1972).
[2] J. Erbacher, *Reduction of the codimension of an isometric immersion,* J. Differential Geometry, **5** (1971), pp. 333–340.
[3] E. Heinz, *Ein elementarer Beweis des Satzes von Radó-Behnke-Stein-Cartan über analytische Funktionen,* Math. Ann., **131** (1956), pp. 258–259.
[4] T. Itoh, *On minimal surfaces in a Riemannian manifold of constant curva-* ture, Math. J. of Okayama Univ., **17** (1974), pp. 19–38.
[5] K. Kenmotsu, *On compact minimal surfaces with non-negative Gaussian curvature in a space of constant curvature* II, Tôhoku Math. J., **27** (1975), pp. 291–301.
[6] H. B. Lawson, *Complete minimal surfaces in S^3*, Ann. of Math., **92** (1970), pp. 335–374.
[7] ——, *Some intrinsic characterizations of minimal surfaces,* J. D'anal. Math., **24** (1971), pp. 151–161.
[8] T. Otsuki, *Minimal submanifolds with m-index 2 and generalized Veronese surfaces,* J. Math. Soc. Japan, **24** (1972), pp. 89–122.
[9] C. G. Ricci, Opere, Vol. 1, Edizione Cremonese, Rome, (1956), p. 411.

Department of Mathematics
Tokyo Institute of Technology
Tokyo, 152
Japan

Minimal Submanifolds and Geodesics
Kaigai Publications, Tokyo, 1978, 143–161

UNIQUENESS AND NON-UNIQUENESS FOR PLATEAU'S PROBLEM—ONE OF THE LAST MAJOR QUESTIONS

JOHANNES C. C. NITSCHE

1. One of the most vexing questions in connection with the classical Plateau problem which remains unanswered today, although it has been a challenge at least since 1939, is concerned with the phenomena of uniqueness and non-uniqueness and with the isolated character of minimal surfaces. It is unknown whether every solution of Plateau's problem, stable or unstable, is isolated (in a suitable topology). Were this the case, only a finite number of disc-type minimal surfaces could be spanned in a given contour. With the exception of those situations where uniqueness is assured beforehand, notably my "4π-theorem" [19], it has also not yet been possible to estimate the number of solutions for Plateau's problem by the geometric properties of their bounding contours. Thus the long range problem remains (see [21], pp. X, 689, 690):

To prove that a reasonable Jordan curve Γ bounds only finitely many solutions of Plateau's problem, and to estimate the number of these solutions by the geometric properties of Γ.

Here the word "reasonable" is used to describe one of the following classes of curves:

i) Γ is an analytic, or sufficiently regular, differential geometric curve.

i) Γ is a polygon.

iii) Γ is an analytic, or sufficiently regular, differential geometric curve. Independent (geometric) considerations imply a priori that all disc-type minimal surfaces spanning Γ are free of branch points on their boundary.

iv) Γ is an analytic, or sufficiently regular, differential geometric curve. Independent considerations imply a priori that all disc-type minimal surfaces spanning Γ are free of branch points in their interior and on their boundary.

For example, a Jordan curve of class $C^{1,\alpha}$, $0 < \alpha < 1$, which lies on the boundary of a convex domain (i.e., an extreme curve) satisfies condition iii); see [21], § 366. There are also various geometric properties which imply condition iv) (see [25], p. 35; [21], §§ 376–386; [23]).

Condition iv) is satisfied for a regular analytic contour possessing a single-valued (parallel or central) projection onto a plane starshaped curve. Of course, by a theorem of T. Radó ([24], p. 10) the solution of Plateau's problem will be unique if Γ allows such a projection onto a convex curve.

Heuristic constructions by P. Lévy and R. Courant which suggest the existence of rectifiable curves bounding continua of minimal surfaces point to the complexity of the issue. For a description see [11], pp. 28–31; [6], pp. 119–122; [17], pp. 396–398; [21], §§ 834–836. It must be noted that the constructions in question have not been and, in fact, may never be substantiated by rigorous proofs.

2. We start with a brief definition of our terms. Let Γ be a rectifiable Jordan curve of length l in Euclidean space and $\{\mathfrak{x} = \mathfrak{z}(s)\,;\, 0 \leq s \leq l\}$ a representation of Γ in terms of its arc length parameter s. We are concerned with vectors $\mathfrak{x}(u, v) = \{x(u, v), y(u, v), z(u, v)\}$ defined in the closure $\bar{P}$ of the unit disc $P = \{u, v\,;\, u^2 + v^2 < 1\}$ which map the boundary ∂P onto Γ. Setting $u + iv = w = \rho e^{i\theta}$ we shall interchangeably use the notations $\mathfrak{x}(u, v)$, or $\mathfrak{x}(w)$, or $\mathfrak{x}(\rho, \theta)$-whichever is most convenient. Denote by $\mathfrak{H} = \mathfrak{H}(\Gamma)$ the set of all vectors $\mathfrak{x}(w) \in C^2(P) \cap C^0(\bar{P})$ which are harmonic in P and which map ∂P onto Γ in a monotonic manner such that three fixed distinct points w_1, w_2, w_3 on ∂P are transformed into three fixed distinct points $\mathfrak{z}_j = \mathfrak{z}(s_j)$, $(j = 1, 2, 3)$, on Γ, respectively. Each element $\mathfrak{x}(u, v)$ of $\mathfrak{H}$ defines a harmonic surface $S = \{\mathfrak{x} = \mathfrak{x}(u, v)\,;\, (u, v) \in \bar{P}\}$ bounded by Γ. In view of the isoperimetric inequality the area of this surface,

$$A[\mathfrak{x}] = \iint_P |\mathfrak{x}_u \times \mathfrak{x}_v|\, dudv\ ,$$

cannot exceed the value $l^2/4\pi$, although the Dirichlet integral

$$D[\mathfrak{x}] = \frac{1}{2}\iint_P (\mathfrak{x}_u^2 + \mathfrak{x}_v^2) dudv\ ,$$

which is subject to the inequality $A[\mathfrak{x}] \leq D[\mathfrak{x}]$, may well be infinite. The equality $A[\mathfrak{x}] = D[\mathfrak{x}]$ holds if, and only if, the additional relations $\mathfrak{x}_u^2 = \mathfrak{x}_v^2$, $\mathfrak{x}_u\mathfrak{x}_v = 0$ identifying u and v as isothermal parameters on S are valid in P. By definition, a *solution of Plateau's problem* for the curve Γ is a generalized minimal surface $S = \{\mathfrak{x} = \mathfrak{x}(u, v)\,;\, (u, v) \in \bar{P}\}$ whose position vector is an element of $\mathfrak{H}$ satisfying precisely these relations. We denote by $\mathfrak{M} = \mathfrak{M}(\Gamma)$ the set of all such vectors. It is a matter of record that the mapping of ∂P onto Γ effected by $\mathfrak{x}(e^{i\theta})$ is a homeomorphism.

The term "generalized" above is used to indicate that the surface S may possess branch points in whose neighborhood S loses its differential geometric character; see [15], p. 232 or [21], § 283. Branch points corresponding to interior points of the parameter domain P are isolated. If Γ belongs to the regularity class $C^{1,1}$, the same holds true for boundary branch points; see [21], § 381.

A solution of Plateau's problem $S = \{\mathfrak{x} = \mathfrak{x}(u, v)\,;\, (u, v) \in \bar{P}\}$ is called *isolated* if there exists a number $\varepsilon > 0$ depending on the properties of Γ and S only such that $\max_{(u,v)\in \bar{P}} |\mathfrak{y}(u, v) - \mathfrak{x}(u, v)| \geq \varepsilon$ for all vectors $\mathfrak{y}(u, v) \in \mathfrak{M}$. Every component of the space $\mathfrak{M}$, that is, every maximal connected closed subset of $\mathfrak{M}$ (in a suitable topology $-C^0(\bar{P}), C^k(\bar{P})$, $H^k(\bar{P})$, etc.) is called a *block* of (generalized) minimal surfaces. A priori it is conceivable that a block of minimal surfaces could be a genuine continuum consisting of more than a single element. Whether this is in fact ever possible has been an intriguing question for the last forty years (see [13], p. 466; [28], p. 854; [6], p. 122; [9], p. 301; [17], p. 394; [30], p. 313; [21], §§ 423, 906).

3. The total curvature $\kappa(\Gamma)$ of the Jordan curve Γ is the supremum of the numbers $\tilde{\kappa}(\Pi)$ for all closed polygons Π inscribed in Γ. Here $\tilde{\kappa}(\Pi)$ is defined as follows: For a closed polygon Π with vertices $\mathfrak{p}_1, \mathfrak{p}_2, \cdots, \mathfrak{p}_m$ $(\mathfrak{p}_m = \mathfrak{p}_0, \mathfrak{p}_{m+1} = \mathfrak{p}_1)$ $\tilde{\kappa}(\Pi)$ is the sum $\sum_{j=1}^{m} \alpha_j$ of the exterior angles α_j, $0 < \alpha_j < \pi$, i.e., the angles between the vectors $\mathfrak{p}_j - \mathfrak{p}_{j-1}$ and $\mathfrak{p}_{j+1} - \mathfrak{p}_j$. For a closed polygon Π the numbers $\kappa(\Pi)$ and $\tilde{\kappa}(\Pi)$ are identical. If Γ is a regular curve of class C^2 then its curvature $k(s) = |\mathfrak{z}''(s)|$ is defined, and $\kappa(\Gamma) = \int_0^l k(s)ds$.

Later on, in section **6**, we shall be interested in simple closed polygons Π whose total curvature $\kappa(\Pi)$ is smaller than 4π. Obviously, every quadrilateral has this property. On the other hand, there are hexagons whose total curvature comes as close to 6π as one pleases. As for pentagons, a brief reflection will convince the reader that *the total curvature of every pentagon is smaller than* 4π. Consider a simple closed pentagon with vertices $\mathfrak{p}_1, \mathfrak{p}_2, \mathfrak{p}_3, \mathfrak{p}_4, \mathfrak{p}_5$ and denote by $\beta_j = \pi - \alpha_j$, $0 < \beta_j < \pi$, its interior angles, i.e., the angles between the vectors $\mathfrak{p}_{j-1} - \mathfrak{p}_j$ and $\mathfrak{p}_{j+1} - \mathfrak{p}_j$ $(j = 1, \cdots, 5)$. Our statement is a consequence of the inequality $\sum \beta_j > \pi$. A similar inequality holds for any simple closed $(2m + 1)$-gon. In particular, *the total curvature of every heptagon is smaller than* 6π.

For the proof*) set $\mathfrak{a}_j = \mathfrak{p}_{j+1} - \mathfrak{p}_j$ $(j = 1, \cdots, 5)$ and $\mathfrak{a}'_j = |\mathfrak{a}_j|^{-1} \mathfrak{a}_j$, $\mathfrak{a}''_j = -\mathfrak{a}'_j$. If the initial points of the vectors $\mathfrak{a}'_j, \mathfrak{a}''_j$ are placed at the origin of the coordinate system, their end points will lie on the unit sphere S_2. The angle β_j is equal to the distance on S_2 between the end points of $\mathfrak{a}'_{j-1}$ and $\mathfrak{a}''_j$ or $\mathfrak{a}''_{j-1}$ and $\mathfrak{a}'_j$, and the sum $\sum \beta_j$ is equal to the sum of the

*) I am grateful for this proof of the inequality $\sum \beta_j > \pi$ which is shorter than my own to my colleague Ian Richards.

distances between the end points of $\mathfrak{a}_1'$ and $\mathfrak{a}_2''$, $\mathfrak{a}_2''$ and $\mathfrak{a}_3'$, $\mathfrak{a}_3'$ and $\mathfrak{a}_4''$, $\mathfrak{a}_4''$ and $\mathfrak{a}_5'$, $\mathfrak{a}_5'$ and $\mathfrak{a}_6'' = \mathfrak{a}_1''$. Since the vectors $\mathfrak{a}_1'$ and $\mathfrak{a}_1''$ have opposite directions, this sum must be larger than or equal to π. Equality can hold only if all vectors $\mathfrak{a}_j$ lie in the same plane which is impossible for a Jordan polygon.

4. The first proof of the isolatedness of some special minimal surfaces was presented 1969 in [17]; see also [20]. In fact, [17] treats a case where $\lambda = 1$ is the first eigenvalue of the associated eigenvalue problem (3) below. Subsequently, R. Böhme and F. Tomi [4] and R. Böhme [3] studied the structure of the space $\mathfrak{M}(\Gamma)$. They proved for a regular analytic contour Γ that $\mathfrak{M}(\Gamma)$ is the union of finitely many blocks in each $C^k(\bar{P})$, $0 \leq k \leq \infty$ ([3], p. 40), and for a contour of class C^∞ that the values of the Dirichlet integral $D[\mathfrak{x}]$ for all $\mathfrak{x} \in \mathfrak{M}$ form a compact set of vanishing Lebesgue measure on the number line ([4], p. 15). F. Tomi [30] showed that a regular analytic Jordan curve bounds only finitely many minimal surfaces of least area. Employing a remark made earlier in this paper, one can use Tomi's method to derive the

Theorem. *An extreme Jordan curve of class C^4 cannot bound infinitely many solutions of Plateau's problem of least area.*

In general, surfaces of least area will not be unique. There are examples of extreme curves bounding at least two such surfaces; see [16] and [21], §§ 388–397. Recently, Tomi [31] has extended his result slightly so a special class of curves satisfying a number of geometric conditions one of which (the "boundary 0-point Radó condition") is evidently equivalent to the requirement that Γ be extreme: If Γ is not a plane curve, and each point of Γ possesses a supporting plane, then Γ must lie on the boundary of its convex hull.

While we thus have some knowledge about minimal surfaces whose areas represent absolute, or relative minima, information regarding the totality of solutions of Plateau's problem, including the unstable ones, is much more difficult to come by. The presence of unstable minimal surfaces poses a real challenge. In the case of non-uniqueness a contour always bounds at least one such surface. There are, of course, those who hold that because unstable minimal surfaces cannot be realized experimentally, the study of their number and their properties is a sterile exercise.

A theorem including all solutions of Plateau's problem has been discovered in [23]:

Theorem. *A regular analytic Jordan curve Γ satisfying condition* iv) *above whose total curvature $\kappa(\Gamma)$ does not exceed the value 6π cannot bound infinitely many solutions of Plateau's problem.*

The assumption of analyticity of Γ can be reduced; it is sufficient to know that Γ belongs to class $C^{3,\alpha}$. It should be noted that the stronger inequality $\kappa(\Gamma) \leq 4\pi$ implies uniqueness; see [19]. Uniqueness holds also for any $C^{2,\alpha}$-curve Γ as long as $\kappa(\Gamma) < 4\pi$.

We shall restrict our comments to one of the main steps of the proof. Let us assume that the curve Γ were capable of bounding infinitely many solutions of Plateau's problem. A compactness argument then shows that at least one of these solutions, say $S^{(0)} = \{\mathfrak{x} = \mathfrak{x}^{(0)}(u, v) ; (u, v) \in \bar{P}\}$, cannot be isolated: Each C^0-neighborhood of $\mathfrak{x}^{(0)}(u, v)$ in $\mathfrak{M}$ contains a vector $\mathfrak{x}(u, v) \in \mathfrak{M}$. In view of the universal boundary regularity of minimal surfaces (see [18] and [21], §§ 336–349) a similar statement is true regarding any C^k-neighborhood, $0 \leq k < \infty$, of $\mathfrak{x}^{(0)}(u, v)$. It is well known (see for instance [4], § 3) that the neighborhood can be restricted so that every surface S in this neighborhood has a representation with the position vector

$$(1) \qquad \mathfrak{x}(u, v) = \mathfrak{x}^{(0)}(u, v) + \zeta(u, v)\mathfrak{X}^{(0)}(u, v) .$$

Here $\mathfrak{X}^{(0)}(u, v)$ denotes the unit normal vector of $S^{(0)}$. The function $\zeta(u, v) \in C^{2,\alpha}(\bar{P})$ vanishes on ∂P and satisfies the inequality $\|\zeta\|_{2,\alpha} \leq \mathscr{C}\|\mathfrak{x} - \mathfrak{x}^{(0)}\|_{2,\alpha}$, where $\mathscr{C}$ is a constant depending on the properties of $S^{(0)}$ only. Note that u and v are isothermal parameters on $S^{(0)}$, but not on S. The condition that $\mathfrak{x}(u, v)$ be the position vector of a minimal surface leads to the following non-linear differential equation for $\zeta(u, v)$:

$$(2) \quad \Delta\zeta + p(u, v)\zeta = \Phi[\zeta] \equiv \Phi_2[\zeta] + \Phi_3[\zeta] + \Phi_3[\zeta] + \Phi_4[\zeta] + \Phi_5[\zeta] .$$

Here each $\Phi_m[\zeta]$ is a homogeneous polynomial of degree m in $\zeta, \zeta_u, \zeta_v, \zeta_{uu}, \zeta_{uv}, \zeta_{vv}$ with coefficients which depend on the vector $\mathfrak{x}^{(0)}(u, v)$ and its derivatives, and $p(u, v) = -2E^{(0)}(u, v)K^{(0)}(u, v) \geq 0$ where $E^{(0)}(u, v) = [\mathfrak{x}^{(0)}(u, v)]^2$ and $K^{(0)}(u, v)$ is the Gaussian curvature of $S^{(0)}$. The discussion of (2), i.e., the search for non-trivial solutions $\zeta(u, v)$ of (2), is closely related to the eigenvalue problem

$$(3) \qquad \Delta\eta + \lambda p(u, v)\eta = 0 \text{ in } P , \qquad \eta = 0 \text{ on } \partial P .$$

This problem has a sequence of eigenvalues $\{\lambda_n\}$ satisfying the inequalities $0 < \lambda_1 < \lambda_2 \leq \lambda_3 \leq \cdots$ and corresponding eigenfunctions $\eta_n(u, v)$ subject to the ortho-normality relations

$$(4) \qquad \iint_P p(u, v)\eta_m\eta_n du dv = \delta_{mn} .$$

Let $k_g(s)$ and $k_n(s)$ denote the geodesic curvature and normal curvature, respectively, of Γ, regarded as a curve on $S^{(0)}$. From the Gauss-Bonnet formula and the relation $k^2 = k_g^2 + k_n^2$ we conclude that

$$(5) \quad 2\pi + \iint_{S^{(0)}} |K^{(0)}|\, do^{(0)} = \int_\Gamma k_g(s)ds \leq \int_\Gamma k(s)ds \equiv \kappa(\Gamma) \leq 6\pi \,.$$

Remember that $S^{(0)}$ is a regular minimal surface.

For a generalized minimal surface $S = \{\mathfrak{x} = \mathfrak{x}(u, v)\,;\, (u, v) \in \bar{P}\}$ with branch points a counterpart to (5) has, under certain implicit assumptions, been derived by S. Sasaki [26]:

$$(6) \quad 1 + \sum (m_\alpha - 1) + \sum M_\beta + \frac{1}{2\pi}\iint_S |K|\, do \leq \frac{1}{2\pi}\kappa(\Gamma) \,.$$

Here $m_\alpha - 1$ denotes the orders of the interior branch points and $2M_\beta$ the orders of the boundary branch points. (The latter must be even; see [18], p. 332.) For a derivation and discussion of (6) see [14] and [21], §§ 377–381. While Sasaki employed (6) to give a new proof of Fenchel's theorem according to which $\kappa(\Gamma) \geq 2\pi$ (see [21], § 26), the workers in minimal surface theory prefer to read the inequality in the opposite direction, thereby obtaining estimates regarding the number and order of branch points on the solutions of Plateau's problem as well as bounds for their total curvature. There is also a useful generalization of (6) for the case that the contour Γ is merely continuous, but has finite total curvature; see [21], § 377:

$$(6') \qquad 1 + \sum (m_\alpha - 1) + \frac{1}{2\pi}\iint_S |K|\, do \leq \frac{1}{2\pi}\kappa(\Gamma) \,.$$

Here boundary branch points cannot generally be defined; but, if $\kappa(\Gamma)$ is finite, S can only possess finitely many interior branch points.

Since $k_n(s)$ cannot vanish identically, unless Γ is a plane curve —a trivial case—, it follows from (5) that

$$(7) \qquad \iint_{S^{(0)}} |K^{(0)}|\, do^{(0)} = \frac{1}{2}\iint_P p(u, v)dudv < 4\pi \,.$$

Since $S^{(0)}$ is assumed to be non-isolated, the differential equation (2) must admit non-trivial solutions $\zeta(u, v)$. This is possible only if $\lambda = 1$ is an eigenvalue of (3). We shall show that $\lambda = 1$ must in fact be the lowest eigenvalue λ_1. For this purpose let us investigate the assumption $1 = \lambda_n$, $n > 1$. The first eigenfunction $\eta_1(u, v)$ does not vanish in P. An inspection of (4) then shows that the eigenfunction $\eta_n(u, v)$ must change its sign in P. Since $\eta_n = 0$ on ∂P and $p(u, v)$ is analytic in a larger disc, $\eta_n(u, v)$ can be continued analytically across ∂P as a solution of (3). It then follows from well known arguments (see for instance [8], Theorems 1, 2) that the gradient $\{\partial\eta_n/\partial u,\, \partial\eta_n/\partial v\}$ has at most isolated zeros on the set

$$\{u, v; \eta_n(u, v) = 0\} \cap \bar{P} \qquad \text{in } \bar{P} .$$

The zeros of $\eta_n(u, v)$ form therefore a finite number of analytic arcs, the nodal lines, which divide a neighborhood of any point where they meet into sectors of equal opening angle. The disc $\bar{P}$ appears as the union of the closures of open sets $Q_1, Q_2, \cdots, Q_k$ $(k \geq 2)$, with mutually empty intersections. Each nodal domain Q_l is connected and bounded by piecewise analytic curves. The function $\eta_n(u, v)$ is zero on ∂Q_l, but does not vanish in Q_l. It follows that $\lambda = 1$ is the smallest eigenvalues of the differential equation $\Delta\eta + \lambda p\eta = 0$ in each domain Q_l. Since

$$\iint_P p(u, v)dudv = \sum_{l=1}^{k} \iint_{Q_l} p(u, v)dudv < 8\pi ,$$

there exists at least one domain, say Q_1, for which

$$\iint_{Q_1} p(u, v)dudv < \frac{8\pi}{k} \leq 4\pi .$$

Then the portion $S_1^{(1)} = \{\mathfrak{x} = \mathfrak{x}^{(0)}(u, v) ; (u, v) \in \bar{Q}_1\}$ of $S^{(0)}$ has total curvature $\iint_{S_1^{(0)}} |K^{(0)}| \, do^{(0)} < 2\pi$. From a theorem of J. L. Barbosa and M. do Carmo [1] one can now conclude that the smallest eigenvalue of the differential equation $\Delta\eta + \lambda p\eta = 0$ in Q_1 is greater than one. This is a contradiction.

Once it is known that $1 = \lambda_1$, it can be shown that $\mathfrak{x}^{(0)}(u, v)$ is an element of a genuine block $\mathfrak{B} \subset \mathfrak{M}$. $\mathfrak{B}$ is a one-dimensional compact topological manifold. This fact leads in turn to another contradiction: The assumption that $S^{(0)}$ is not isolated must be dismissed. For details see [23], [30].

5. A goal of the theorems in the previous sections is the conclusion that a specific contour Γ, or a well defined class of contours Γ, possess the finiteness property, i.e., that no Γ is capable of bounding infinitely many solutions of Plateau's problem. There is also a *generic* approach. R. Böhme and A. J. Tromba [5] have announced the following result:

The space of all Jordan curves of class C^∞ (i.e., all C^∞ embeddings of ∂P into $\mathbf{R}^3$, or $\mathbf{R}^n$, equipped with the C^∞ topology) contains an open and dense set of curves which have the finiteness property.

There is also a H^k Sobolev space version of this theorem.

Absolutely no information is available about the properties of the closed and nowhere dense exceptional set and about the type of curves it might contain. In a recent paper of F. Morgan [12] it is shown that "almost every" $C^{2,\alpha}$ Jordan curve bounds a unique area minimizing solution of Plateau's problem. This is related to A. J. Tromba's result [32] that the set of C^k curves, $k \geq 7$, bounding a unique surface of least area, has a dense interior. (There appear to be some discrepancies. For instance, Theorem 8 seems to be contradicted*) by the portion of Enneper's surface spanned into the curve

$$(8) \qquad \Gamma_r = \begin{Bmatrix} x = r\cos\theta - \frac{1}{3}r^3\cos 3\theta \\ y = -r\sin\theta - \frac{1}{3}r^3\sin 3\theta\,;\ 0 \leq \theta \leq 2\pi \\ z = r^2\cos 2\theta \end{Bmatrix}$$

if the parameter r lies in the interval $1.69 < r < \sqrt{3}$; see [16] and [21], § 394.) It is not likely that the curves bounding a unique disc-type minimal surface of least area form an open set in any reasonable topology, but no examples confirming this belief are as yet available. The statement would follow from [22] if it could be shown that the curve Γ_1, i.e., the curve (8) for $r = 1$, bounds a unique disc-type minimal surface ([21], § 891). Each curve Γ_r lies on Enneper's minimal surface which is free of self intersections for $0 < r < \sqrt{3}$; see [21], § 90. It is proved in [22] that as r passes increasingly through the value $r = 1$, the curves Γ_r acquire the capability of bounding, in addition to a suitable portion of Enneper's surface, two further minimal surfaces which appear in a continuous bifurcation process.

Further remarks regarding generic uniqueness, in part based on the heuristic constructions of Lévy and Courant mentioned in section **1** (especially the "bridge theorem" —[11], p. 19; [6], p. 121; [9]— which has not yet been proved) can be found in [10], pp. 87–94. Also forthcoming will be a general discussion by A. J. Tromba [33].

6. Assume it to be known that a curve Γ cannot bound infinitely many solutions of Plateau's problem. It would then be desirable to estimate the actual number of solutions in terms of the geometric properties of Γ. To-day the only known explicit estimate of this nature is contained in the uniqueness theorem [19]: *A regular analytic Jordan curve Γ of total curvature $\kappa(\Gamma) \leq 4\pi$ (or a $C^{2,\alpha}$ curve Γ satisfying $\kappa(\Gamma) < 4\pi$) bounds a unique disc-type minimal surface. This surface is regular up to its boundary.* We shall prove here the corresponding theorem for polygons, referring to the definition of the total curvature to section **3**.

Theorem. *A simple closed polygon Π of total curvature $\kappa(\Pi) < 4\pi$ bounds a unique solution of Plateau's problem.*

Let us mention two special consequences of this theorem.

a) *Every quadrilateral bounds a unique solution of Plateau's problem.* This statement has a long history. It was already formulated by H. A. Schwarz; see [27], p. 111. Schwarz never published his promised proof.

*) Unless "minimal" is replaced by "absolutely area minimizing".

A simple demonstration combines a theorem of T. Radó ([24], p. 10) with the observation that every quadrilateral possesses a suitable parallel projection onto a plane convex quadrilateral; see [21], § 400. In recent years another proof was given by R. Garnier [7]. Garnier's proof is complex but well worth studying in view of its potential applicability to more general situations. Our theorem now provides still another proof.

b) The remarks in section **3** enable us to draw as second consequence a new uniqueness theorem: *Every pentagon bounds a unique solution of Plateau's problem.* There are octagons which bound at least three disc-type minimal surfaces. It would be interesting to clarify the situation for hexagons and heptagons; see [21], § 896.

7. As we shall see in section **9** below, our theorem will be a consequence of the fact that every disc-type minimal surface of least area bounded by Π represents a strict relative minimum for the area functional. Precisely:

Let $S = \{\mathfrak{x} = \mathfrak{x}(u, v)\,;\ (u, v) \in \bar{P}\}$ be a solution of Plateau's problem of least area bounded by Π. There is a number $d > 0$, depending on S, such that every surface $T = \{\mathfrak{x} = \mathfrak{y}(u, v)\,;\ (u, v) \in \bar{P}\}$, whose position vector $\mathfrak{y}(u, v)$ is an element of $\mathfrak{H}(\Pi)$ subject to the condition $\max_{\bar{P}} |\mathfrak{y}(u, v) - \mathfrak{x}(u, v)| < d$, satisfies the inequalities $D[\mathfrak{y}] \geq A[\mathfrak{y}] > A[\mathfrak{x}] = D[\mathfrak{x}]$.

It is obviously sufficient to establish this statement for every solution of Plateau's problem in a d-neighborhood of S. The proof consists of several steps. We proceed to sketch these steps.

a) Let $S = \{\mathfrak{x} = \mathfrak{x}(u, v)\,;\ (u, v) \in Q\}$ be an open minimal surface whose total curvature has the value

$$(9)\qquad \iint_S |K|\, do = \iint_P |K(u, v)|\, E(u, v) du dv \leq 2\pi - \mu\,, \qquad \mu > 0\,.$$

Let $D \subset Q$ be a domain in the (u, v)-plane bounded by piecewise smooth arcs. There is a contant $\nu = \nu(\mu) > 0$ such that the first eigenvalue λ_1 of problem (3) for the domain D, i.e.,

$$\Delta\eta + \lambda p(u, v)\eta = 0 \text{ in } D\,, \qquad \eta = 0 \text{ on } \partial D\,,$$

satisfies the inequality $\lambda_1 \geq 1 + \nu$. In other words: For every function $\zeta(u, v) \in C^{0,1}(\bar{D})$ vanishing on ∂D we have

$$(10)\qquad \iint_D (\zeta_u^2 + \zeta_v^2) du dv \geq (1 + \nu) \iint_D p\zeta^2 du dv\,.$$

This statement constitutes a quantitative strengthening of a theorem of J. L. Barbosa and M. do Carmo [1]. A simple proof will be published elsewhere.

b) It is necessary to derive an estimate for the areas of nearby surfaces. Let $S = \{\mathfrak{x} = \mathfrak{x}(u, v)\,;\,(u, v) \in \bar{Q}\}$ be a minimal surfaces of class $C^2(\bar{Q})$ for which u and v are isothermal parameters so that $E(u, v) \equiv \mathfrak{x}_u^2 = \mathfrak{x}_v^2 \equiv G(u, v)$, $F(u, v) \equiv \mathfrak{x}_u \mathfrak{x}_v = 0$. Denote by $L(u, v), M(u, v)$ and $N(u, v)$ the coefficients of the second fundamental form of S and by $K(u, v)$ its Gaussian curvature. Consider a nearby $C^2(\bar{Q})$ surface $T = \{\mathfrak{x} = \mathfrak{y}(u, v) = \mathfrak{x}(u, v) + \zeta(u, v)\mathfrak{X}(u, v)\,;\,(u, v) \in \bar{Q}\}$ spanned by the boundary of S, i.e., $\zeta(u, v) = 0$ for $(u, v) \in \partial Q$. Then

$$|\mathfrak{y}_u \times \mathfrak{y}_v|^2 = E^2\Big\{1 + \frac{1}{E}[\zeta_u^2 + \zeta_v^2 - p\zeta^2] + \frac{2\zeta}{E^2}[L\zeta_u^2 + 2M\zeta_u\zeta_v + N\zeta_v^2] + K^2\zeta^4 - \frac{K\zeta^2}{E}(\zeta_u^2 + \zeta_v^2)\Big\}\,.$$

With the help of the inequality

$$|L\zeta_u^2 + 2M\zeta_u\zeta_v + N_v^2| \leq E\sqrt{|K|}(\zeta_u^2 + \zeta_v^2)$$

we obtain

$$|\mathfrak{y}_u \times \mathfrak{y}|^2 \geq E^2\Big\{1 + \frac{1}{E}[(1 - |\zeta|\sqrt{|K|})^2(\zeta_u^2 + \zeta_v^2) - p\zeta^2]\Big\}\,.$$

Here $p = -2EK \geq 0$, as in section **4**. We now apply the inequality

$$\sqrt{1 + a + b} \geq 1 + \frac{1 - \delta}{2}a + \frac{1}{2}b - \frac{1}{2}b^2\,,$$

which is valid whenever $0 \leq a \leq 2\delta$, $|b| \leq \delta$, where δ is an arbitrary number in the interval $0 < \delta < 1$, and find

$$|\mathfrak{y}_u \times \mathfrak{y}_v| \geq E + \frac{1 - \delta}{2}(1 - |\zeta|\sqrt{|K|})^2(\zeta_u^2 + \zeta_v^2) - \frac{1}{2}(1 + \delta)p\zeta^2$$

whenever

$$(11) \qquad |\zeta|\sqrt{|K|} \leq 1\,, \qquad \zeta_u^2 + \zeta_v^2 \leq 2\delta E,\; p\zeta^2 \leq \delta E\,.$$

Thus

$$(12) \qquad \begin{aligned} A[\mathfrak{y}] \geq A[\mathfrak{x}] &+ \frac{1 - \delta}{2} \min_{\bar{Q}} (1 - |\zeta|\sqrt{|K|})^2 \iint_Q (\zeta_u^2 + \zeta_v^2)dudv \\ &- \frac{1 + \delta}{2}\iint_Q p\zeta^2 dudv\,. \end{aligned}$$

c) Let now $S = \{\mathfrak{x} = \mathfrak{x}(u, v)\,;\,(u, v) \in \bar{P}\}$ be a solution of Plateau's problem for the polygon Π of m vertices. From the investigations of

Weierstrass, Darboux, Garnier a.o. (see especially [7], pp. 140–144; further also [2]) we have precise information about the behavior of S near a vertex of Π. For a discussion it will be convenient to work temporarily with the upper half $\{u, v\,;\, u > 0\}$ of the w-plane as parameter domain P, so that the position vector $\mathfrak{x}(u, v)$ maps the real axis $u = 0$ onto Π. Consider three consecutive vertices $\mathfrak{p}_{j-i}, \mathfrak{p}_j, \mathfrak{p}_{j+1}$ of Π and define the interior angle β_j of Π at $\mathfrak{p}_j$ as in section **3**. We may assume that the vertex $\mathfrak{p}_j$ corresponds to the point $w = 0$ while the vertices $\mathfrak{p}_{j-1}$ and $\mathfrak{p}_{j+1}$ are images of points w_{j-1} and w_{j+1} on the negative real axis and the positive real axis, respectively. Choose the (x, y, z)-coordinate system in such a way that $\mathfrak{p}_{j-1} = (h\cos\beta, h\sin\beta, 0)$, $\mathfrak{p}_j = (0, 0, 0)$, $\mathfrak{p}_{j+1} = (k, 0, 0)$ with $\beta = \beta_j$ and $h > 0$, $k > 0$. The analytic functions $\Phi(w)$ and $\Psi(w)$ in the Weierstrass representation of S,

$$
\begin{aligned}
x(u, v) &= \operatorname{Re} \int_0^w (\Phi^2 - \Psi^2)dw \\
x(u, v) &= \operatorname{Re} -i \int_0^w (\Phi^2 + \Psi^2)dw \qquad (13)\\
z(u, v) &= \operatorname{Re} 2 \int_0^w \Phi\Psi\, dw
\end{aligned}
$$

have near $w = 0$ expansions

$$
\begin{aligned}
\Phi(w) &= w^{-\lambda}(a_0 + a_1 w + \cdots) \\
\Psi(w) &= w^{\lambda}(b_0 + b_1 w + \cdots)
\end{aligned}
, \qquad \lambda = \frac{1}{2} - \frac{\beta}{2\pi} \qquad (14)
$$

with suitable coefficients a_n, b_n: $\pi a_0^2/\beta \equiv a > 0$, $a_0 b_0 \equiv -ib$, $b \neq 0$ real, etc. From (13) and (14) asymptotic expressions for the components of the position vector $\mathfrak{x}$ are obtained:

$$
\begin{aligned}
x(\rho, \theta) &= a\rho^{\beta/\pi} \cos\left(\frac{\beta\theta}{\pi}\right) + \cdots \\
y(\rho, \theta) &= a\rho^{\beta/\pi} \sin\left(\frac{\beta\theta}{\pi}\right) + \cdots \qquad (14')\\
z(\rho, \theta) &= 2b\rho \sin\theta + \cdots .
\end{aligned}
$$

The unit normal vector $\mathfrak{X}$ of S near $\mathfrak{p}_j$ has the form

$$
\mathfrak{X}(\rho, \theta) = \{0, 0, 1\} - \frac{2b\pi}{a\beta}\rho^{1-\beta/\pi}\left\{\sin\left(1 - \frac{\beta}{\pi}\right)\theta, \cos\left(1 - \frac{\beta}{\pi}\right)\theta, 0\right\} + \cdots .
$$

The asymptotic expansions for other geometric quantities are

$$E(\rho,\theta) = \frac{a^2\beta^2}{\pi^2}\rho^{2\beta/\pi-2} + \cdots$$

$$(15) \qquad K(\rho,\theta) = -\frac{4\pi^2 b^2(\pi-\beta)^2}{a^4\beta^4}\rho^{2-4\beta/\pi} + \cdots$$

$$p(\rho,\theta) = \frac{8b^2(\pi-\beta)^2}{a^2\beta^2}\rho^{2-2\beta/\pi} + \cdots .$$

d) Assume that $\tilde{S} = \{\mathfrak{x} = \tilde{\mathfrak{x}}(u,v)\,;\,(u,v)\in\bar{P}\}$ is another solution of Plateau's problem for Π which is close to S. It can be arranged that the vertex $\mathfrak{p}_j = \{0,0,0\}$ is the image of $w=0$ also under the mapping by $\tilde{\mathfrak{x}}$. The other vertices $\mathfrak{p}_l$ of Π will be images of points $\tilde{w}_l$ on the real axis close to the corresponding points w_l, but in general different from them. The analytic functions $\tilde{\Phi}(w)$ and $\tilde{\Psi}(w)$ associated with $\tilde{S}$ will have asymptotic expansions similar to those for $\Phi(w)$ and $\Psi(w)$. In fact, a slight deformation of the surface S into the surface $\tilde{S}$ which implies a slight translation of the points w_l will result in only a slight change of the coefficients a_n, b_n.

A direct, albeit lengthy computation shows the following. If the distance $d = \max_{\bar{P}}|\tilde{\mathfrak{x}}(u,v) - \mathfrak{x}(u,v)| \equiv \|\tilde{\mathfrak{x}} - \mathfrak{x}\|_0^{\bar{P}}$ between the surfaces S and $\tilde{S}$ is sufficiently small, then a portion of $\tilde{S}$ near the vertex $\mathfrak{p}_j$ can be represented in the form (1), i.e., if d is sufficiently small, say $d \le d_0$, then there are constants $\rho_0 > 0$ and $\mathscr{C}_0 > 0$, depending on d_0 and S alone, such that a portion of $\tilde{S}$ near $\mathfrak{p}_j$ has a representation with the position vector

$$(16) \qquad \mathfrak{y}(u,v) = \mathfrak{x}(u,v) + \zeta(u,v)\mathfrak{X}(u,v)\,, \qquad u^2+v^2 < \rho_0^2\,.$$

Here the function $\zeta(u,v)$ is analytic with exception of the point $w=0$, vanishes on the real axis and satisfies the inequalities

$$(16') \qquad |\zeta| \le \mathscr{C}_0 d\rho\,, \quad |\zeta_u| \le \mathscr{C}_0 d\,, \quad |\zeta_v| \le \mathscr{C}_0 d$$

for $u^2+v^2=\rho^2<\rho_0^2$, $v>0$. u and v are not isothermal parameters on $\tilde{S}$.

e) We now return to the unit disc $P = \{u,v\,;\,u^2+v^2<1\}$ as parameter domain and denote by $w_j = e^{i\theta_j}$ the points on ∂P whose images under the mapping by the position vector $\mathfrak{x}(u,v)$ are vertices of Π. A reflection across the sides of Π leads to an analytic continuation of S. Thus the vector $\mathfrak{x}(u,v)$ is analytic in $\bar{P}$ except at the points w_j. Since the total curvature $\kappa(\Pi)$ is less than 4π, it follows from (6) and (6′) that the surface S cannot have branch points, either in its interior or on the open sides of Π. Then, in view of (15_1), there must be a constant $E_0 > 0$ such that $E(u,v) \ge E_0$ for $(u,v)\in P$.

For a given $\varepsilon > 0$ denote by P_ε the domain

$$P_\varepsilon = \{w\,; |w| < 1, |w - w_j| > \varepsilon \text{ for } j = 1, 2, \cdots, m\} .$$

Assume that S is not isolated. It then follows from the regularity results (especially [21], § 348) as in section **4** that there is a sequence of solutions of Plateau's problem whose position vectors converge to $\mathfrak{x}(u, v)$ in $\bar{P}_\varepsilon$ in the $C^{3,\alpha}$-norm. Let $\tilde{S} = \{\mathfrak{x} = \tilde{\mathfrak{x}}(u, v)\,; (u, v) \in \bar{P}\}$ be such a solution and set $\bar{d} = \|\tilde{\mathfrak{x}} - \mathfrak{x}\|_{3,\alpha}^{P_\varepsilon}$. For sufficiently small values of $\bar{d}$, say for $\bar{d} \leq d_1(\varepsilon)$, a large portion of $\tilde{S}$, roughly the portion corresponding to the domain P_ε, can be represented in the form

$$\text{(17)} \qquad \mathfrak{x} = \mathfrak{y}(u, v) = \mathfrak{x}(u, v) + \zeta(u, v)\mathfrak{X}(u, v) , \qquad (u, v) \in \bar{P}_\varepsilon$$

$$\text{(17')} \qquad |\zeta| \leq \mathscr{C}_1\bar{d} , \quad |\zeta_u| \leq \mathscr{C}_1\bar{d} , \quad |\zeta_v| \leq \mathscr{C}_1\bar{d}$$

for $(u, v) \in \bar{P}_\varepsilon$ and $\zeta = 0$ on $\partial P_\varepsilon \cap \partial P$. The constant $\mathscr{C}_1$ depends on ε and on the properties of S.

We set $\varepsilon_1 = \rho_0/2$ and $d_2 = \min(d_0, d_1(\varepsilon_1))$ and summarize: If $d \equiv \|\tilde{\mathfrak{x}} - \mathfrak{x}\|_0^{\bar{P}} < d_2$ and $\bar{d} \equiv \|\tilde{\mathfrak{x}} - \mathfrak{x}\|_{3,\alpha}^{P_{\varepsilon_1}} < d_2$, then the surface $\tilde{S}$ has a representation with the position vector

$$\text{(18)} \qquad \mathfrak{y}(u, v) = \mathfrak{x}(u, v) + \zeta(u, v)\mathfrak{X}(u, v) , \qquad (u, v) \in \bar{P}$$

where

$$\text{(18')} \qquad \begin{aligned} &\zeta = 0 && \text{for } (u, v) \in \partial P_{\varepsilon_1} \cap \partial P \\ &|\zeta| \leq \mathscr{C}_1\bar{d},\ |\zeta_u| \leq \mathscr{C}_1\bar{d},\ |\zeta_v| \leq \mathscr{C}_1\bar{d} && \text{for } (u, v) \in \bar{P}_{\varepsilon_1} \\ &|\zeta| \leq \mathscr{C}_0 d\,|w - w_j|,\ |\zeta_u| \leq \mathscr{C}_0 d,\ |\zeta_v| \leq \mathscr{C}_0 d && \\ & && \text{for } |w - w_j| < 2\varepsilon_1 . \end{aligned}$$

f) For a number ε in the interval $0 < \varepsilon \leq \varepsilon_1$ we now choose a cut-off function $\eta^{(\varepsilon)}(u, v) \in \mathscr{C}^\infty(\bar{P})$ which is equal to 1 for $(u, v) \in \bar{P}_{2\varepsilon}$ and equal to 0 for $(u, v) \in P \backslash P_\varepsilon$ and which satisfies the inequality $0 \leq \eta^{(\varepsilon)}(u, v) \leq 1$ in $\bar{P}$. There is a constant $\mathscr{C}_2$ independent of ε such that

$$\text{(19)} \qquad |\eta_u^{(\varepsilon)}(u, v)| \leq \frac{\mathscr{C}_2}{\varepsilon} , \quad |\eta_v^{(\varepsilon)}(u, v)| \leq \frac{\mathscr{C}_2}{\varepsilon} , \quad (u, v) \in \bar{P} .$$

Set $\zeta^{(\varepsilon)}(u, v) = \eta^{(\varepsilon)}(u, v)\zeta(u, v)$ and consider the surface $S^{(\varepsilon)}$ with the position vector

$$\mathfrak{y}^{(\varepsilon)}(u, v) = \mathfrak{x}(u, v) + \zeta^{(\varepsilon)}(u, v)\mathfrak{X}(u, v) .$$

Denote by $A[\mathfrak{x}\,; D]$ the area of the portion $\{\mathfrak{x} = \mathfrak{x}(u, v)\,; (u, v) \in D \subset P\}$ of the surface S.

Let δ be a number in the interval $0 < \delta < 1$ whose precise value shall be left to a later determination. It follows from (12) that

$$A[\mathfrak{y}^{(\varepsilon)}; P_\varepsilon] \geq A[\mathfrak{x}; P_\varepsilon] + \frac{1-\delta}{2}\sigma \iint_{P_\varepsilon} [(\zeta_u^{(\varepsilon)})^2 + (\zeta_v^{(\varepsilon)})^2]dudv$$
$$- \frac{1+\delta}{2} \iint_{P_\varepsilon} p(\zeta^{(\varepsilon)})^2 dudv .$$

Here we have set

$$\sigma = \min_{P_\varepsilon} (1 - |\zeta^{(\varepsilon)}|\sqrt{|K|})^2 .$$

The conditions for the validity of the inequality are those of (11). Since $E(u, v) \geq E_0 > 0$, the second and third condition of (11) can be fulfilled if d and $\bar{d}$ are chosen sufficiently small, say $d, \bar{d} \leq d_3(\delta) \leq d_2$. As for the first condition of (11), the inequalities (15_2) and ($18'_{2,3}$) imply the existence of a constant $\mathscr{C}_3$ with the property that

$$\sqrt{|K(u, v)|}\,|\zeta(u, v)| \leq \begin{cases} \mathscr{C}_3 \bar{d} & \text{for } (u, v) \in P_{\varepsilon_1} \\ \mathscr{C}_3 d & \text{for } (u, v) \in P_{2\varepsilon_1} . \end{cases}$$

If we choose $d, \bar{d} \leq d_4(\delta) \equiv \min (d_3(\delta), \delta/\mathscr{C}_3)$, then $\sqrt{|K(u, v)|}\,|\zeta(u, v)| \leq \delta$ for all $(u, v) \in P$.

Recall that the total curvature of Π has a value $\kappa(\Pi) = 4\pi - \mu$ where $\mu > 0$, so that in view of (6′),

$$\iint_S |K|\, do \leq 2\pi - \mu .$$

It can then be concluded from (10) that

$$\iint_{P_\varepsilon} [(\zeta_u^{(\varepsilon)})^2 + (\zeta_v^{(\varepsilon)})^2]dudv \geq (1 + \nu) \iint_{P_\varepsilon} p(\zeta^{(\varepsilon)})^2 dudv ,$$

where $\nu = \nu(\mu) > 0$ is the constant from section **7a**. We now obtain

$$A[\mathfrak{y}^{(\varepsilon)}; P_\varepsilon] \geq A[\mathfrak{x}; P_\varepsilon]$$
$$+ \frac{1}{2}[(1-\delta)^3(1+\nu) - (1+\delta)] \iint_{P_\varepsilon} p(\zeta^{(\varepsilon)})^2 dudv .$$

Denoting by $\delta = \delta_0$, $0 < \delta_0 < 1$, the root of the equation

$$(1-\delta)^3(1+\nu) - (1+\delta) = \frac{\nu}{2} ,$$

we find

$$A[\mathfrak{y}^{(\varepsilon)}; P_\varepsilon] \geq A[\mathfrak{x}; P_\varepsilon] + \frac{\nu}{4}\iint_{P_\varepsilon} p(\zeta^{(\varepsilon)})^2 dudv$$

$$\geq A[\mathfrak{x}; P_\varepsilon] + \frac{\nu}{4}\iint_{P_{2\varepsilon}} p\zeta^2 dudv$$

for all $\varepsilon \leq \varepsilon_1$ and all $d, \bar{d} \leq d_4(\delta_0)$. On the other hand,

$$A[\mathfrak{y}^{(\varepsilon)}; P_\varepsilon] = A[\mathfrak{y}; P_{2\varepsilon}] + A[\mathfrak{y}^{(\varepsilon)}; P_\varepsilon \backslash P_{2\varepsilon}] .$$

The set $P_\varepsilon \backslash P_{2\varepsilon}$ is the union of the m domains $\{w; |w| < 1, \varepsilon \leq |w - w_j| \leq 2\varepsilon\}$. For $\varepsilon \leq \rho \equiv |w - w_j| \leq 2\varepsilon$ the relations ($18'_3$) and (19) imply the estimates

$$|\zeta_u^{(\varepsilon)}(u, v)|, |\zeta_v^{(\varepsilon)}(u, v)| \leq \mathscr{C}_0(1 + 2\mathscr{C}_2)d .$$

It now follows from (15) and from the formulas in section **7b** that

$$|\mathfrak{y}_u^{(\varepsilon)} \times \mathfrak{y}_v^{(\varepsilon)}| \leq |\mathfrak{x}_u \times \mathfrak{x}_v| + \mathscr{C}_4 d\rho^{\beta/\pi - 1} .$$

The constant $\mathscr{C}_4$ depends on ε_1, d_4, μ and on the properties of S, but not on ε. Integration yields

$$A[\mathfrak{y}^{(\varepsilon)}; P_\varepsilon \backslash P_{2\varepsilon}] \leq A[\mathfrak{x}; P_\varepsilon \backslash P_{2\varepsilon}] + \mathscr{C}_5 d\varepsilon^{1+\beta/\pi}$$

where $\mathscr{C}_5 = 3m\pi\mathscr{C}_4$.

Combining all inequalities, we find

$$A[\mathfrak{y}] \geq A[\mathfrak{y}; P_{2\varepsilon}]$$

$$\geq A[\mathfrak{x}] + \frac{\nu}{4}\iint_{P_{2\varepsilon}} p\zeta^2 dudv - A[\mathfrak{x}; P\backslash P_{2\varepsilon}] - \mathscr{C}_5 d\varepsilon^{1+\beta/\pi}$$

for all $\varepsilon \leq \varepsilon_1$, and all $d, \bar{d} \leq d_4$. In the limit $\varepsilon \to 0$ this inequality reduces to

(20) $$A[\mathfrak{y}] \geq A[\mathfrak{x}] + \frac{\nu}{4}\iint_P p\zeta^2 dudv .$$

Our statement from the beginning of section **7** is proved: *Every solution of Plateau's problem of least area represents a strict relative minimum for the Dirichlet functional in the space $\mathfrak{H}(\Pi)$.*

8. Let $S = \{\mathfrak{x} = \mathfrak{x}(u, v); (u, v) \in \bar{P}\}$ be a solution of Plateau's problem for the m-gon Π. For $0 < r < 1$ denote by γ_r the curve $\{\mathfrak{x} = \mathfrak{x}(r \cos \theta, r \sin \theta); 0 \leq \theta \leq 2\pi\}$, i.e., the image on S of the concentric circle $u^2 + v^2 = r^2$, and by $\kappa(\gamma_r)$ its total curvature.

For each $\varepsilon > 0$ there is a $\delta = \delta(\varepsilon) > 0$ such that $\kappa(\gamma_r) \leq \kappa(\Pi) + \varepsilon$ whenever $1 - \delta < r < 1$.

For the proof we shall use again the special coordinate systems and the notations introduced in section **7c**. Choose a fixed point $w = u_1$ on the real axis between the points w_{j-1} and w_j and a fixed point $w = u_2$ on the real axis between the points w_j and w_{j+1}. We shall be interested in the total curvature of an arc on S which corresponds to the segment $\sigma_v = \{w = u + iv\,;\, u_1 \leq u \leq u_2\}$ parallel to the real axis, for small positive values of v. This total curvature is equal to

$$\kappa(\sigma_v) = \int_{u_1}^{u_2} |\mathfrak{x}_u(u, v) \times \mathfrak{x}_{uu}(u, v)|\,|\mathfrak{x}_u(u, v)|^{-2} du\ .$$

A detailed computation using the asymptotic expansions of section **7c** shows that in a vicinity of the point $u = v = 0$

$$\mathfrak{x}_u \times \mathfrak{x}_{uu} = \frac{a^2\beta^2(\pi - \beta)}{\pi^3}\rho^{2\beta/\pi - 3} \sin\theta\{0, 0, 1\} + \mathfrak{M}_1(\rho)$$

and

$$|\mathfrak{x}_u| = \frac{a\beta}{\pi}\rho^{\beta/\pi - 1} + M_2(\rho)\ .$$

Here $\mathfrak{M}_1(\rho)$ and $M_2(\rho)$ can be estimated in the form

$$|\mathfrak{M}_1(\rho)| \leq \mathscr{C}_1(\rho^{2\beta/\pi - 2} + \rho^{-1})$$
$$|M_2(\rho)| \leq \mathscr{C}_2(\rho^{\beta/\pi} + \rho^{1-\beta/\pi})\ .$$

The fact that we have to include two different exponents, each of which can dominate, stems from the two different expansions of the functions $\Phi(w)$ and $\Psi(w)$ in (14). From the above,

$$|\mathfrak{x}_u \times \mathfrak{x}_{uu}|\,|\mathfrak{x}_u|^{-2} = \left(1 - \frac{\beta}{\pi}\right)\rho^{-1} \sin\theta + M_3(\rho)$$
$$= \left(1 - \frac{\beta}{\pi}\right)\frac{v}{u^2 + v^2} + M_3(\rho)$$

where

$$|M_3(\rho)| \leq \mathscr{C}_3(1 + \rho^{1-2\beta/\pi})\ .$$

Let $\rho_0 < 1$ be so chosen that the above asymptotic expansions are valid in $u^2 + v^2 < \rho_0^2$. Then for $0 < u_0 < \rho_0/2$, $0 < v < \rho_0/2$,

$$\int_{-u_0}^{u_0} |\mathfrak{x}_u \times \mathfrak{x}_{uu}| |\mathfrak{x}_u|^{-2} du = 2\left(1 - \frac{\beta}{\pi}\right) \arctan\left(\frac{u_0}{v}\right) + M_4(u_0)$$

with

$$|M_4(u_0)| \leq \mathscr{C}_4(u_0 + u_0^{2-2\beta/\pi}) .$$

We now choose u_0 so small that $u_0 \leq \min(-u_1, u_2)$ and $|M_4(u_0)| \leq \varepsilon/4m$ (m is the number of vertices of Π). After fixing u_0 a positive number v_0 is selected for which $\arctan(u_0/v_0) \geq \pi/2 - \varepsilon/8m$. Then, for all v in $0 < v < v_0$,

$$\left| \int_{u_1}^{u_2} - \int_{u_1}^{-u_0} - \int_{u_0}^{u_2} (|\mathfrak{x}_u \times \mathfrak{x}_{uu}| |\mathfrak{x}_u|^{-2}) du - (\pi - \beta) \right| \leq \frac{\varepsilon}{2m} .$$

Since the position vector of S is analytic away from the corners of S, we have

$$\lim_{v \to 0} \int_{u_1}^{-u_0} |\mathfrak{x}_u(u, v) \times \mathfrak{x}_{uu}(u, v)| |\mathfrak{x}_u(u, v)|^{-2} du = 0$$

etc. Thus there is a positive number $v_1 \leq v_0$ such that

$$\left| \int_{u_1}^{u_2} |\mathfrak{x}_u(u, v) \times \mathfrak{x}_{uu}(u, v)| |\mathfrak{x}_u(u, v)|^{-2} du - (\pi - \beta) \right| \leq \frac{\varepsilon}{m}$$

for all v satisfying $0 < v < v_1$. Now observe that $\pi - \beta$ is precisely the total curvature of the part of Π between the vertices $\mathfrak{p}_{j-1}$ and $\mathfrak{p}_{j+1}$. An addition of the above inequality for all vertices of Π leads to our statement.

9. We are now finally in the position to prove the theorem formulated in section **6**. Let Π be a simple closed polygon whose total curvature $\kappa(\Pi)$ is smaller than 4π and assume that Π bounds two district solutions of Plateau's problem, $S_1 = \{\mathfrak{x} = \mathfrak{x}_1(u, v) \in \bar{P}\}$ and $S_2 = \{\mathfrak{x} = \mathfrak{x}_2(u, v) ; (u, v) \in \bar{P}\}$; assume also that at least one of them, say S_1, is area minimizing. Two cases have to be distinguished.

a) The areas of the surfaces S_1 and S_2 are different: $D[\mathfrak{x}_1] < D[\mathfrak{x}_2]$. Set $d = D[\mathfrak{x}_2] - D[\mathfrak{x}_1]$. Denote by γ_r the image of the circle $u^2 + v^2 = r^2$ on S_2 under the mapping by the vector $\mathfrak{x}_2(u, v)$. From section **8** we know that there is a number r_0, $0 < r_0 < 1$, such that γ_r is an analytic Jordan curve of total curvature $\kappa(\gamma_r) < 4\pi$ for $r_0 \leq r < 1$. For a suitable number r_1, $r_0 \leq r_1 < 1$, the area of the ring-type surface $T = \{\mathfrak{x} = \mathfrak{x}_2(u, v) ; r_2^2 \leq u^2 + v^2 \leq 1\}$ is less than $d/3$.

The curve γ_{r_1} bounds the disc-type minimal surface $\{\mathfrak{x} = \mathfrak{x}_2(u, v) ; u^2 + v^2 \leq r_1^2\}$. The area of this surface is greater than $D[\mathfrak{x}_2] - d/3$. γ_{r_1} also bounds a surface which is composed of S_1 and T. The area of the latter is smaller than $D[\mathfrak{x}_1] + d/3 < D[\mathfrak{x}_2] - d/3$. A solution of Plateau's problem of least area for γ_{r_1} which must exist and whose area connot exceed $D[\mathfrak{x}_1] + d/3$ must be distinct from the portion of S_2 bounded by γ_{r_1}. This is a contradiction to the uniqueness theorem [19].

b) The areas of the surfaces S_1 and S_2 are equal: $D[\mathfrak{x}_1] = D[\mathfrak{x}_2]$. From section **7** we know that both S_1 and S_2 represent strict relative minima for the Dirichlet functional in the space $\mathfrak{H}(\Pi)$, so that $\mathfrak{x}_1$ and $\mathfrak{x}_2$ are separated by a wall of positive height. An application of Morse theory ([29], also [6], pp. 223–243) now assures us that the polygon Π bounds a third minimal surface S_3 whose area is larger than that of S_1 and S_2. The reasoning of section **9a** leads again to a contradiction.

Our main theorem is proved.

Acknowledgement. The preceding research was supported in part by the National Science Foundation through Grant No. MCS 77–00950.

Bibliography

[1] Barbosa, J. L. & M. do Carmo: *On the size of a stable minimal surface in* R^3, Amer. J. Math. **98** (1976), 515–528.

[2] Beeson, M.: *The behavior of a minimal surface in a corner*, Arch. Rat. Mech. Anal. **65** (1977), 379–394.

[3] Böhme, R.: *Die Zusammenhangskomponenten der Lösungen analytischer Plateauprobleme*, Math. Z. **133** (1973), 31–40.

[4] Böhme, R. & F. Tomi: *Zur Struktur der Lösungsmenge des Plateauproblems*, Math. Z. **133** (1973), 1–29.

[5] Böhme, R. & A. J. Tromba: *The number of solutions to the classical Plateau problem is generically finite*, Research Announcement, Bull. Amer. Math. Soc. **83** (1977), 1043–1044.

[6] Courant, R.: Dirichlet's principle, conformal mapping, and minimal surfaces, Interscience, New York, 1950.

[7] Garnier, R.: *Sur un théorème de Schwarz*, Comment. Math. Helv. **25** (1951), 140–172.

[8] Hartman, P. & A. Wintner: *On the local behavior of solutions of nonparabolic partial differential equations*, Amer. J. Math. **75** (1953), 449–476.

[9] Kruskal, M.: *The bridge theorem for minimal surfaces*, Comm. P. Appl. Math. **7** (1954), 297–316.

[10] Lawson, H. B.: Lectures on minimal submanifolds, Impa, Rio de Janeiro.

[11] Lévy, P.: *Le problème de Plateau*, Mathematica **23** (1947–48), 1–45.

[12] Morgan, F.: *Almost every curve in* R^3 *bounds a unique area minimizing surface*, Preprint. Princeton University, 1977.

[13] Morse, M. & C. Tompkins: *The existence of minimal surfaces of general critical types*, Ann. of Math. (2) **40** (1939), 443–472; (2) **42** (1941), 334.

[14] Nitsche, J. C. C.: *On the total curvature of a closed curve*, MR 25 #492 (1963), 104.

[15] Nitsche, J. C. C.: *On new results in the theory of minimal surfaces*, Bull. Amer. Math. Soc. **71** (1965), 195–270.

[16] Nitsche, J. C. C.: *Contours bounding at least three solutions of Plateau's*

problem, Arch. Rat. Mech. Anal. **30** (1968), 1–11.
[17] Nitsche, J. C. C.: *Concerning the isolated character of solutions of Plateau's problem,* Math. Z. **109** (1969), 393–411.
[18] Nitsche, J. C. C.: *The boundary behavior of minimal surfaces. Kellogg's theorem and branch points on the boundary,* Inventiones math. **8** (1969), 313–333; **9** (1970), 270.
[19] Nitsche, J. C. C.: *A new uniqueness theorem for minimal surfaces,* Arch. Rat. Mech. Anal. **52** (1973), 319–329.
[20] Nitsche, J. C. C.: *Local uniqueness for Plateau's problem,* Symposia Math., Vol. 14, pp. 403–411, Academic Press, London-New York, 1974.
[21] Nitsche, J. C. C.: Vorlesungen über Minimalflächen, Springer, Berlin-Heidelberg-New York, 1975.
[22] Nitsche, J. C. C.: *Non-uniqueness for Plateau's problem. A bifurcation process,* Ann. Acad. Sci. Fenn., Ser. A. I. Math., **2** (1976), 361–373.
[23] Nitsche, J. C. C.: *Contours bounding at most finitely many solutions of Plateau's problem,* To appear in I. N. Vekua anniversary issue.
[24] Radó, T.: *Contributions to the theory of minimal surfaces,* Acta. Litt. Scient. Univ. Szeged **6** (1932), 1–20.
[25] Radó, T.: On the problem of Plateau, Springer, Berlin, 1932.
[26] Sasaki, S.: *On the total curvature of a closed curve,* Japan. J. Math. **29** (1959), 118–125.
[27] Schwarz, H.: Gesammelte Mathematische Abhandlungen, Vol. 1, Springer, Berlin, 1890.
[28] Shiffman, M.: *The Plateau problem for non-relative minima,* Ann. of Math. (2) **40** (1939), 834–854.
[29] Shiffman, M.: *Unstable minimal surfaces with any rectifiable boundary,* Proc. Nat. Acad. Sci. USA **28** (1942), 103–108.
[30] Tomi, F.: *On the local uniquesness of the problem of least area,* Arch. Rat. Mech. Anal. **52** (1973), 312–318.
[31] Tomi, F.: On the finite solvability of Plateau's problem, Springer Lecture Notes in Mathematics, #597 (1977), 679–695.
[32] Tromba, A. J.: The set of curves of uniqueness for Plateau's problem has a dense interior, Springer Lecture Notes in Mathematics, #597 (1977), 696–706.
[33] Tromba, A. J.: On the number of minimal surfaces spanning a wire, Preprint, to appear as Memoir Amer. Math. Soc.

UNIVERSITY OF MINNESOTA
MINNEAPOLIS, MN 55455
U.S.A.

Minimal Submanifolds and Geodesics
Kaigai Publications, Tokyo, 1978, 163–172

PROPERTIES OF SOLUTIONS OF THE MINIMAL SURFACE EQUATION IN HIGHER CODIMENSION

ROBERT OSSERMAN

The minimal surface equation arose classically as the Euler-Lagrange equations associated with the problem of minimizing the area of a surface given as the graph of a function $f(x, y)$. If f is defined in a bounded domain D, then the area of the graph is given by

$$A = \int_D W , \tag{1}$$

where

$$W = \sqrt{1 + |\nabla f|^2} . \tag{2}$$

If the surface so defined minimizes area among all surfaces given as graphs of functions over D having the same boundary values as f, then the Euler-Lagrange equations must be satisfied by f:

$$\frac{\partial}{\partial x}\left(\frac{1}{W}\frac{\partial f}{\partial x}\right) + \frac{\partial}{\partial y}\left(\frac{1}{W}\frac{\partial f}{\partial y}\right) = 0 , \tag{3}$$

or equivalently:

$$(1 + f_y^2)f_{xx} - 2f_xf_yf_{xy} + (1 + f_x^2)f_{yy} = 0 . \tag{4}$$

Equation (4) is the usual form for the *minimal surface equation* in two variables. It is a second-order quasilinear elliptic equation. The fundamental problem for this class of equations is the *Dirichlet problem*: let D be a bounded domain; given a continuous function

$\varphi: \partial D \to \boldsymbol{R}$, find $f: \bar{D} \to \boldsymbol{R}$ such that

i) f is continuous in $\bar{D}$,

ii) $f|_{\partial D} = \varphi$,

iii) $f|_D$ satisfies (4) .

The basic questions concerning the Dirichlet problem are those of existence, uniqueness, and regularity of solutions. The results are the following:

A. Existence. *Let D be a bounded domain. Then there exists a solution f of the Dirichlet problem corresponding to an arbitrary continuous function φ on the boundary of D if and only if D is convex.*

The proof that convexity of D is sufficient was given in 1930 by Radó ([15], pp. 795–796; see also [16], p. 101.) That the convexity of D is also necessary is due to Finn [5] in 1965.

B. Uniqueness. *There can be at most one solution f of Dirichlet's problem corresponding to a given continuous function φ on ∂D.*

This result follows immediately from the fact that the difference of any two solutions of (4) satisfies a maximum principle. (See, for example, [3], p. 323.)

C. Regularity. The regularity of solutions to the minimal surface equation has been treated in many places. (See, for example, Radó [16], Ch. IV, Nitsche [10] VII. 3, Harvey-Lawson [6].) We shall restrict our attention here to just one aspect of the problem, that of the removability of isolated singularites. One has the theorem of Bers [1] from 1951: *Every solution of* (4) *in* $0 < x^2 + y^2 < \varepsilon^2$ *extends continuously to the origin. The extended function is smooth at the origin, and satisfies* (4) *in the full disk*: $x^2 + y^2 < \varepsilon^2$.

Our object here is to consider these same questions of existence, uniqueness and regularity for solutions of the system of equations satisfied by a (vector) function whose graph is a minimal submanifold of arbitrary dimension and codimension in some euclidean space. We shall use the following notation:

$$D \text{ is a bounded domain in } \boldsymbol{R}^n$$
$$f: D \to \boldsymbol{R}^k$$
$$F: D \to \boldsymbol{R}^{n+k}; \; F(x) = (x, f(x))$$
$$M \text{ is the graph of } f; \; M = F(D)) \, .$$

M is *minimal* if its first variation of area is zero for all variations keeping the boundary fixed. Geometrically this is equivalent to the vanishing of the mean curvature vector. Analytically it may be expressed by a (vector) differential equation that must be satisfied by f. The Dirichlet problem has the same form as before, except that we are given a vector function $\varphi: \partial D \to \boldsymbol{R}^k$ and must find a vector function $f: \bar{D} \to \boldsymbol{R}^k$ satisfying the given equation in D.

Concerning the form of the equation and the properties of solutions, it is preferable to carry out the generalization one step at a time, holding either the dimension or codimension fixed, before going to the most general case. We are thus led to consider four cases.

I. $n = 2, k = 1$. This is the classical case discussed above, and the corresponding differential equation is (4).

II. n arbitrary, $k = 1$. Here M is a *minimal hypersurface* defined by a real function f of n variables. The corresponding equation is a direct generalization of (3); namely

$$\sum_{i=1}^{n} \frac{\partial}{\partial x_i}\left(\frac{1}{W}\frac{\partial f}{\partial x_i}\right) = 0 \tag{5}$$

where W is again given by (2).

III. $n = 2, k$ arbitrary. In this case M is a two-dimensional surface in $\boldsymbol{R}^{k+2}$, and the differential equation satisfied by f is ([11], p. 24)

$$(1 + |f_y|^2)f_{xx} - 2(f_x \cdot f_y)f_{xy} + (1 + |f_x|^2)f_{yy} = 0 \,. \tag{6}$$

IV. n arbitrary, k arbitrary. This is the general case, and the differential equation is ([12], p. 1099)

$$\sum_{i,j=1}^{n} g^{ij}\frac{\partial^2 f}{\partial x_i \partial x_j} = 0\,, \tag{7}$$

where

$$((g^{ij})) = ((g_{ij}))^{-1}\,, \qquad g_{ij} = \frac{\partial F}{\partial x_i}\cdot\frac{\partial F}{\partial x_j}\,.$$

For each of these cases, we may pose the questions of existence, uniqueness, and regularity that were disucssed above for the classical case, case I. We can record the results in the form of a table:

	I	II	III	IV
A	√			
B	√			
C	√			

where again, A: existence for Dirichlet problem
B: uniqueness for Dirichlet problem
C: removability of isolated singularities.

A check means that one has an affirmative answer in the given case.

We now proceed to the next case, II: n arbitrary, $k = 1$.

A. Theorem (Jenkins-Serrin [7]; see also Bombieri [2] § 7). *Let D be a bounded domain in $\boldsymbol{R}^n$ with C^3-boundary. The Dirichlet problem for equation* (5) *has a solution for all continuous $\varphi\colon \partial D \to \boldsymbol{R}$, if and only if the mean curvature of ∂D with respect to the inner normal is non-negative at every point.*

B. *There can be at most one solution f of Dirichlet's problem for* (5) *corresponding to a given continuous function φ on ∂D.*

The proof is exactly the same as in the special case $n = 2$.

C. *Every solution of* (5) *in $0 < |x| < \varepsilon$ extends continuously to the origin. The extended solution is smooth at the origin and satisfies* (5) *in the full ball $|x| < \varepsilon$.*

This result is due to Finn [4]. As in the case $n = 2$, much stronger results of this type are known. The best such result is given in a recent paper of Simon [17]. An interesting feature of these results is that unlike the situation for the Laplacian, one does not have to assume boundedness of the solution in order to conclude that a singularity is removable.

In summary, there are no surprises if one generalizes to arbitrary dimension, *provided* we stay in codimension one:

	I	II	III	IV
A	√	√		
B	√	√		
C	√	√		

Case III: $n = 2$, k arbitrary.

A. Exactly the same result holds for arbitrary k as in the classical case, $k = 1$: *a solution exists for arbitrary continuous (vector-valued) boundary functions if and only if D is convex.* The classical argument carries over without difficulty. (For details, see [11], Theorem 7.2).

B. Uniqueness is not nearly as clear. We postpone the discussion until later.

C. One sees quickly that in higher codimension there *can* be isolated singularities. In fact, since a complex analytic curve in $\boldsymbol{C}^2$ may be viewed as a real two-dimensional minimal surface in $\boldsymbol{R}^4$, any isolated singularity of an analytic function, such as

$$w = 1/z \tag{8}$$

provides an example. Of course, the solution can never be bounded in the neighborhood of such a singularity, since the singularity would then be removable, by Riemann's theorem. However, there do exist bounded solutions of (6) with isolated singularities. An example ([13], p. 287) is

$$f(x, y) = \sqrt{1 + \frac{4}{x^2 + y^2}}\,(y, -x)\ . \tag{9}$$

Both (8) and (9) are defined in the whole plane except for the origin; (8) tends to zero at infinity and blows up at the origin, while (9) tends to infinity at infinity but is bounded at the origin. Since the vector $f(x, y)$ in (9) tends to a different limit along each ray approaching the origin, this solution cannot be extended to be continuous at the origin. Recently it has been shown that slight variants of the examples (8) and (9) both appear as special cases of a one-parameter family of solutions of (6), all defined in the whole plane except the origin, all having a singularity at the origin, and all but (8) being bounded in a neighborhood of the origin. We shall give details of that construction below.

In view of these results it would appear that the extension of Bers' theorem on isolated singularities to higher codimension is hopeless. However, that is not quite the case, since one has the following result.

Theorem (Osserman [14]). *Let f be a solution of* (6) *defined in* $0 < x^2 + y^2 < \varepsilon^2$, *and suppose that all the components of f except for at most one extend continuously to the origin. Then f extends continuously to the origin, is smooth there, and satisfies* (6) *in the full disk.*

This theorem contains as special cases both Bers' theorem and the following result, proved independently by Harvey and Lawson [6].

Corollary. *If $f(x, y)$ is continuous in $x^2 + y^2 < \varepsilon^2$ and is a solution of* (6) *in* $0 < x^2 + y^2 < \varepsilon^2$, *then f is a solution in the full disk.*

Thus one has removability of singularities under suitable restrictions. We may summarize the results described so far:

	I	II	III	IV
A	√	√	√	
B	√	√	?	
C	√	√	$\frac{1}{2}$	

This was the situation as it appeared in 1975. To my knowledge, there were no results on either existence or uniqueness for the Dirichlet problem in case IV, where the dimension was greater than two, and the codimension greater than one. Concerning isolated singularities, there was a very pretty analysis of one case, due to Gulliver and Spruck (unpublished).

They looked for solutions to equation (7) of the form

$$f(x) = g(|x|)x \tag{10}$$

where

$$x = (x_1, \cdots, x_n) , \qquad r = |x| = \sqrt{x_2^1 + \cdots + x_n^2} .$$

Thus the dimension n may be arbitrary, and the codimension k equals the dimension. Quite surprisingly, Gulliver and Spruck were able to carry out a complete analysis of this case, obtaining an ordinary differential equation for the function g, and bringing that equation into a form that could be explicitly integrated. Their results are as follows.

First of all, the coefficients g_{ij} for the graph of f take the form

$$g_{ij} = \delta_{ij} + \frac{\partial f}{\partial x_i} \cdot \frac{\partial f}{\partial x_j} = (1 + g^2)(\delta_{ij} + bx_i x_j) , \tag{11}$$

where

$$b = b(r) = \frac{g'(rg' + 2g)}{r(1 + g^2)} . \tag{12}$$

The inverse matrix $((g^{ij}))$ is then of the form

$$g^{ij} = \frac{1}{1 + g^2}(\delta_{ij} - cx_i x_j) , \qquad c = c(r) = \frac{b}{1 + r^2 b} . \tag{13}$$

Substituting in (7), one finds

$$(1 + g^2) \sum g^{ij} \frac{\partial^2 f}{\partial x_i \partial x_j} = \left[g'\left(\frac{n + 1}{r} - 2cr \right) + g''(1 - cr^2) \right] x .$$

It is convenient to introduce also the function

$$G(r) = rg(r) ,$$

Then

$$1 + r^2 b = \frac{1 + (G')^2}{1 + g^2} .$$

Using the expression for c in (13), one arrives at

$$\sum g^{ij} \frac{\partial^2 f}{\partial x_i \partial x_j} = \frac{1}{r(1 + br^2)(1 + g^2)} \left[G'' + (n - 1) \frac{1 + (G')^2}{1 + g^2} g' \right] x ,$$

and the equation (7) reduces to

$$\frac{G''}{1+(G')^2} + (n-1)\frac{g'}{1+g^2} = 0\,. \tag{14}$$

It follows that

$$\tan^{-1} G' + (n-1)\tan^{-1} g = c_1 \tag{15}$$

for some constant c_1. If we introduce the complex curve given parametrically by

$$F(t) = t + iG(t)\,,$$

then (14) reduces to

$$\arg \frac{d}{dt}[e^{-ic_1}F(t)^n] = 0\,,$$

so that

$$\operatorname{Im}\{e^{-ic_1}F(t)^n\} = c_2\,. \tag{16}$$

Thus, all solutions to our original problem are obtained by taking the inverse image of horizontal lines under the complex map $w \to w^n$, rotating through some angle, and choosing a branch which projects simply onto the real axis.

If $c_2 = 0$, then $F(t)$ represents a straight line through the origin, so that $G(t) = kt$, and $g = k$, for some constant k. It follows that $f(x) = kx$, so that we obtain the trivial solution to (7) where each component is a linear function, and where there is no singularity at the origin.

In the contrary case, $c_2 \neq 0$, the curve $F(t)$ does not pass through the origin. There are two cases. For certain values of c_1 there will be a branch of the curve $F(t)$ projecting simply onto the positive real axis and asymptotic to the positive imaginary axis. The corresponding solution $f(x)$ is defined everywhere except at the origin, and tends to infinity at the origin. For other values of c_1 there will be a branch of $F(t)$ intersecting the positive imaginary axis, and that will give a solution $f(x)$ defined everywhere except at the origin, and *bounded* at the origin. In fact, from (10)

$$|f(x)| = g(|x|)\,|x| = G(|x|) \to G(0) \qquad \text{as } x \to 0$$

and $G(0)$ is defined by the intersection of the curve $F(t)$ with the imaginary axis.

Example 1. $n = 2$, $c_1 = 0$, $c_2 = 2$. Then (16) reduces to $tG(t) = 1$, and (10) becomes

$$f(x) = \frac{x}{|x|^2} .$$

This is exactly the example (8) except for a reflection in the x_1, x_2-plane.

Example 2. $n = 2$, $c_1 = \pi/2$, $c_2 = 4$. Then one finds $G(t)^2 = t^2 + 4$, and

$$f(x) = \sqrt{1 + \frac{4}{r^2}}\, x .$$

Up to a rotation in the x_1, x_2-plane, this is exactly the example (9).

Example 3. $n = 2$, $0 < c_1 < \pi/2$, $c_2 > 0$. This gives a two-parameter family of solutions $f(x)$ defined everywhere except at the origin, bounded in a neighborhood of the origin, but not continuously extendable to the origin. They all have the further property that

$$\lim_{x \to 0} |f(x)| = G(0) > G(|x|) = |f(x)| \qquad \text{for } 0 < |x| < \varepsilon .$$

whereas the reverse inequality holds in Example 2.

Example 4. For arbitrary n, we obtain a solution of (15) in parametric form:

$$(17) \qquad g(t) = \cot\theta , \quad \text{where} \quad t = \frac{\sin\theta}{(\cos n\theta)^{1/n}} , \quad 0 < \theta < \frac{\pi}{2n} .$$

Since $t \to 0$ as $\theta \to 0$, $t \to \infty$ as $\theta \to \pi/2n$, and t is monotone increasing in θ, the corresponding solution $f(x) = g(|x|)x$ is defined everywhere except at the origin, where it is bounded:

$$|f(x)| = |x|\, g(|x|) = \frac{\cos\theta}{(\cos n\theta)^{1/n}} \to 1 \qquad \text{as } |x| \to 0 .$$

As a final comment on this family of solutions, we may note that the graph of $f(x)$ is bounded by a sphere of radius $G(0)$ lying in the n-dimensional subspace of $\boldsymbol{R}^{2n}$ which projects onto the origin: $(x_1, \cdots, x_n) = 0$. Thus, the solution $f(x)$ extends continuously to the origin if and only if $G(0) = 0$. But as we observed above, that only happens for the entire solutions $f(x) = kx$. In view of the result cited earlier for $n = 2$, and arbitrary codimension, one might be led to conjecture that an isolated singularity where the solution extends continuously is always removable. However, that turns out to be false. That fact together with the answers to the remaining open questions on our chart, is contained in a recent paper of Lawson and Osserman [9]. It turns out that *all* the results that are valid in the classical cases fail when one goes to arbitrary dimension and codimension.

Specifically, the results are these.

IV. A ([9], Theorem 6.1). *Let D be the unit ball in $\boldsymbol{R}^n$, and $\partial D = S^{n-1}$. Let $\varphi_0 : S^{n-1} \to S^{n-2} \subset \boldsymbol{R}^{n-1}$ be any C^2-map which is not homotopic to zero as a map into S^{n-2}. Then $\exists R_0 > 0$ such that there is no solution to the Dirichlet problem in D for the boundary function $\varphi = R\varphi_0$ whenever $R \geq R_0$.*

The simplest illustration is provided by the Hopf map $\eta : S^3 \to S^2$, given by the restriction to the unit sphere of the map

$$\eta_1 = |z_1|^2 - |z_2|^2, \qquad \eta_2 + i\eta_3 = 2z_1\bar{z}_2 \tag{18}$$

where $z_1 = x_1 + ix_2$, $z_2 = x_3 + ix_4$. This gives an example where the dimension $n = 4$, the codimension $k = 3$, and no solution to the Dirichlet problem exists for the simplest domain D: the unit ball, and the smoothest boundary values: quadratic polynomials.

B ([9], Theorem 5.1). *Let D be the unit disk in $\boldsymbol{R}^n$. There exists a real analytic map $\varphi : \partial D \to \boldsymbol{R}^2$ with the property that there exist at least three distinct solutions to the Dirichlet problem in D for the boundary function φ.*

Thus, uniqueness fails as soon as the codimension is greater than one, even for the simplest case where $n = 2$.

C ([9], Theorem 7.1). *There exist solutions to* (7) *in the whole space except the origin, extending continuously to the origin, but not differentiable there.*

An example of such a solution in the case $n = 4$ and $k = 3$ is the cone

$$f(x) = \frac{\sqrt{5}}{2} |x| \, \eta\left(\frac{x}{|x|}\right), \tag{19}$$

where η is the Hopf map (18).

It is interesting to note that not only is the cone defined by (18) a minimal submanifold, but it is absolutely area-minimizing. That appeared as a special case in recent work of Harvey and Lawson. For further details, see the paper by Lawson in the present volume.

We can summarize all the results indicated above in the form of the completed chart:

	I	II	III	IV
A	√	√	√	×
B	√	√	×	×
C	√	√	½	×

References

[1] L. Bers, *Isolated singularities of minimal surfaces,* Ann. of Math. **53** (1951), 364–386.

[2] E. Bombieri, Theory of minimal surfaces and a counterexample to the Bernstein, Theorem in high dimensions, Lecture Notes, Courant Institute, New York University, 1970.

[3] R. Courant and D. Hilbert, Methods of Mathematical Physics, Vol. II, Interscience, New York, 1962.

[4] R. Finn, *On partial differential equations (whose solutions admit no isolated singularities)*, Scripta Math. **26** (1961), 107–115.

[5] R. Finn, *Remarks relevant to minimal surfaces and surfaces of prescribed mean curvature,* J. D'Anal. Math., **14** (1965), 139–160.

[6] R. Harvey and H. B. Lawson, Jr., *Extending minimal varieties,* Inventiones Math. **28** (1975), 209–226.

[7] H. Jenkins and J. Serrin, *The Dirichlet Problem for the minimal surface equation in higher dimensions,* J. Reine Angew. Math. **229** (1968), 170–187.

[8] H. B. Lawson, Jr., Lectures on Minimal Submanifolds, I. M. P. A., Rua Luiz de Camões 68, Rio de Janeiro, 1973.

[9] H. B. Lawson, Jr. and R. Osserman, *Non-existence, non-uniqueness and irregularity of solutions to the minimal surface system,* Acta Math. **139** (1977), 1–17.

[10] J. C. C. Nitsche, Vorlesungen über Minimalflächen, Springer-Verlag, New York 1975.

[11] R. Osserman, A Survey of Minimal Surfaces, Van Nostrand, N. Y., 1969.

[12] R. Osserman, *Minimal varieties,* Bull. A. M. S., **75** (1969), 1092–1120.

[13] R. Osserman, *Some properties of solutions to the minimal surface system for arbitrary codimension,* pp. 283–291, Proc. of Symp. in Pure Math., Vol. XV, "Global Analysis", A. M. S., Providence, 1970.

[14] R. Osserman, *On Bers' Theorem on isolated singularities,* Indiana Univ. Math. J., **23** (1973), 337–342.

[15] T. Radó, *The problem of the least area and the problem of Plateau,* Math. Z. **32** (1930), 763–796.

[16] T. Radó, On the Problem of Plateau, Springer-Verlag, Berlin 1933. Reprint 1971.

[17] L. Simon, *On a theorem of de Giorgi and Stampacchia,* Math. Z. **155** (1977), 199–204.

STANFORD UNIVERSITY
STANFORD, CA 94305
U.S.A.

Minimal Submanifolds and Geodesics
Kaigai Publications, Tokyo, 1978, 173–192

A CERTAIN PROPERTY ON GEODESICS OF THE FAMILY OF RIEMANNIAN MANIFOLDS $O_n^2(I)$

TOMINOSUKE OTSUKI

§ 0. Introduction

The present author proved in [4] and [6] the fact that any complete minimal hypersurface in the unit $(n+1)$-sphere S^{n+1} with 2 principal curvatures, one of which is simple, is congruent to a hypersurface of S^{n+1} in $\boldsymbol{R}^{n+2} = \boldsymbol{R}^n \times \boldsymbol{R}^2$, whose orthogonal projection into $\boldsymbol{R}^2$ (with coordinates u and v) is a curve whose support function $x(t)$ is a solution of the nonlinear differential equation of order 2:

$$\text{(E)} \qquad nx(1-x^2)\frac{d^2x}{dt^2} + \left(\frac{dx}{dt}\right)^2 + (1-x^2)(nx^2-1) = 0\,.$$

He also proved that i) there are countably and infinitely many compact minimal hypersurfaces of this type immersed in S^{n+1}, ii) $S^{n-1}(\sqrt{(n-1)/n}) \times S^1(\sqrt{1/n})$ corresponds to the solution $x(t) \equiv 1/\sqrt{n}$, and iii) the existence of imbedded minimal hypersurfaces of this type other than the above Clifford torus is equivalent to the existence of solutions of (E) with period 2π.

Any non-constant solution $x(t)$ of (E) such that $x^2 + x'^2 < 1$ is always periodic and its period T is given by

$$\text{(0.1)} \qquad T = 2\int_{a_0}^{a_1} \{1 - x^2 - C(x^{-2}-1)^{1/n}\}^{-1/2} dx\,,$$

where $a_0 = \min x(t)$, $a_1 = \max x(t)$ and $0 < a_0 < 1/\sqrt{n} < a_1 < 1$, and

$$\text{(0.2)} \qquad C = (a_0^2)^{1/n}(1-a_0^2)^{1-1/n} = (a_1^2)^{1/n}(1-a_1^2)^{1-1/n}$$

is the integral constant of (E) with $0 < C < A = (1/n)^{1/n}(1-1/n)^{1-1/n}$.

(E) can be considered for any real constant $n > 1$ and any nonconstant solution $x(t)$ (with $x^2 + x'^2 < 1$) has the same property above and the curve C with $x(t)$ as its support function is a geodesic in the 2-dimensional Riemannian manifold O_n^2 with the metric ([6]):

$$\text{(0.3)} \quad ds^2 = (1-u^2-v^2)^{n-2}\{(1-v^2)du^2 + 2uvdudv + (1-u^2)dv^2\}\,.$$

Using this fact, the period T can be also given by the improper integral ([10]):

$$(0.4) \qquad T = \sqrt{nc}\int_{x_0}^{x_1} \frac{dx}{x\sqrt{(n-x)\{x(n-x)^{n-1}-c\}}} \cdot$$

where $x_0 = na_0^2$, $x_1 = na_1^2$ and

$$(0.5) \qquad c = (nC)^n = x_0(n-x_0)^{n-1} = x_1(n-x_1)^{n-1} ,$$

$$(0.6) \qquad 0 < x_0 < 1 < x_1 < n .$$

On the other hand, considering T as a function of c, the following facts were proved by the present author:

(i) $\pi < T$ (in [4]),

(ii) $\lim_{c\to 0} T = \pi$ and $\lim_{c\to B} T = \sqrt{2}\,\pi$, where $B = (n-1)^{n-1}$ (in [4]),

(iii) $T < \sqrt{2}\,\pi$ (in [8] and [9]),

(iv) T is monotone increasing with respect to c (in [10]).

The facts (iii) and (iv) were originally conjectured by virtue of a numerical observation done by M. Urabe for the solutions of (E). According to the data offered by him, it was also conjectured at the same time in order to prove (iii) that, considering T as a function of $\sigma = (a_1 - 1/\sqrt{n})/(1 - 1/\sqrt{n})$ and n ($\geqq 2$), T is monotone decreasing with respect to n, because in the case of $n = 2$ (iii) can be easily proved.

In the present paper, we shall consider T as a function of x_0 and n and investigate its property from the same point of view as above.

§ 1. Preliminaries

Using the notation in [10], we consider a Riemann surface $\mathscr{F}$ in C^2 (with coordinates z and w) given by the equation

$$(1.1) \qquad z(n-z)^{n-1} - w^2 = c ,$$

where c is a real constant given by

$$(1.2) \qquad c = x_0(n-x_0)^{n-1} , \qquad 0 < x_0 < 1 .$$

Setting $b = \sqrt{(n-1)^{n-1} - c}$, we take a piecewise smooth curve γ on $\mathscr{F}$, whose projections γ_z and γ_w into the z-plane and the w-plane are illustrated as in Fig. 1, with the condition that all the roots except x_0 and x_1 of the equation $z(n-z)^{n-1} - c = 0$ are outside of γ_z.

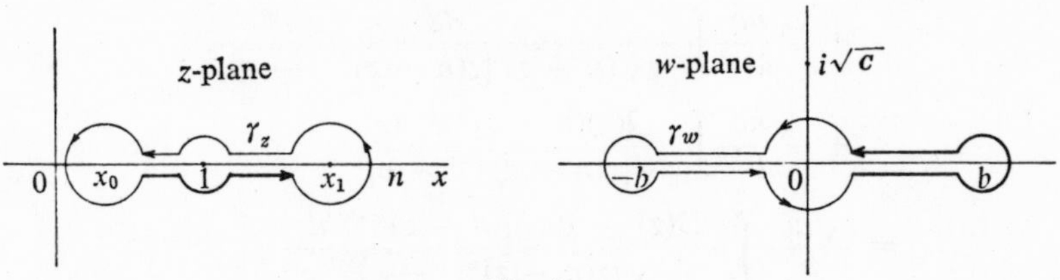

Fig. 1

Then, $T = T(x_0, n)$ given by (0.4) can be written by the integral along γ as follows:

$$T = -\frac{\sqrt{nc}}{2}\int_\gamma \frac{dz}{z\sqrt{(n-z)\{z(n-z)^{n-1}-c\}}}\,. \tag{1.3}$$

If we consider n as a fixed constant and c as a variable constant in (1.3), we have

$$\frac{\partial T}{\partial c} = -\frac{1}{4}\sqrt{\frac{n}{c}}\int_c \frac{(n-z)^{n-3/2}dz}{\sqrt{\{z(n-z)^{n-1}-c\}^3}}\,, \tag{1.4}$$

as is shown in [10]. We have easily

$$\frac{\partial c}{\partial n} = c\left\{\log(n-x_0) + \frac{n-1}{n-x_0}\right\}.$$

Setting

$$\lambda(z) := \log(n-z) + \frac{n-1}{n-z}\,, \tag{1.5}$$

from (1.3) we get

$$\begin{aligned}\frac{\partial T(x_0, n)}{\partial n} &= \frac{\partial T}{\partial c}\frac{\partial c}{\partial n} - \frac{1}{4}\sqrt{\frac{c}{n}}\int_\gamma \frac{dz}{z\sqrt{(n-z)}\{z(n-z)^{n-1}-c\}} \\ &\quad + \frac{\sqrt{nc}}{4}\int_\gamma \frac{1}{z\sqrt{(n-z)\{z(n-z)^{n-1}-c\}}} \\ &\qquad\qquad \times \left[\frac{1}{n-z} + \frac{z(n-z)^{n-1}\lambda(z)}{z(n-z)^{n-1}-c}\right]dz \\ &= -\frac{\sqrt{nc}}{4}\int_\gamma \frac{\lambda(x_0)(n-z)^{n-3/2}dz}{\sqrt{\{z(n-z)^{n-1}-c\}^3}} \\ &\quad - \frac{1}{4}\sqrt{\frac{c}{n}}\int_\gamma \frac{dz}{z\sqrt{(n-z)\{z(n-z)^{n-1}-c\}}}\end{aligned}$$

$$+\frac{\sqrt{nc}}{4}\int_{\gamma}\frac{dz}{z\sqrt{(n-z)^3}\{z(n-z)^{n-1}-c\}}$$
$$+\frac{\sqrt{nc}}{4}\int_{\gamma}\frac{\lambda(z)(n-z)^{n-3/2}dz}{\sqrt{\{z(n-z)^{n-1}-c\}^3}}$$
$$=\frac{\sqrt{nc}}{4}\int_{\gamma}\frac{\{\lambda(z)-\lambda(x_0)\}(n-z)^{n-3/2}dz}{\sqrt{\{z(n-z)^{n-1}-c\}^3}}$$
$$+\sqrt{\frac{c}{n}}\int_{\gamma}\frac{dz}{\sqrt{(n-z)^3}\{z(n-z)^{n-1}-c\}}\,.$$

Hence, setting

$$J_1(\gamma):=\int_{\gamma}\frac{\{\lambda(z)-\lambda(x_0)\}(n-z)^{n-3/2}dz}{\sqrt{\{z(n-z)^{n-1}-c\}^3}} \tag{1.6}$$

and

$$J_2(\gamma):=\int_{\gamma}\frac{dz}{\sqrt{(n-z)^3}\{z(n-z)^{n-1}-c\}}\,, \tag{1.7}$$

we have

$$\frac{\partial T(x_0,n)}{\partial n}=\frac{1}{4}\sqrt{\frac{c}{n}}\{nJ_1(\gamma)+J_2(\gamma)\} \tag{1.8}$$

and

$$J(\gamma):=nJ_1(\gamma)+J_2(\gamma)=\int_{\gamma}\frac{F(z)dz}{\sqrt{(n-z)^3\{z(n-z)^{n-1}-c\}^3}}\,, \tag{1.9}$$

where

$$F(z):=z(n-z)^{n-1}-c+n(n-z)^n\{\lambda(z)-\lambda(x_0)\}\,, \tag{1.10}$$

which is analytic and has only a singular point at $z=n$.

We can easily see that

$$J_2(\gamma)=-2\int_{x_0}^{x_1}\frac{dx}{\sqrt{(n-x)^3}\{x(n-x)^{n-1}-c\}}<0\,. \tag{1.11}$$

§ 2. Some lemmas on the function $\lambda(z)$

Lemma 2.1. *The complex analytic function $\lambda(z)-\lambda(x_0)$ in z has $z=x_0$ as a zero point of order* 1.

Proof. It is evident that $z = x_0$ is a zero point for $\lambda(z) - \lambda(x_0)$. From (1.5) we have

$$\frac{d}{dz}\lambda(z) = -\frac{1}{n-z} + \frac{n-1}{(n-z)^2} = \frac{z-1}{(n-z)^2} \neq 0 \quad \text{at } z = x_0\,, \tag{2.1}$$

which implies the lemma. Q.E.D.

Lemma 2.2. *The function* $\lambda(z) - \lambda(x_0)$ *has only one zero point* x^* *in the intervals* $(0, x_0)$ *and* (x_0, n) *on the real axis of the z-plane, which is of order* 1 *and* $1 < x^* < x_1$.

Proof. For the real variable x, we see from (2.1) that $\lambda(x)$ is monotone decreasing in $(0, 1)$ and increasing in $(1, n)$, and take its minimum value in $(0, n)$ at $x = 1$, i.e.

$$\lambda(1) = \log(n-1) + 1\,.$$

On the other hand, we have

$$\lim_{x\to n-0} \lambda(x) = \lim_{x\to n-0} \frac{(n-x)\log(n-x) + n - 1}{n-x}$$
$$= \lim_{x\to n-0} \frac{n-1}{n-x} = +\infty\,.$$

Hence, we see that there exists only one zero pointt x^* of $\lambda(z) - \lambda(x_0)$ in the intervals $(0, x_0)$ and (x_0, n), which is of order 1 and $1 < x^* < n$.

Now, we shall prove that $x^* < x_1$. Hereafter, we consider x_1 and x^* as functions of x_0 $(0 < x_0 < 1)$. It is clear that

$$\lim_{x_0\to 1} x_1 = \lim_{x_0\to 1} x^* = 1$$

and

$$\lim_{x_0\to 0} x_1 = n > \lim_{x_0\to 0} x^* = \delta_0 > 1\,.$$

Hence, for sufficiently small x_0 we have $x^* < x_1$,

Let us assume that there exists a real number ξ such that

$$0 < \xi < 1\,, \qquad x^*(\xi) = x_1(\xi)$$

and

$$x^*(x_0) < x_1(x_0) \qquad \text{for } 0 < x_0 < \xi\,.$$

Then, by Roll's theorem there exists a real number η such that

$$\xi < \eta < 1 \quad \text{and} \quad \left.\frac{dx^*}{dx_0}\right|_{x_0=\eta} = \left.\frac{dx_1}{dx_0}\right|_{x_0=\eta} .$$

On the other hand, from the equalities

$$x_0(n - x_0)^{n-1} = x_1(n - x_1)^{n-1} \quad \text{and} \quad \lambda(x_0) = \lambda(x^*)$$

we obtain easily

$$(2.2) \qquad \frac{dx_1}{dx_0} = \frac{(1 - x_0)(n - x_0)^{n-2}}{(1 - x_1)(n - x_1)^{n-2}} \quad \text{and} \qquad \frac{dx^*}{dx_0} = \frac{(x_0 - 1)(n - x^*)^2}{(x^* - 1)(n - x_0)^2} .$$

Hence we have

$$\frac{dx^*}{dx_0} = \left(\frac{n - x_1}{n - x_0}\right)^n \frac{dx_1}{dx_0} \qquad \text{at } x_0 = \xi .$$

Since we have

$$0 < \frac{n - x_1}{n - x_0} < 1 ,$$

it must be

$$\left.\frac{dx_1}{dx_0}\right|_{x_0=\xi} < \left.\frac{dx^*}{dx_0}\right|_{x_0=\xi} < 0 .$$

Hence we may suppose that $x_1(\eta) < x^*(\eta)$. Then, we have from (2.2)

$$\frac{(n - x_0)^{n-2}}{(x_1 - 1)(n - x_1)^{n-2}} = \frac{(n - x^*)^2}{(x^* - 1)(n - x_0)^2} \qquad \text{at } x_0 = \eta$$

and hence

$$\begin{aligned}(n - \eta)^n &= \frac{x_1(\eta) - 1}{x^*(\eta) - 1}(n - x^*(\eta))^2(n - x_1(\eta))^{n-2} \\ &< (n - x^*(\eta))^2(n - x_1(\eta))^{n-2} < (n - x_1(\eta))^n .\end{aligned}$$

which is a contradiction, since we have

$$0 < \eta < 1 < x_1(\eta) < n .$$

Thus, we see that $1 < x^*(x_0) < x_1(x_0)$ for $0 < x_0 < 1$. Q.E.D.

§ 3. A formula on $J(\gamma)$

First of all, we take a real constant a such that $1 < a < x_1$ and replace γ by another piecewise smooth curve $\gamma' + \gamma''$ through a piecewise smooth homotopy on the Riemann surface $\mathscr{F}$ as in Fig. 2. The end points of γ' and γ'' are (a, h) and $(a, -h)$ of C^2, where $h = \sqrt{a(n-a)^{n-1} - c}$.

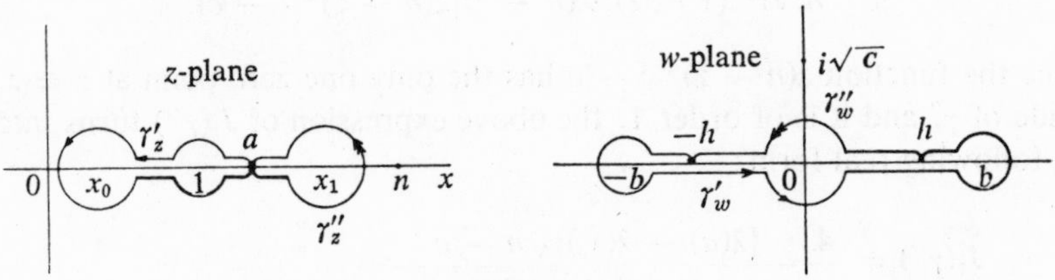

Fig. 2

We can take this homotopy since it avoids the roots of the equation $z(n-z)^{n-1} - c = 0$ and $z = n$. Then, by analyticity of the integrand of $J_1(\gamma)$, we have

$$J_1(\gamma) = J_1(\gamma' + \gamma'') = J_1(\gamma') + J_1(\gamma'') .$$

On the other hand, on $\mathscr{F}$ we have the differential equality:

$$n(1-z)(n-z)^{n-2}dz = 2wdw$$

and hence

$$\frac{(n-z)^{n-3/2}dz}{\sqrt{\{z(n-z)^{n-1} - c\}^3}} = \frac{2\sqrt{n-z}dw}{n(1-z)w^2} .$$

Thus, we have

$$\begin{aligned}
J_1(\gamma'') &= \int_{\gamma''} \frac{\{\lambda(z) - \lambda(x_0)\}(n-z)^{n-3/2}dz}{\sqrt{\{z(n-z)^{n-1} - c\}^3}} \\
&= \frac{2}{n}\int_{\gamma''} \frac{\{\lambda(z) - \lambda(x_0)\}\sqrt{n-z}}{1-z} \cdot \frac{dw}{w^2} \\
&= -\frac{2}{n}\left[\frac{\{\lambda(z) - \lambda(x_0)\}\sqrt{n-z}}{1-z} \cdot \frac{1}{w}\right]_{\partial\gamma''} \\
&\quad + \frac{2}{n}\int_{\gamma''} \frac{1}{w}\left[\frac{1}{\sqrt{(n-z)^3}} + \{\lambda(z) - \lambda(x_0)\}\left\{\frac{\sqrt{n-z}}{(1-z)^2}\right.\right. \\
&\qquad \left.\left. - \frac{1}{2(1-z)\sqrt{n-z}}\right\}\right]dz
\end{aligned}$$

$$= \frac{4}{n}\cdot\frac{\{\lambda(a)-\lambda(x_0)\}\sqrt{n-a}}{(a-1)\sqrt{a(n-a)^{n-1}-c\}}}$$
$$-\frac{2}{n}\int_{\gamma''}\frac{dz}{\sqrt{(n-z)^3\{z(n-z)^{n-1}-c\}}}$$
$$+\frac{1}{n}\int_{\gamma''}\frac{\{\lambda(z)-\lambda(x_0)\}(2n-1-z)dz}{(1-z)^2\sqrt{(n-z)\{z(n-z)^{n-1}-c\}}}\,.$$

Since the function $z(n-z)^{n-1}-c$ has the only one zero point at $z=x_1$ inside of γ''_z and it is of order 1, the obove expression of $J_1(\gamma'')$ turns into the following real form:

$$J_1(\gamma'') = \frac{4}{n}\frac{\{\lambda(a)-\lambda(x_0)\}\sqrt{n-a}}{(a-1)\sqrt{a(n-a)^{n-1}-c}}$$
$$(3.1)\qquad +\frac{4}{n}\int_a^{x_1}\frac{dx}{\sqrt{(n-x)^3\{x(n-x)^{n-1}-c\}}}$$
$$-\frac{2}{n}\int_a^{x_1}\frac{\{\lambda(x)-\lambda(x_0)\}(2n-1-x)dx}{(x-1)^2\sqrt{(n-x)\{x(n-x)^{n-1}-c\}}}\,.$$

Next, noticing the fact that $\lambda(z)-\lambda(x_0)$ has the zero point at $z=x_0$ of order 1, we obtain easily

$$J_1(\gamma') = \int_{\gamma'}\frac{\{\lambda(z)-\lambda(x_0)\}(n-z)^{n-3/2}dz}{\sqrt{\{z(n-z)^{n-1}-c\}^3}}$$
$$(3.2)\qquad = -2\int_{x_0}^{a}\frac{\{\lambda(x)-\lambda(x_0)\}(n-x)^{n-3/2}dx}{\sqrt{\{x(n-x)^{n-1}-c\}^3}}\,.$$

Hence, by (1.9), (1.11), (3.1) and (3.2) we have

$$J(\gamma) = n\{J_1(\gamma')+J_1(\gamma'')\}+J_2(\gamma)$$
$$= \frac{4\{\lambda(a)-\lambda(x_0)\}\sqrt{n-a}}{(a-1)\sqrt{a(n-a)^{n-1}-c}}$$
$$+4\int_a^{x_1}\frac{dx}{\sqrt{(n-x)^3\{x(n-x)^{n-1}-c\}}}$$
$$-2\int_a^{x_1}\frac{\{\lambda(x)-\lambda(x_0)\}(2n-1-x)dx}{(x-1)^2\sqrt{(n-x)\{x(n-x)^{n-1}-c\}}}$$
$$-2n\int_{x_0}^{a}\frac{\{\lambda(x)-\lambda(x_0)\}(n-x)^{n-3/2}dx}{\sqrt{\{x(n-x)^{n-1}-c\}^3}}$$
$$-2\int_{x_0}^{x_1}\frac{dx}{\sqrt{(n-x)^3\{x(n-x)^{n-1}-c\}}}\,,$$

i.e.

$$
(3.3)\qquad
\begin{aligned}
J(\gamma) = -2\Bigg[\int_{x_0}^{a} &\frac{F(x)dx}{\sqrt{(n-x)^3\{x(n-x)^{n-1}-c\}^3}} \\
&- \int_{a}^{x_1} \frac{(2n-1-x)G(x)dx}{(x-1)^2\sqrt{(n-x)\{x(n-x)^{n-1}-c\}}} \\
&- \frac{2\sqrt{n-a}\{\lambda(a)-\lambda(x_0)\}}{(a-1)\sqrt{a(n-a)^{n-1}-c}}\Bigg],
\end{aligned}
$$

where

$$(3.4)\qquad F(x) := x(n-x)^{n-1} - c + n(n-x)^n\{\lambda(x) - \lambda(x_0)\},$$

$$(3.5)\qquad G(x) := \frac{(x-1)^2}{(n-x)(2n-1-x)} - \{\lambda(x) - \lambda(x_0)\}.$$

Especially, using x^* in Lemma 2.2, we have

$$
(3.6)\qquad
\begin{aligned}
J(\gamma) = -2&\int_{x_0}^{x^*} \frac{F(x)dx}{\sqrt{(n-x)^3\{x(n-x)^{n-1}-c\}^3}} \\
+ 2&\int_{x^*}^{x_1} \frac{(2n-1-x)G(x)dx}{(x-1)^2\sqrt{(n-x)\{x(n-x)^{n-1}-c\}}}.
\end{aligned}
$$

Lemma 3.1. *The function $F(x)$ is monotone increasing in $[x_0, x^*]$ and monotone decreasing in $[0, x_0]$ and $[x^*, n)$. It is positive in $(0, x_0)$ and (x_0, x_1) and real analytic in $(0, n)$.*

Proof. We have easily

$$(3.7)\qquad F'(x) = -n^2(n-x)^{n-1}\{\lambda(x) - \lambda(x_0)\}.$$

Noticing the properties of $\lambda(x)$ described in Lemma 2.2, we obtain the first part of the lemma. From the definition we have

$$F(x_0) = 0, \qquad F(x_1) = n\{\lambda(x_1) - \lambda(x_0)\}(n-x_1)^n > 0$$

and

$$F(n) = -c < 0, \qquad F'(n) = 0. \qquad \text{Q.E.D.}$$

Lemma 3.2. *The function $G(x)$ is monotone increasing in $(0, n)$ and tends to $+\infty$ as $x \to n$. It is positive in $[x_0, n)$ and real analytic in $(0, n)$, and $G'(1) = 0$.*

Proof. By (1.5) and (3.5), we have

$$G(x) = -\frac{2n-3+x}{2n-1-x} - \log(n-x) + \lambda(x_0)\ ,$$

which implies

$$\lim_{x\to n-0} G(x) = +\infty\ .$$

We obtain easily

(3.8) $$G'(x) = \frac{(x-1)^2}{(n-x)(2n-1-x)^2}\ ,$$

which shows that $G(x)$ is monotone increasing in $(0, n)$. The rest of the lemma is clear. Q.E.D.

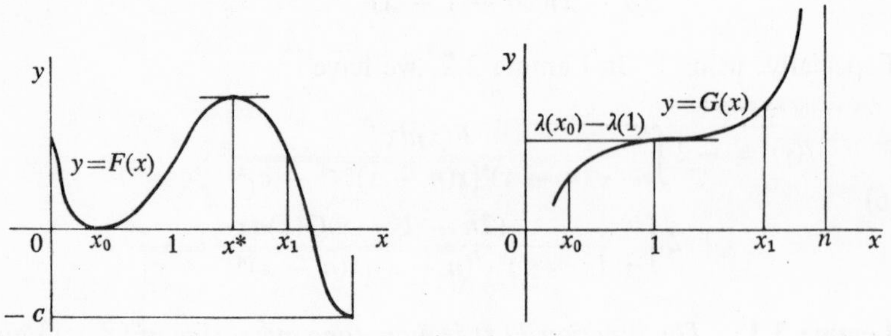

Fig. 3

§ 4. Certain expressions of $J(\gamma)$ by proper integrals

In this section, we shall make some preparations to see the sign of the integral:

(4.1) $$J(\gamma) = nJ_1(\gamma) + J_2(\gamma) = \int_\gamma \frac{F(z)dz}{\sqrt{(n-z)^3}\{z(n-z)^{n-1} - c\}}\ .$$

Setting

(4.2) $$\psi(x) := x(n-x)^{n-1}\ ,$$

we have

(4.3) $$\psi'(x) = n(1-x)(n-x)^{n-2} \quad \text{and}$$
$$\psi''(x) = -n(n-1)(2-x)(n-x)^{n-3}\ .$$

Since $G(x)$ is continuous in $[x_0, x_1]$ and $\psi'(x_1) \neq 0$, we obtain from (3.3)

$$
(4.4)\qquad J(\gamma) = -2 \lim_{a\to x_1-0}\left[\int_{x_0}^{a} \frac{F(x)dx}{\sqrt{(n-x)^3\{\psi(x)-c\}^3}} - \frac{2\sqrt{n-a}\{\lambda(a)-\lambda(x_0)\}}{(a-1)\sqrt{\psi(a)-c}}\right].
$$

On the other hand, by the definition of $F(x)$ we have

$$
\begin{aligned}
\frac{\sqrt{n-x}\{\lambda(x)-\lambda(x_0)\}}{(x-1)\sqrt{\psi(x)-c}} &= \frac{F(x)-\psi(x)+c}{n(x-1)(n-x)^{n-1/2}\sqrt{\psi(x)-c}} \\
&= \frac{F(x)}{n(x-1)(n-x)^{n-1/2}\sqrt{\psi(x)-c}} \\
&\quad - \frac{\sqrt{\psi(x)-c}}{n(x-1)(n-x)^{n-1/2}}
\end{aligned}
$$

and

$$
\lim_{a\to x_1-0} \frac{\sqrt{\psi(a)-c}}{n(a-1)(n-a)^{n-1/2}} = 0 .
$$

Since we have

$$
\lim_{a\to x_1} \frac{B-\psi(a)}{b^2} = 1 ,
$$

where

$$
(4.5)\qquad B = (n-1)^{n-1} \quad\text{and}\quad b^2 = B-c, b>0 ,
$$

we obtain

$$
(4.6)\qquad J(\gamma) = -2 \lim_{a\to x_1-0}\left[\int_{x_0}^{a} \frac{F(x)dx}{\sqrt{(n-x)^3\{\psi(x)-c\}^3}} - \frac{2F(a)\{B-\psi(a)\}}{nb^2(a-1)(n-a)^{n-1/2}\sqrt{\psi(a)-c}}\right].
$$

Now, noticing the second term in the brackets of (4.6), we define an auxiliary function

$$
(4.7)\qquad L(x) := \begin{cases} \dfrac{F(x)\{B-\psi(x)\}}{(x-1)(n-x)^{n-1/2}} & \text{for } 0<x<n,\ x\neq 1 \\ 0 & \text{for } x=1 . \end{cases}
$$

Lemma 4.1. *The function*

$$(4.8)\qquad \mu(x) := \begin{cases} \dfrac{B - \psi(x)}{(x-1)^2} & \text{for } 0 < x < n,\ x \neq 1 \\ \dfrac{n(n-1)^{n-2}}{2} & \text{for } x = 1 \end{cases}$$

is positive, decreasing and real analytic in $(0, n)$.

Proof. From (4.2) and (4.3) we have

$$\psi(1) = (n-1)^{n-1} = B\,, \quad \psi'(1) = 0\,, \quad \psi''(1) = -n(n-1)^{n-2}\,,$$

which shows that $\mu(x)$ is real analytic on $(0, n)$. Since $\psi(x) < B$ for $x \neq 1$, $0 < x < n$, $\mu(x)$ is positive.

Next, we have easily

$$(4.9)\qquad \mu'(x) = -\frac{2B - \{(n-2)x^2 + n\}(n-x)^{n-2}}{(x-1)^3} \qquad \text{for } x \neq 1\,.$$

Setting

$$(4.10)\qquad g(x) := 2B - \{(n-2)x^2 + n\}(n-x)^{n-2}\,,$$

we have

$$g'(x) = n(n-2)(x-1)^2(n-x)^{n-3}\,,$$

which shows that $g'(x) > 0$ for $x \neq 1$, $x < n$. Since $g(1) = 0$, it must be

$$g(x) < 0 \qquad \text{for } x < 1 \text{ and } g(x) > 0 \text{ for } 1 < x < n\,.$$

Hence, we have

$$\mu'(x) = -\frac{g(x)}{(x-1)^3} < 0 \qquad \text{for } x < n \text{ and } x \neq 1\,,$$

which implies that $\mu(x)$ is monotone decreasing in $(0, n)$. Q.E.D.

Using the function $\mu(x)$, we have

$$L(x) = \frac{(x-1)F(x)\mu(x)}{(n-x)^{n-1/2}}\,,$$

which shows that $L(x)$ is real analytic in $(0, n)$ and

$$(4.11)\qquad \begin{cases} L(x_0) = 0 \\ L(x) < 0 \text{ in } (x_0, 1) \text{ and } L(x) > 0 \text{ in } (1, x_1]\,. \end{cases}$$

With $L(x)$, (4.6) can be written as

$$(4.6') \quad J(\gamma) = -2 \lim_{a \to x_1 - 0} \left[\int_{x_0}^{a} \frac{F(x)dx}{\sqrt{(n-x)^3\{\psi(x)-c\}^3}} - \frac{2}{nb^2} \cdot \frac{L(a)}{\sqrt{\psi(a)-c}} \right].$$

Remark. The reason of replacing the second term in the brackets of (4.4) with the one of (4.6′) is that the function

$$\frac{\sqrt{n-x}\{\lambda(x)-\lambda(x_0)\}}{x-1}$$

has a singular point at $x = 1$ (a pole of order 1 as a complex analytic function).

Now, we have

$$\left(\frac{L(x)}{\sqrt{\psi(x)-c}}\right)' = \frac{2\{\psi(x)-c\}L'(x) + n(x-1)(n-x)^{n-2}L(x)}{2\sqrt{\{\psi(x)-c\}^3}}$$

and, using $L(x_0) = 0$, we have

$$\begin{aligned}
&\int_{x_0}^{a} \frac{F(x)dx}{\sqrt{(n-x)^3\{\psi(x)-c\}^3}} - \frac{2}{nb^2} \cdot \frac{L(a)}{\sqrt{\psi(a)-c}} \\
&= \int_{x_0}^{a} \frac{F(x)dx}{\sqrt{(n-x)^3\{\psi(x)-c\}^3}} - \frac{2}{nb^2} \int_{x_0}^{a} \left(\frac{L(x)}{\sqrt{\psi(x)-c}}\right)' dx \\
&= \int_{x_0}^{a} \frac{1}{b^2\sqrt{(n-x)^3\{\psi(x)-c\}^3}} \Big[b^2F(x) \\
&\quad - \frac{2}{n}(n-x)^{3/2}\{\psi(x)-c\}L'(x) - (x-1)(n-x)^{n-1/2}L(x) \Big] dx.
\end{aligned}$$

The expression in the brackets of the last equality can be written as

$$\begin{aligned}
&b^2F(x) - \frac{2}{n}(n-x)^{3/2}\{\psi(x)-c\}L'(x) - \{B-\psi(x)\}F(x) \\
&= \{\psi(x)-c\}\left[F(x) - \frac{2}{n}(n-x)^{3/2}L'(x)\right].
\end{aligned}$$

On the other hand, by (3.4) and 4.7) we have

$$\begin{aligned}
L'(x) = \frac{(B-\psi)F}{(x-1)(n-x)^{n-1/2}} \Big[&- \frac{n^2(n-x)^{n-1}(\lambda-\lambda_0)}{F} \\
&+ \frac{n(x-1)(n-x)^{n-2}}{B-\psi} - \frac{1}{x-1} + \frac{2n-1}{2(n-x)} \Big],
\end{aligned}$$

where $\lambda_0 = \lambda(x_0)$, and hence

$$F - \frac{2}{n}(n-x)^{3/2}L' = -F + \frac{2n(n-x)(B-\psi)(\lambda-\lambda_0)}{x-1} + \frac{\{4n-1-(2n+1)x\}(B-\psi)F}{n(x-1)^2(n-x)^{n-1}},$$

Now, setting

$$(4.12)\qquad \begin{aligned} M(x) := {} & \frac{\{4n-1-(2n+1)x\}\{B-x(n-x)^{n-1}\}}{n(x-1)^2(n-x)^{n-1}}F(x) - F(x) \\ & + \frac{2n(n-x)}{x-1}\{B-x(n-x)^{n-1}\}\{\lambda(x)-\lambda(x_0)\}, \end{aligned}$$

we obtain the formula:

$$(4.13)\qquad \begin{aligned} & \int_{x_0}^{a} \frac{F(x)dx}{\sqrt{(n-x)^3\{\psi(x)-c\}^3}} - \frac{2}{nb^2}\frac{L(a)}{\sqrt{\psi(a)-c}} \\ & = \frac{1}{b^2}\int_{x_0}^{a}\frac{M(x)dx}{\sqrt{(n-x)^3\{\psi(x)-c\}}}. \end{aligned}$$

Hence, from the above argument we obtain the following

Proposition 1. *We have*

$$(4.14)\qquad J(\gamma) = -\frac{2}{b^2}\int_{x_0}^{x_1}\frac{M(x)dx}{\sqrt{(n-x)^3\{x(n-x)^{n-1}-c\}}}.$$

where $M(x)$ is real analytic in $(0, n)$.

Proposition 2. *The function $M(x)$ is positive in $(x_0, 1)$ and negative in $(1, x^*)$ and $M(x_0)=M(1)=0$.*

In order to prove this proposition, we prove the following lemmas.

Lemma 4.2. *The function*

$$(4.15)\qquad h(x) := \{4n-1-(2n+1)x\}\mu(x) - n(n-x)^{n-1}$$

is monotone decreasing in $[0, p_n]$, and $h(x) > 0$ in $[0, 1)$ and $h(x) < 0$ in $(1, n)$, where p_n is determined in Lemma 4.3.

Proof. Differentiating $h(x)$ and using Lemma 4.1, we have

$$h'(x) = -(2n+1)\mu(x) - \frac{\{4-1-(2n+1)x\}g(x)}{(x-1)^3} + n(n-1)(n-x)^{n-2}$$

$$= \frac{1}{(x-1)^3}[-(2n+1)(x-1)\{B-\psi(x)\}$$
$$-\{4n-1-(2n+1)x\}\{2B-((n-2)x^2+n)(n-x)^{n-2}\}$$
$$+n(n-1)(x-1)^3(n-x)^{n-2}]$$
$$= \frac{1}{(x-1)^3}[\{-3(2n-1)+(2n+1)x\}B$$
$$+(n-x)^{n-2}\{3n^2-n(n+5)x+3(n^2-n+1)x^2$$
$$-(n-1)(n+1)x^2\}] .$$

Setting

$$\rho(x) := \{-3(2n-1)+(2n+1)x\}B$$
$$+(n-x)^{n-2}\{3n^2-n(n+5)x+3(n^2-n+1)x^2 \tag{4.16}$$
$$-(n-1)(n+1)x^3\},$$

we have

$$h'(x) = \frac{\rho(x)}{(x-1)^3} . \tag{4.17}$$

Now, we investigate in partcular the function $\rho(x)$.

Lemma 4.3. *The function $\rho(x)$ is monotone decreasing in $[0, p_n]$, where p_n is the minimum of the roots of the quadratic equation of x:*

$$n(11n-4)-(8n^2+n+2)x+(n+1)^2x^2 = 0 , \tag{4.18}$$

and $1 < p_n < 2$, and

$$\rho(x) > 0 \quad \text{in } [0,1) \text{ and } \rho(x) < 0 \text{ in } (1, p_n] . \tag{4.19}$$

Proof. From (4.16) we obtain easily

$$\rho'(x) = (2n+1)B-(n-x)^{n-3}[n^2(4n-1)-n(7n^2-2n+1)x$$
$$+3n^2(2n-1)x^2-(n-1)(n+1)^2x^3] ,$$

and

$$\rho''(x) = n(n-1)(n-x)^{n-4}[n(11n-4)-(19n^2-3n+2)x$$
$$+3(3n^2+n+1)x^2-(n+1)^2x^3] .$$

Setting

$$\tau(x) := n(11n-4) - (19n^2 - 3n + 2)x + 3(3n^2 + n + 1)x^2 - (n+1)^2x^3 , \tag{4.20}$$

we see easily that $\tau(1) = 0$, hence we obtain

$$\tau(x) = (1-x)\{n(11n-4) - (8n^2+n+2)x + (n+1)^2x^2\} .$$

The discriminant of the quadratic equation (4.18) is

$$\begin{aligned}(8n^2 + n + 2)^2 &- 4n(11n-4)(n+1)^2 \\ &= 20n^4 - 56n^3 + 21n^2 + 20n + 4 \\ &= (n-2)^2(20n^2 + 24n + 37) + 72(n-2) > 0 .\end{aligned}$$

Therefore (4.18) has positive two roots p_n and q_n $(p_n < q_n)$. Since we have easily

$$2 < \frac{8n^2 + n + 2}{2(n+1)^2} < n \qquad \text{for } n > 2,$$

and

$$\begin{aligned}&[n(11n-4) - (8n^2+n+2)x + (n+1)^2x^2]_{x=1} \\ &\quad = (n-1)(4n+1) > 0 , \\ &[n(11n-4) - (8n^2+n+2)x + (n+1)^2x^2]_{x=2} \\ &\quad = -n(n-2) < 0 ,\end{aligned}$$

and

$$\begin{aligned}&[n(11n-4) - (8n^2+n+2)x + (n+1)^2x^2]_{x=n} \\ &\quad = n(n-1)(n-2)(n-3) ,\end{aligned}$$

it must be $1 < p_n < 2$.

Furthermore, we have easily $\rho'(1) = 0$, hence we obtain from the above facts

$$\rho'(x) < 0 \qquad \text{for} \quad 0 < x < p_n,\ x \neq 1 ,$$

from which we see that $\rho(x)$ is monotone decreasing in $[0, p_n]$. Then, we have easily $\rho(1) = 0$, and hence

$$\begin{cases}\rho(x) > 0 & \text{for } 0 \leq x < 1 , \\ \rho(x) < 0 & \text{for } 1 < x < p_n ,\end{cases} \qquad \text{Q.E.D.}$$

Now, going back to the verificatian of Lemma 4.2, from (4.17) and (4.19), we obtain

$$h'(x) < 0 \qquad \text{for } 0 < x < 1 \text{ and } 1 < x < p_n .$$

Furthermore, investigating the sign of $\rho''(x)$, $\rho'(x)$ and $\rho(x)$ in the interval (p_n, n) separately for the cases $n \geqq 3$ and $2 < n < 3$ and using the facts

$$\rho'(n) = (2n + 1)B > 0 ,$$
$$\rho(n) = (n - 1)(2n - 3)B > 0$$

and

$$h(1) = 0 , \qquad h(n) = -(2n - 1)(n - 1)^{n-2} < 0 ,$$

we obtain

$$h(x) > 0 \qquad \text{in } [0, 1) \text{ and } h(x) < 0 \text{ in } (1, n) . \qquad \text{Q.E.D.}$$

Proof of Proposition 2. By means of (4.8), (4.12), $M(x)$ can be written as

$$M(x) = \frac{h(x)F(x)}{n(n-x)^{n-1}} + 2n(x-1)(n-x)\mu(x)\{\lambda(x) - \lambda(x_0)\} , \tag{4.21}$$

from which we see easily that Proposition 2 is true by Lemmas 2.2, 3.1, 4.1 and 4.2. Q.E.D.

Remark 1. From (1.8), (1.9), (3.6), Lemma 3.1 and Lemma 3.2, we have

$$\begin{aligned} \frac{\partial T(x_0, n)}{\partial n} = -\frac{1}{2}\sqrt{\frac{c}{n}}\Bigg[\int_{x_0}^{x^*} \frac{F(x)dx}{\sqrt{(n-x)^3}\{x(n-x)^{n-1} - c\}^3} \\ - \int_{x^*}^{x_1} \frac{(2n - 1 - x)G(x)dx}{(x-1)^2\sqrt{(n-x)}\{x(n-x)^{n-1} - c\}}\Bigg] . \end{aligned} \tag{4.22}$$

$$\int_{x_0}^{x^*} \frac{F(x)dx}{\sqrt{(n-x)^3}\{x(n-x)^{n-1} - c\}^3} > 0 \quad \text{and}$$

$$\int_{x^*}^{x_0} \frac{(2n - 1 - x)G(x)dx}{(x-1)^2\sqrt{(n-x)}\{x(n-x)^{n-1} - c\}} > 0 .$$

This fact shows us that it is very difficult to know the sign of $\partial T(x_0, n)/\partial n$. Propositions 1 and 2 also show us that in order to know the sign of $\partial T(x_0, n)/\partial n$ a through investigation of $M(x)$ is needed.

Remark 2. Here, we shall give a clue from which we make a bet on the sign of $\partial T(x_0, n)/\partial n$. According to the data given by M. Urabe, the values of $T(x_0, 2)$, $T(x_0, 4)$ and $T(x_0, 8)$ for $x_0 = 1/2$ and $1/4$ are given approximately in the table as follows:

n	a_0	a_1	$\sigma = \dfrac{\sqrt{n}\,a_1 - 1}{\sqrt{n} - 1}$	$T_0 = T/2\pi$	
2	$1/2$	$\sqrt{3}/2$	0.543	0.696	
4	$1/2\sqrt{2}$	0.642	0.284	0.699	for $x_0 = \dfrac{1}{2}$
8	$1/4$	0.462	0.167	0.701	
2	$\sqrt{2}/4$	$\sqrt{14}/4$	0.779	0.672	
4	$1/4$	0.738	0.476	0.686	for $x_0 = \dfrac{1}{4}$
8	$\sqrt{2}/8$	0.546	0.298	0.688	

From these data, it seems to us that, if we regards the period T as a function of x_0 and n, it may be increasing with respect to n (≥ 2). This conjecture is rather contrary to the one described in Introduction in considering T as a function of $\sigma = (a_1 - 1/\sqrt{n})/(1 - 1/\sqrt{n})$ and n.

In Part (II) to continue the present part we shall develope the argumenet on $T(x_0, n)$ from this point of view.

References

[1] S. S. Chern, M. do Carmo and S. Kobayashi, *Minimal submanifolds of a sphere with second fundamental form of constant length,* Functional Analysis and Related Fields, Springer-Verlag, 1970, 60–75.

[2] S. Furuya, *On periods of periodic solutions of a certain nonlinear differential equation,* Japan-United States Seminar on Ordinary Differential and Functional Equations, Lecture Notes in Mathematics, Springer-Verlag, **243** (1971), 320–323.

[3] W. Y. Hsiang and H. B. Lawson, Jr., *Minimal submanifolds of low cohomogeneity,* J. Differ. Geometry, **5** (1970), 1–38.

[4] T. Otsuki, *Minimal hypersurfaces in a Riemannian manifold of constant curvature,* Amer. J. Math., **92** (1970), 145–173.

[5] T. Otsuki, *On integral inequalities related with a certain nonlinear differential equation,* Proc. Japan Acad., **48** (1972), 9–12.

[6] T. Otsuki, *On a 2-dimensional Riemannian manifold,* Differential Geometry, in honor of K. Yano, Kinokuniya, Tokyo, 1972, 401–414.

[7] T. Otsuki, *On a family of Riemannian manifolds defined on an m-disk,* Math. J. Okayama Univ., **16** (1973), 85–97.

[8] T. Otsuki, *On a bound for period of solutions of a certain nonlinear differential equation* (I), J. Math. Soc. Japan, **26** (1974) 206–233.

[9] T. Otsuki, *On a bound for periods of solutions of a certain nonlinear differential equation* (II), Funkcialaj Ekvacioj, **17** (1974), 193–205.

[10] T. Otsuki, *Geodesics of O_n^2 and an analysis on a related Riemann surface,* Tôhoku Math. J., **28** (1976), 411–427.

[11] M. Maeda and T. Otsuki, *Models of the Riemannian manifolds O_n^2 in the Lorentzian 4-space*, J. Differ. Geometry, **9** (1974), 97–108.

[12] M. Urabe, *Computations of periods of a certain nonlinear autonomous oscilations,* Study of algorithms of numerical computations, Sûrikaiseki Kenkyûsho Kôkyû-roku, **149** (1972), 111–129 (Japanese).

TOKYO INSTITUTE OF TECHNOLOGY
TOKYO, 152 JAPAN
PRESENTLY
SCIENCE UNIVERSITY OF TOKYO
TOKYO, 162 JAPAN

Minimal Submanifolds and Geodesics
Kaigai Publications, Tokyo, 1978, 193–207

CUT LOCI OF COMPACT SYMMETRIC SPACES

TAKASHI SAKAI

§ 1. Introduction

Let M be a compact riemannian manifold of dimension n and fix a point o of M. Let γ_X be a geodesic emanating from o with the unit initial direction $X \in T_oM$. Then we define a cut point of o along γ_X as the last point on γ_X to which the geodesic arc minimizes the distance. The locus C of all cut points of o is called the *cut locus* of o. Now let Exp : $T_oM \rightarrow M$ be the exponential mapping. Lifting γ_X and the cut point of o along γ_X via Exp, we get a half-ray from 0 and the tangent cut point of o along γ_X which may be expressed in the form $\bar{t}_0(X)X, \bar{t}_0(X) \in \boldsymbol{R}^+$. The locus $\tilde{C}$: $= \{\bar{t}_0(X)X ; X \in T_oM$ unit vector$\}$ is called the *tangent cut locus* of o. By definition we have $C = \operatorname{Exp} \tilde{C}$. Since $\bar{t}_0(X)$ depends continuously on X, a domain $\mathscr{S} : = \{tX ; 0 \leqq t < \bar{t}_0(X) ; X \in T_oM$ unit vector$\}$ is homeomorphic to an n-cell, on which Exp is a diffeomorphism, and its boundary $\tilde{C}$ is mapped onto C. Thus M is obtained from C by attaching an n-cell via Exp and cut locus contains the essence of the topology of M. $\mathscr{S}$ is a maximal domain on which a normal coordinate system around o is valid and the estimate of its size (i.e., injectivity radius) played an essential role in celebrated pinching theorems due to W. Klingenberg, M. Berger and others. ([3], [5], [9], [12]).

On the other hand the structure of the cut locus is interesting in connection with the singularities of the exponential mapping. In this note we will be concerned with the cut locus of a point in a compact riemannian symmetric space and report some of the results obtained in [15], [16], [17] and [19].

For a riemannian symmetric space the following theorem due to R. Crittenden is well-known.

Theorem ([3], [6], [16]). *For a compact simply connected riemannian symmetric space, the* (*tangent*) *cut locus of a point coincide with the* (*tangent*) *first conjugate locus.*

Firstly we give an example and explain our motivation. Let (E^{2d}, α) be a $2d$-dimensional real symplectic vector space with symplectic form α, i.e., α is a non-degenerate skew-symmetric 2-form on E. A d-dimensional subspace of E on which α vanishes identically is called a Lagrangean subspace of E. We denote by $\Lambda(E)$ the space of all Lagrangean subspaces of E. Then $\Lambda(E)$ has the structure of a compact connected manifold, in fact a regular algebraic variety, of $d(d+1)/2$ dimension. Now we put for $\lambda_0 \in \Lambda(E)$, $\Lambda^k(\lambda_0): = \{\lambda \in \Lambda(E); \dim(\lambda \cap \lambda_0) = k\}$ which is a submanifold of codimension $k(k+1)/2$ in $\Lambda(E)$. Then $\Sigma(\lambda_0): = \bigcup_{k=1}^d \Lambda^k(\lambda_0)$ is an algebraic subvariety with the singular part $\bigcup_{k=2}^d \Lambda^k(\lambda_0)$ and defines an oriented cycle of codimension 1, whose Poincarè dual $[\quad] \in H^1(\Lambda(E); \boldsymbol{Z})$ defines the Maslov-Arnold index ([1]). That is, for a closed curve γ in $\Lambda(E)$, $[\gamma]$ is the intersection number of γ with the cycle $\Sigma(\lambda_0)$.

This index plays an important role in the index theorem of geodesics and the theory of Fourier integral operator ([1], [7], [8], [11]). First we give a differential geometric characterization of $\Sigma(\lambda_0)$. Choosing an adapted basis $\{e_1, \cdots, e_d, f_1, \cdots, f_d\}$ of E such that $\alpha(e_i, e_j) = \alpha(f_i, f_j) = 0$, $\alpha(e_i, f_j) = -\delta_{ij}$, we can show that the real respresentation of the unitary group $\boldsymbol{U}(d)$ acts transitively on $\Lambda(E)$ and that $\Lambda(E)$ is diffeomorphic to $\boldsymbol{U}(d)/\boldsymbol{O}(d)$ ([1]). Now note that $\boldsymbol{U}(d)/\boldsymbol{O}(d)$ has a structure of a compact riemannian symmetric space, where the symmetry is defined by the complex conjugate in $\boldsymbol{U}(d)$. Then we have

Theorem ([15]). *With respect to the above riemannian structure on $\Lambda(E)$, $\Sigma(\lambda_0)$ is the cut locus of $\lambda_0^\perp$ in $\Lambda(E)$, where $\lambda_0^\perp$ is the Lagrangean complement of λ_0.*

Moreover we can determine all closed geodesics on $\Lambda(E)$ and calculate their Maslov-Arnold index. Note that $\pi_1(\boldsymbol{U}(d)/\boldsymbol{O}(d)) \cong \boldsymbol{Z}$. Then the following problems naturally arise.

(I) *Characterize the tangent cut locus of a point in a compact riemannian symmetric space.*

In fact for a simply connected case Crittenden's theorem gives a characterization. We shall give a characterization for the general case in § 3.

Next in the above example the cut locus $\Sigma(\lambda_0) = \bigcup_{k=1}^d \Lambda^k(\lambda_0)$ has a natural stratification. Thus we have the second problem.

(II) *Study the structure of the cut locus of a point in a compact riemannian symmetric space.*

Here we must mention that recently M. Buchner has shown that the cut locus of a point in a compact analytic riemannian manifold is triangulable [4].

On problem (II) we shall describe the structure of the cut locus and its singularities in terms of algebraic structure of compact riemannian symmetric space.

In § 4 we shall treat the simply connected case following [17]. Now very recently M. Takeuchi has given a structure theorem for the cut locus and conjugate locus of a point in general (not necessarily simply connected) compact riemannain symmetric space using his theory of fundamental group of compact riemannain symmetric space. On the other hand, we may give another proof of his result based on Crittenden's theorem and our result given in § 4. We shall present it in § 5 for completeness. The author would like to express his sincere thanks to Professor Takeuchi for informing him of the result.

§ 2. Preliminaries

Let (G, K) be a compact riemannian symmetric pair which consists of the following: (1) a compact connected Lie group G and a closed subgroup K of G, (2) an involutive automorphism σ of G such that $(G_\sigma)_o \subset K \subset G_\sigma$, where G_σ is the subgroup of fixed points of σ and $(G_\sigma)_o$ denotes the identity component of G_σ and (3) a G-invariant riemannian structure on $M: = G/K$.

Then $M = G/K$ is a compact riemannian symmetric space and conversely every compact riemannian symmetric space may be expressed in this form. Let $\mathfrak{g}, \mathfrak{k}$ be Lie albegras of G, K respectively. Then the differential $D\sigma: \mathfrak{g} \to \mathfrak{g}$ is an involutive automorphism and $\mathfrak{k}$ is the $(+1)$-eigenspace of $D\sigma$. If we put $\mathfrak{m}: = \{X \in \mathfrak{g}; D\sigma(X) = -X\}$, we have a vector space direct sum $\mathfrak{g} = \mathfrak{k} + \mathfrak{m}$. Next let $\pi: G \to G/K$ be the canonical projection and $o: = \pi(e)$ be the origin of $M = G/K$. Then we may identify the tangent space T_oM with $\mathfrak{m}$ via $D\pi$. Moreover we may assume that G-invariant riemannian structure on M is induced from an Ad G-invariant and $D\sigma$-invariant inner product $\langle\ ,\ \rangle$ on $\mathfrak{g}$.

Now let $\mathfrak{a}$ be a Cartan subalgebra of (G, K) and take a maximal abelian subalgebra $\mathfrak{t}$ of $\mathfrak{g}$ containing $\mathfrak{a}$. Then it is well-known that $\mathfrak{m} = \mathrm{Ad}\, K_o\mathfrak{a}$, where K_o denotes the identity component of K. Let $\Sigma(G)$ (resp. $\Sigma(G, K)$) be the root space of G (resp. (G, K)), i.e., $\Sigma(G): = \{\alpha \in \mathfrak{t}\backslash\{0\}; \tilde{\mathfrak{g}}_\alpha \neq \{0\}\}$ (resp. $\Sigma(G, K): = \{\gamma \in \mathfrak{a}\backslash\{0\}; \mathfrak{g}^c_\gamma \neq \{0\}\}$), where $\tilde{\mathfrak{g}}_\alpha: = \{X \in \mathfrak{g}^c; [H, X] = 2\pi\sqrt{-1}\langle\alpha, H\rangle X$ for every $H \in \mathfrak{t}\}$ (resp. $\mathfrak{g}^c_\gamma: = \{X \in \mathfrak{g}^c; [H, X] = 2\pi\sqrt{-1}$ $\langle\alpha, H\rangle X$ for every $H \in \mathfrak{a}\}$). Then we have the root space decomposition $\mathfrak{g}^c = \tilde{\mathfrak{g}}_0 + \sum_{\alpha \in \Sigma(G)} \tilde{\mathfrak{g}}_\alpha (\tilde{\mathfrak{g}}_0 = \mathfrak{t}^c)$ and $\mathfrak{g}^c = \mathfrak{g}^c_0 + \sum_{\gamma \in \Sigma(G,K)} \mathfrak{g}^c_\gamma$. If we put $\Sigma_0(G): = \Sigma(G) \cap \mathfrak{b}$, where $\mathfrak{b} = \mathfrak{t} \cap \mathfrak{k}$, then we get $\mathfrak{t} = \mathfrak{b} \oplus \mathfrak{a}$ and $\mathfrak{g}^c_\gamma = \sum_{\bar{\alpha}=\gamma} \tilde{\mathfrak{g}}_\alpha$, where $\bar{\alpha}$ denotes the orthogonal projection of $\alpha \in \mathfrak{t}$ to $\mathfrak{a}$. Especially we have $\Sigma(G, K) = \{\bar{\alpha}; \alpha \in \Sigma(G)\backslash\Sigma_0(G)\}$. Next let $\Sigma^+(G)$ (resp. $\Sigma^+(G, K)$) be the set of positive roots with respect to an appropriate linear order on $\mathfrak{t}$. Then we have $\Sigma^+(G, K) = \{\bar{\alpha}; \alpha \in \Sigma^+(G)\backslash\Sigma_0(G)\}$. Now put $\mathfrak{k}_\gamma: = \mathfrak{k} \cap (\mathfrak{g}^c_\gamma + \mathfrak{g}^c_{-\gamma})$, $\mathfrak{m}_\gamma: = \mathfrak{m} \cap (\mathfrak{g}^c_\gamma + \mathfrak{g}^c_{-\gamma})$ for $\gamma \in \Sigma^+(G, K)$ and let $\mathfrak{k}_0$ be the centralizer of $\mathfrak{a}$. We have the decomposition $\mathfrak{k} = \mathfrak{k}_0 + \sum_{\gamma \in \Sigma^+(G,K)} \mathfrak{k}_\gamma$, $\mathfrak{m} = \mathfrak{m}_0 + \sum_{\gamma \in \Sigma^+(G,K)} \mathfrak{m}_\gamma$. Then for every $\alpha \in \Sigma^+(G) - \Sigma_0(G)$, there exist $S_\alpha \in \mathfrak{k}$, $T_\alpha \in \mathfrak{m}$ such that the following hold:

(1) for every $\gamma \in \Sigma^+(G, K)$, $\{S_\alpha : \bar{\alpha} = \gamma\}$, $\{T_\alpha : \bar{\alpha} = \gamma\}$ form a basis of $\mathfrak{k}_\gamma$, $\mathfrak{m}_\gamma$ respectively.

(2) $[H, S_\alpha] = 2\pi\langle\alpha, H\rangle T_\alpha$, $[H, T_\alpha] = -2\pi\langle\alpha, H\rangle T_\alpha$ for $H \in \mathfrak{a}$.

Now in a riemannian symmetric space M, the geodesic $\gamma_X : t \to \operatorname{Exp} tX$ emanating from o with the initial direction $X \in \mathfrak{m}$ is given by $\operatorname{Exp} tX = \pi \exp X$, where exp denotes the exponential mapping of Lie group. For details of above facts see [10], [18].

We shall give a lemma on the differential $D\operatorname{Exp}$ of the exponential mapping $\operatorname{Exp} : \mathfrak{m} \to M$.

Lemma 1 ([17]). *Let H_0 be an element of $\mathfrak{a}$ and consider $Y = H + \sum_{\alpha \in \Sigma^+(G)\setminus\Sigma_0(G)} a_\alpha T_\alpha \in \mathfrak{m}(H \in \mathfrak{a})$ as a tangent vector to $\mathfrak{m}$ at H_0. Then we have*

$$D\operatorname{Exp}_{H_0} Y = D\tau_{\exp H_0}(H + \textstyle\sum_{\langle\alpha, H_0\rangle = 0} a_\alpha T_\alpha + \sum_{\langle\alpha, H_0\rangle \neq 0} a_\alpha \sin 2\pi\langle\bar{\alpha}, H_0\rangle / \{2\pi\langle\bar{\alpha}, H_0\rangle\} \cdot T_\alpha) ,$$

where τ_g denotes the left translation of M by an element $g \in G$.

Recall that $X \in \mathfrak{m}\setminus\{0\}$ (resp. $\operatorname{Exp} X$) is called a tangent conjugate point (resp. conjugate point) to o along $\gamma_{X/\|X\|}$ iff Exp is singular at X. By the above lemma X is a tangent conjugate point iff $2\langle\gamma, X\rangle \in Z\setminus\{0\}$ for some $\gamma \in \Sigma(G, K)$. Especially we get

Corollary 2. *Let $X \in \mathfrak{a}$ be a unit vector. Then the first tangent conjugate point $t_0(X)X$ of o along γ_X is given by*

$$t_0(X) = \operatorname{Min}_{\alpha \in \Sigma^+(G)\setminus\Sigma_0(G)} 1/\{2\,|\langle\alpha, X\rangle|\} .$$

§ 3. A characterization of tangent cut locus

Now we shall consider the first problem (I). Let (G, K) be a compact riemannian symmetric pair and $\mathfrak{a}$ be a Cartan subalgebra of (G, K). Since $\mathfrak{m}\,(\cong T_o M) = \bigcup_{k \in K_0} \operatorname{Ad} k\mathfrak{a}$ and $\operatorname{Ad} k$ acts on $\mathfrak{m}$ as an isometry, for problem (I), it suffices to study the tangent cut points of o in $\mathfrak{a}$. Now we define $\Gamma(G, K) := \{X \in \mathfrak{a} ; \operatorname{Exp} X = o\}$. Then $\Gamma(G, K)$ is a lattice in $\mathfrak{a}$. We shall consider the flat torus $\mathfrak{a}/\Gamma(G, K)$. For a unit vector X in $\mathfrak{a}$ define

$$\bar{t}_0(X) := \operatorname{Min}_{A \in \Gamma(G,K)\setminus\{0\}} \frac{\langle A, A\rangle}{2\,|\langle X, A\rangle|} .$$

Then it is easy to see that $\tilde{C}_\mathfrak{a} := \{\bar{t}_0(X)X ; X \text{ is a unit vector in } \mathfrak{a}\}$ is the tangent cut locus of o in a flat torus $\mathfrak{a}/\Gamma(G, K)$. Then we have

Theorem 3 ([16]). *Let (G, K) be a compact riemannian symmetric pair. Then the tangent cut locus $\tilde{C}$ of o in $M = G/K$ is determined by*

the tangent cut locus $\tilde{C}_{\mathfrak{a}} := \{\bar{t}_0(X)X\,;\, X \in \mathfrak{a},\ \|X\| = 1\}$ *of* o *in a flat torus* $\mathfrak{a}/\Gamma(G, K)$. *In fact, we have* $\tilde{C} = \mathrm{Ad}\, K_o \tilde{C}_{\mathfrak{a}}$.

In stead of giving a proof of this theorem (see [16]), I'll explain it by the above mentioned example $\boldsymbol{U}(d)/\boldsymbol{O}(d)$. For simplicity assume that $d = 2$. Then rank $\boldsymbol{U}(2)/\boldsymbol{O}(2) = 2$ and dim $\boldsymbol{U}(2)/\boldsymbol{O}(2) = 3$. Define an inner product $\langle\ ,\ \rangle$ on $\mathfrak{g} = \mathfrak{u}(2)$ by

$$\langle X, Y\rangle := -\tfrac{1}{2}\,\mathrm{tr}\, XY\,, \qquad X, Y \in \mathfrak{u}(2)$$

and matrices $A_1, A_2 \in \mathfrak{u}(2)$ by

$$A_1 := \begin{pmatrix} \sqrt{-2} & 0 \\ 0 & 0 \end{pmatrix} \qquad A_2 := \begin{pmatrix} 0 & 0 \\ 0 & \sqrt{-2} \end{pmatrix}.$$

Then $\{A_1, A_2\}$ is orthonormal and spans a Cartan subalgebra $\mathfrak{a}$. An easy calculation shows that the lattice $\Gamma(\boldsymbol{U}(2), \boldsymbol{O}(2))$ is generated by $A_i/\sqrt{2}$, $i = 1, 2$. The tangent conjugate locus in $\mathfrak{a}$ is denoted in Diagram 1 by dotted lines.

Now from Theorem 3, the tangent cut locus $\tilde{C}_{\mathfrak{a}}$ in $\mathfrak{a}$ is given by the square $P_1P_2P_3P_4$ in Diagram 1, where the identification under the exponential mapping is shown by arrows. Note that the line $y = x$ is the center $\mathfrak{c}$ of $\mathfrak{u}(2)$. Now by Theorem 3, the whole tangent cut locus $\tilde{C}$ is given by $\tilde{C} = \mathrm{Ad}\, \boldsymbol{SO}(2)\tilde{C}_{\mathfrak{a}}$. Since Ad $\boldsymbol{SO}(2)$ fixes the center $\mathfrak{c}$ pointwise, the whole tangent cut locus of o is obtained by rotating the above square $P_1P_2P_3P_4$ aroud the axis $\mathfrak{c}$. Now under the exponential map the great circle obtained by rotating a conjugate point P_2 around $\mathfrak{c}$ is mapped into one point $\mathrm{Exp}\, P_1 = \mathrm{Exp}\, P_2 = \mathrm{Exp}\, P_3 = \mathrm{Exp}\, P_4$. Thus we have the Diagram 2 which shows the shape of cut locus and its singularity.

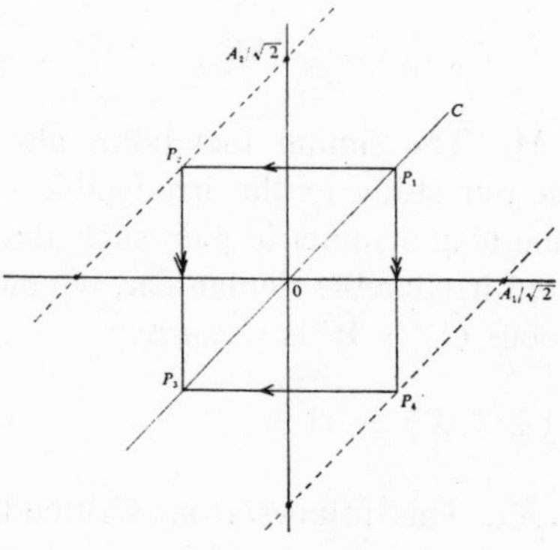

Diagram 1

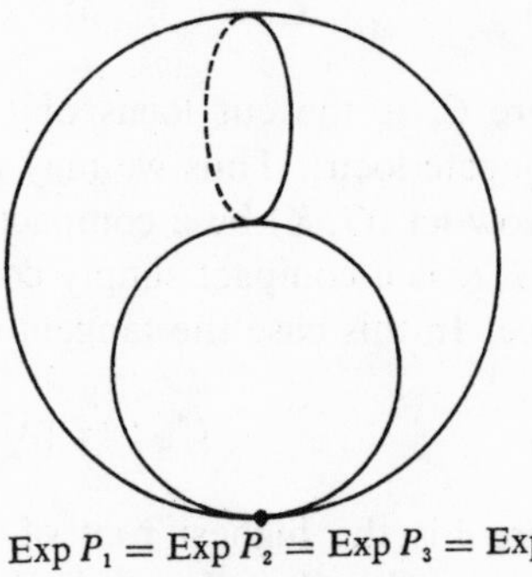

$\mathrm{Exp}\, P_1 = \mathrm{Exp}\, P_2 = \mathrm{Exp}\, P_3 = \mathrm{Exp}\, P_4$

Diagram 2

Remark. If $\Gamma(G, K)$ has an orthonormal basis, then the tangent cut locus of o in $\mathfrak{a}$ is particularly easy to be seen, this happens in case of so-called symmetric R-spaces (see H. Naitoh [13]).

Now for general case we must investigate what identifications occur in the tangent cut locus under the exponential mapping. This will be done in the next section in terms of the root system of (G, K).

§ 4. Structure theorem (simply connected case)

Let (G, K) be a compact riemannian symmetric pair and $\Sigma(G, K)$ be the set of roots of (G, K). Introducing an appropriate linear order in $\mathfrak{a}$ we have the fundamental root system $\pi(G, K) := \{\gamma_1, \cdots, \gamma_l\}$, which consists of simple roots (i.e., positive roots which can not be expressed as the sum of two positive roots). Then for every positive root γ, we have $\gamma = \sum m_i \gamma_i$ with non-negative integer m_i.

Next connected components of $\mathfrak{a} \backslash \bigcup_{\gamma \in \Sigma(G,K)} \gamma^\perp, \gamma^\perp := \{X \in \mathfrak{a}; \langle \gamma, X \rangle = 0\}$, are called *Weyl chambers*. Especially a Weyl chamber $W := \{X \in \mathfrak{a}; \langle \gamma, X \rangle > 0$ for all $\gamma \in \pi(G, K)\}$ will be called a *positive* Weyl chamber. Now let $W(G, K) = N(G, K)/W(G, K)$ be the Weyl group of (G, K), where $N(G, K) := \{k \in K; \mathrm{Ad}\, k(\mathfrak{a}) = \mathfrak{a}\}$ and $Z(G, K) := \{k \in K; \mathrm{Ad}\, k|_{\mathfrak{a}} = \mathrm{id}_{\mathfrak{a}}\}$. Then $W(G, K)$ acts simply transitively on the set of Weyl chambers. Especially we have $\mathfrak{m} = \mathrm{Ad}\, K\overline{W}$.

Now we shall consider a simply connected compact riemannian symmetric space M. Then by Crittenden's theorem the cut locus of o coincides with the first conjugate locus. Since M is simply connected, M is decomposed into the riemannian product of irreducible factors $M = M_1 \times \cdots \times M_t$, where M_i $(i = 1, \cdots, t)$ is a compact simply connected irreducible riemannian symmetric space. Then it is easy to see that the cut locus C of $o = (o_1, \cdots, o_t)$ in M is given by

$$(*) \qquad C = \bigcup_{i=1}^{t} M_1 \times \cdots \times C_i \times \cdots \times M_t ,$$

where C_i is the cut locus of o_i in M_i. The similar fact holds also for conjugate locus. Thus we may reduce our study to the irreducible case.

Now let (G, K) be a compact riemannian symmetric pair such that $M = G/K$ is a compact simply connected irreducible riemannian symmetric space. In this case the tangent cut locus $\tilde{C}_{\overline{W}}$ in $\overline{W}$ is given by

$$\tilde{C}_{\overline{W}} := \{X \in \overline{W}; 2\langle \delta, X \rangle = 1\} ,$$

where δ is the highest root of $\Sigma(G, K)$. This follows from Crittenden's theorem, Corollary 2 and the fact that if we write $\delta = \sum m_i \gamma_i$, then m_i is positive and not less than the i-th component of all positive roots.

Now we shall give a cell decomposition of $\tilde{C}_{\overline{W}}$. For that purpose put $\pi^{\#} := \pi(G, K) \cup \{\delta\}$, and for $\Delta \in \pi^{\#}$ put

$$S_\Delta := \{X \in \overline{W}\,;\, 2\langle\gamma, X\rangle \begin{array}{ll} > 0 & \text{for } \gamma \in \pi \cap \Delta \\ < 1 & \text{for } \gamma \in \{\delta\} \cap \Delta \\ = 0 & \text{for } \gamma \in \pi \backslash \Delta \\ = 1 & \text{for } \gamma \in \{\delta\} \backslash \Delta\} \,. \end{array}$$

Note that $S_\Delta \neq \varnothing$ iff $\Delta \neq \varnothing$ and $S_\Delta \subset \tilde{C}_{\overline{W}}$ iff $\delta \notin \Delta$. If we put $\mathscr{D} := \{\Delta \subset \pi^{\#}\,;\, \Delta \neq \varnothing,\, \delta \notin \Delta\}$, then $\{S_\Delta\}_{\Delta \in \mathscr{D}}$ gives a cell decomposition of $\tilde{C}_{\overline{W}}$. Note that $\Delta'\ (\neq \varnothing) \subset \Delta \in \mathscr{D}$ implies $\Delta' \in \mathscr{D}$. Now define $Z_\Delta := \{k \in K\,;$ $\operatorname{Exp} \operatorname{Ad} k\, X = \operatorname{Exp} X$ for all $X \in S_\Delta\}$. Then we have

Lemma 4. *Let $\Phi_\Delta : K/Z_\Delta \times S_\Delta \to M$ be a differentiable map defined by $\Phi_\Delta(kZ_\Delta, X) = \operatorname{Exp} \operatorname{Ad} k\, X$. $(\Delta \subset \pi^{\#})$. Then*

(1) *Φ_Δ is an everywhere regular injective mapping.*

(2) *$C_\Delta := \operatorname{Im} \Phi_\Delta$ is an embedded submanifold of M and the topology of C_Δ induced via Φ_Δ coincides with the relative topology of M.*

Next the following lemma is easy to see.

Lemma 5. (1) *$C_\Delta \cap C_{\Delta'} \neq \varnothing$ iff $\Delta = \Delta'$*

(2) $\overline{C}_\Delta = \bigcup_{\Delta' \subset \Delta} C_{\Delta'}$.

Now we have our main theorem ([17]).

Theorem 6. *Let M be a compact simply connected irreducible riemannian symmetric space. Then the cut locus C of o in M has the following structure: $C = \bigcup_{\Delta \in \mathscr{D}} C_\Delta$ which is the finite disjoint union such that*

(1) *C_Δ is an embedded submanifold of M diffeomorphic to $K/Z_\Delta \times S_\Delta$.*

(2) $\overline{C}_\Delta = \bigcup_{\Delta' \subset \Delta} C_{\Delta'}$.

Example. $P_d(\boldsymbol{C}) := \boldsymbol{U}(d+1)/\boldsymbol{U}(1) \times \boldsymbol{U}(d)$ (complex projective space).

In this case rank $P_d(\boldsymbol{C}) = 1$. Thus the tangent cut locus $C_{\overline{W}}$ in $\overline{W}$ reduces to a point $\Delta = \{H_0\}$.

We can see by an easy calculation that $Z_\Delta = \boldsymbol{U}(1) \times \boldsymbol{U}(d-1) \times \boldsymbol{U}(1)$. Then we have

$$\begin{aligned} C = C_\Delta \cong K/Z_\Delta &= \boldsymbol{U}(1) \times \boldsymbol{U}(d)/\boldsymbol{U}(1) \times \boldsymbol{U}(d-1) \times \boldsymbol{U}(1) \\ &= \boldsymbol{U}(d)/\boldsymbol{U}(d-1) \times \boldsymbol{U}(1) \cong P_{d-1}(\boldsymbol{C}) \,. \end{aligned}$$

This agrees with the well-known fact.

Remark. The notation in Theorem 6, which is slightly different from our original paper [17], is due to Takeuchi [19]. An exactly similar result holds also for the whole conjugate locus.

In fact put $\mathscr{D}_0 := \{\Delta \subsetneq \pi^{\#}\,;\, \Delta \neq \varnothing\}$. Then $\{S_\Delta\}_{\Delta \in \mathscr{D}_0}$ gives a cell decomposition of the boundary ∂F of the fundamental cell $F(G, K) := \{X \in W\,;$

$2\langle\delta, X\rangle < 1\}$. Note that the full conjugate locus M_s is given by $M_s = \operatorname{Exp} \operatorname{Ad} K\, \partial F$. Put $\mathscr{D}' := \{\Delta \in \mathscr{D}_0 ; \delta \in \Delta\}$. If $\Delta \in \mathscr{D}'$, then injectivity of Φ_Δ in Lemma 4 follows in this case directly from Crittenden's theorem and regularity of Φ_Δ follows from Lemma 1 by an easy calculation. Moreover Crittenden's theorem shows also that $Z_\Delta = \{k \in K ; \operatorname{Ad} k\, X = X \text{ for all } X \in S_\Delta\}$ in this case. Thus we get

Corollary 7. *Let M be as in Theorem 6. Then the whole conjugate locus M_s of o in M has the following structure*: $M_s = \bigcup_{\Delta \in \mathscr{D}_0} C_\Delta$ *which is the finite disjaint union such that*

(1) *C_Δ is an embedded submanifold of M diffeomorphic to $K/Z \times S_\Delta$.*

(2) $\bar{C}_\Delta = \bigcup_{\Delta' \subset \Delta} C_{\Delta'}$.

Next we shall consider the general compact simply connected riemannian symmetric space M. Then M is the riemannian product of irreducible factors $M = M_1 \times \cdots \times M_t$. Note that M may be expressed in the form $M = G/K$, where (G, K) is a compact riemannian symmetric pair such that $G = G_1 \times \cdots \times G_t$, $K = K_1 \times \cdots \times K_t$, $\sigma = \sigma_1 \times \cdots \times \sigma_t$ and $M_i = G_i/K_i$ is a simply connected irreducible factor $(i = 1, \cdots, t)$. Then the root space $\Sigma = \Sigma(G, K)$ decomposes into mutually orthogonal components $\bigcup_{i=1}^t \sum (G_i, K_i)$. Let δ_i be the highest root of the irreducible component $\Sigma_i := \Sigma(G_i, K_i)$. Now put

$$\pi_i^\# := \pi_i(G, K) \cup \{\delta_i\},\ \pi^\# := \bigcup_{i=1}^t \pi_i^\#,\ \Sigma^\# := \bigcup_{i=1}^t \{\delta_i\}\ .$$

And we define for $\Delta \subset \pi^\#$ as above

$$S_\Delta := \{X \in \overline{W} ; 2\langle\gamma, X\rangle \begin{array}{ll} > 0 & \text{if } \gamma \in \Delta \cap \pi(G, K) \\ < 1 & \text{if } \gamma \in \Delta \cap \Sigma^\# \\ = 0 & \text{if } \gamma \in \pi(G, K)\backslash\Delta \\ = 1 & \text{if } \gamma \in \Sigma^\#\backslash\Delta\}\ . \end{array}$$

Then $S_\Delta \neq \emptyset$ iff $\Delta \cap \pi_k^\# \neq \emptyset$ $(k = 1, \cdots, t)$. Put $\mathscr{D}_o := \{\Delta \subsetneq \pi^\# ; \Delta \cap \pi_k^\# \neq \emptyset\ (k = 1, \cdots, t)\}$ and $\mathscr{D} := \{\Delta \in \mathscr{D}_0 ; \Sigma^\#\backslash\Delta \neq \emptyset\}$. Then $C_\Delta, \Phi_\Delta, Z_\Delta$ etc. are defined exactly in the same manner as the above. Note that S_Δ may be expressed in the form $S_\Delta = S_{\Delta_1} \times \cdots \times S_{\Delta_t}$ where $\Delta_i \subset \pi_i^\#$ $(i = 1, \cdots, t)$. On the other hand it is easy to see $Z_\Delta = Z_{\Delta_1} \times \cdots \times Z_{\Delta_t}$ and we have $C_\Delta = C_{\Delta_1} \times \cdots \times C_{\Delta_t}$. Then $(*)$ gives the following.

Theorem 8. *Let $M = G/K$ be a compact simply connected riemannian symmetric space. Then the cut locus C (resp. the whole conjugate locus M_s) of o in M has the following structure*: $C = \bigcup_{\Delta \in \mathscr{D}} C_\Delta$ *(resp.* $M_s = \bigcup_{\Delta \in \mathscr{D}_0} C_\Delta$*) which is the finite disjoint union such that*

(1) *C_Δ is an embedded submanifold of M diffeomorphic to $K/Z_\Delta \times S_\Delta$.*

(2) $\bar{C}_\Delta = \bigcup_{\Delta' \subset \Delta} C_\Delta$.

§ 5. General case (Takeuchi's result)

In this section we give alternative proofs of Takeuchi's results (Theorems 12, 16) based on Theorem 8 and Crittenden's theorem. Let (G, K) be a compact riemannian symmetric pair and $(\mathfrak{g}, \mathfrak{k}, s = D\sigma)$ be the corresponding orthogonal involutive Lie algebra (*oila*). Since the root system and the Weyl group of the pair (G, K) are determined only by $(\mathfrak{g}, \mathfrak{k}, s)$, we shall denote them by $\Sigma(\mathfrak{g}, \mathfrak{k})$, $W(\mathfrak{g}, \mathfrak{k})$ respectively in this section. Since $\mathfrak{g}$ is a compact Lie algebra, we have the decomposition $\mathfrak{g} = \mathfrak{c} \oplus \mathfrak{g}'$, where $\mathfrak{c}$ is the center of $\mathfrak{g}$ and $\mathfrak{g}' := [\mathfrak{g}, \mathfrak{g}]$ is the semi-simple part of $\mathfrak{g}$. If we put $\mathfrak{c}_{\mathfrak{k}} := \mathfrak{c} \cap \mathfrak{k}$, $\mathfrak{c}_{\mathfrak{m}} := \mathfrak{c} \cap \mathfrak{m}$, $\mathfrak{k}' := \mathfrak{k} \cap \mathfrak{g}'$ and $\mathfrak{m}' := \mathfrak{m} \cap \mathfrak{g}'$, we have easily the following.

$$\mathfrak{c} = \mathfrak{c}_{\mathfrak{k}} \oplus \mathfrak{c}_{\mathfrak{m}}, \ \mathfrak{g}' = \mathfrak{k}' + \mathfrak{m}', \ \mathfrak{k} = \mathfrak{c}_{\mathfrak{k}} \oplus \mathfrak{k}', \ \mathfrak{m} = \mathfrak{c}_{\mathfrak{m}} + \mathfrak{m}' .$$

Thus we have the orthogonal decomposition of *oila*:

$$(\mathfrak{g}, \mathfrak{k}, s) = (\mathfrak{c}, \mathfrak{c}_{\mathfrak{k}}, s|_{\mathfrak{c}}) \oplus (\mathfrak{g}', \mathfrak{k}', s|_{\mathfrak{g}}) ,$$

where $(\mathfrak{c}, \mathfrak{c}_{\mathfrak{k}}, s|_{\mathfrak{c}})$ is an euclidean *oila* (i.e. $[\mathfrak{c}_{\mathfrak{m}}, \mathfrak{c}_{\mathfrak{m}}] = 0$) and $(\mathfrak{g}', \mathfrak{s}', s|_{\mathfrak{g}'})$ is semi-simple. Let $\tilde{G}_o$ be a compact simply connected semi-simple Lie group with the Lie algebra $\mathfrak{g}'$ and $\tilde{K}_o$ be the connected Lie subgroup of $\tilde{G}_o$ with the Lie algebra $\mathfrak{k}'$. Note that $\tilde{K}_o$ $(= (G_o)_\sigma)$ is closed. Then $\tilde{M}_o = \tilde{G}_o/\tilde{K}_o$ is a compact simply connected riemannian symmetric space associated with $(\mathfrak{g}', \mathfrak{k}', s|_{\mathfrak{g}'})$. On the other hand $C_{\mathfrak{m}}$ with the flat riemannian structure associates with the euclidean *oila* $(\mathfrak{c}, \mathfrak{c}_{\mathfrak{k}'}, s|_{\mathfrak{c}})$. Thus the universal convering space $\tilde{M}$ of M is given by $\tilde{M} = C_{\mathfrak{m}} \times \tilde{M}_0$.

Now let $\tilde{M}_r := \{x \in \tilde{M} ; x$ is not a conjugate point to the origin o along any geodesic connecting o and $x\}$ $(= \tilde{M} \backslash \tilde{M}_s)$ be the set of regular elements of $\tilde{M}$. Then we have clearly $\tilde{M}_r = C_{\mathfrak{m}} \times (\tilde{M}_o)_r$, Again let $F(\mathfrak{g}, \mathfrak{k}) := \{H \in \mathfrak{a} ; 0 < 2\langle\gamma, H\rangle < 1$ for all $\gamma \in \Sigma^+(\mathfrak{g}, \mathfrak{k})\}$ be a fundamental cell. Then, since $\Sigma(\mathfrak{c}, \mathfrak{c}_{\mathfrak{k}}) = \varnothing$ we get $F(\mathfrak{g}, \mathfrak{k}) = \mathfrak{c}_{\mathfrak{m}} \times F(\mathfrak{g}', \mathfrak{k}')$. Firstly we shall give a preliminary lemma.

Lemma 9. *$\tilde{M}_r$ is diffeomorphic to $K/Z(G, K) \times F(\mathfrak{g}, \mathfrak{k})$, where*

$$Z(G, K) := \{k \in K ; \mathrm{Ad}\, k|_{\mathfrak{a}} = \mathrm{id}_{\mathfrak{a}}\} .$$

Proof. Define a differentiable mapping $\Phi : K_o/Z(\tilde{G}_o, \tilde{K}_o) \times F(\mathfrak{g}', \mathfrak{k}') \to (\tilde{M}_o)_r$ by $\Phi(\tilde{k}Z(\tilde{G}_o, \tilde{K}_o), X) := \widetilde{\mathrm{Exp}}\, \mathrm{Ad}\, \tilde{k}\, X$, where $Z(\tilde{G}_o, \tilde{K}_o) := \{\tilde{k} \in \tilde{K}_o ; \mathrm{Ad}\, \tilde{k}|_{\mathfrak{a}'} = \mathrm{id}_{\mathfrak{a}'}\}$. Then by Crittenden's theorem and Lemma 1 Φ is easily seen to be a diffeomorphism. Next it is well-known that $\tilde{K}_o/Z(\tilde{G}_o, \tilde{K}_o)$ is diffeomorphic to $K/Z(G, K)$. For later use we shall give the outline of its proof. Let G' (resp. K') be the connected Lie subgroup of G (resp. K) with Lie algebra $\mathfrak{g}'$ (resp. $\mathfrak{k}'$). Then the covering homomorphism $\pi: \tilde{G}_o \to G'$ induces a diffeomorphism $\tilde{K}_o/Z(\tilde{G}_o, \tilde{K}_o) \cong K'/Z(G', K')$. Next let K_o be the identity component of K and C_K ($\subset Z(G, K)$) be the connected Lie subgroup of K_o with the Lie algebra $\mathfrak{c}_{\mathfrak{k}}$. Then we get $K_o = C_K K'$ and the inclusion $K' \subsetneq K_o$ induces a diffeomorphism $K'/Z(G', K') \cong K_o/Z(G, K_o)$. Finally note that $K = K_o(K \cap A)$, where $A = \exp \mathfrak{a}$ ($\subset Z(G, K)$), holds. The inclusion $K_o \subsetneq K$ induces a diffeomorphism $K_o/Z(G, K_o) \cong K/Z(G, K)$. Summing up, we get $\tilde{M}_r = C_{\mathfrak{m}} \times (M_o)_r \cong C_{\mathfrak{m}} \times F(\mathfrak{g}', \mathfrak{k}') \times K/Z(G, K) \cong K/Z(G, K) \times F(\mathfrak{g}, \mathfrak{k})$. q.e.d.

Next we will give our interpretation of Takeuchi's theory of fundamental group of compact riemannian symmetric space ([19], [20]). Let $p: \tilde{M} \to M$ be the universal covering projection. Since p is a local isometry we have $\tilde{M}_r = p^{-1}(M_r)$ and every deck transformation of p maps $\tilde{M}_r$ onto itself. By Lemma 9 we see that the covering projection $p: \tilde{M}_r \to M_r$ may be identified with $p: K/Z(G, K) \times F(\mathfrak{g}, \mathfrak{k}) \to M_r$ defined by $p(kZ(G, K), X) := \operatorname{Exp} \operatorname{Ad} k\, X$. Then for every deck transformation φ of p, $\varphi(kZ(G, K), X) = (k'Z(G, K), X')$, $X, X' \in F(\mathfrak{g}, \mathfrak{k})$, $k, k' \in K$ implies that $\operatorname{Exp} \operatorname{Ad} k\, X = \operatorname{Exp} \operatorname{Ad} k'\, X'$, i.e., $\operatorname{Exp} \operatorname{Ad} k'^{-1}k\, X = \operatorname{Exp} X'$. Since $X, X' \in F(\mathfrak{g}, \mathfrak{k})$, $\operatorname{Ad} k'^{-1}k$ maps $\mathfrak{a}$ onto itself and there exists $\varphi_W \in W(\mathfrak{g}, \mathfrak{k})$ such that $\operatorname{Ad} k'^{-1} k|_{\mathfrak{a}} = \varphi_W$. Next by definition we get $X' = \varphi_W X + \varphi_\Gamma$ for some $\varphi_\Gamma \in \Gamma(G, K)$. Since $W(\mathfrak{g}, \mathfrak{k}) \cong W(\mathfrak{g}', \mathfrak{k}')$ is a finite group and $\Gamma(G, K)$ is discrete in $\mathfrak{a}$, φ_W and φ_Γ depend only on φ and do not depend on the choice of $(kZ(G, K), X)$. If we put $\bar{\varphi} := t(\varphi_\Gamma) \circ \varphi_W$, where $t(\cdot)$ denotes the translation by vector $\cdot$, we get $X' = \bar{\varphi} X$. Thus we have defined a map from the fundamental group $\pi_1(M; o) \ni \varphi \to \bar{\varphi} \in \tilde{W}_*(G, K) := \{\bar{\varphi} \in \Gamma(G, K) W(\mathfrak{g}, \mathfrak{k}); \bar{\varphi} F(\mathfrak{g}, \mathfrak{k}) = F(\mathfrak{g}, \mathfrak{k})\}$. Then it is easy to see that this is in fact an isomorphism between these groups.

We may write $\varphi(kZ(G, K), X) = (kZ(G, K)\varphi_W^{-1}, \bar{\varphi} X)$, where $s \in W(\mathfrak{g}, \mathfrak{k})$ acts on $K/Z(G, K)$ from the right by $kZ(G, K)s = khZ(G, K)$ if $s = \operatorname{Ad} h|_{\mathfrak{a}}$.

Remark. In the course of the proof of Lemma 9, we get also a diffeomorphism $\tilde{M}_r \cong \tilde{K}_o/Z(\tilde{G}_o, \tilde{K}_o) \times F(\mathfrak{g}, \mathfrak{k})$. For a deck transformation φ of $p: \tilde{M}_r \to M_r$ such that $\varphi(\tilde{h}Z(\tilde{G}_o, \tilde{K}_o), X) = (\tilde{k}'Z(\tilde{G}_o, \tilde{K}_o), X')$, we have also $\varphi_W = \operatorname{Ad} \tilde{k}'^{-1}\tilde{k}$ and $X' = \varphi_W X + \varphi_\Gamma$.

Now we shall consider the action of deck transformations of p on $\tilde{M}_s$. Recall that $\tilde{M}_s = C_{\mathfrak{m}} \times (\tilde{M}_o)_s$ holds and by Theorem 8 for $\tilde{M}_o$ we have $\tilde{M}_s = \bigcup_{\Delta \in \mathscr{D}_0} \tilde{C}_\Delta$, where $\tilde{C}_\Delta$ for $\tilde{M}$ is defined as in p. 199 and p. 200.

Lemma 10. *Let φ be a deck transformation of p. Then we have*

(1) $\varphi\tilde{C}_{\Delta_1} \cap \tilde{C}_{\Delta_2} \neq \emptyset$ *iff* $\varphi\tilde{C}_{\Delta_1} = \tilde{C}_{\Delta_2}$.

(2) $\bar{\tilde{C}}_{\Delta_1} \cap \varphi\tilde{C}_{\Delta_2} \neq \emptyset$ *iff* $\bar{\tilde{C}}_{\Delta_1} \supset \varphi\tilde{C}_{\Delta_2}$.

Proof. Let $\Delta \in \mathscr{D}_0$. For any $x \in \tilde{C}_\Delta$, there exists a unique $\Delta' \in \mathscr{D}_0$ such that $\varphi(x) \in \tilde{C}_{\Delta'}$. We may write $x = \widetilde{\mathrm{Exp}}\,\mathrm{Ad}\,\tilde{k}\,X$, $\tilde{k} \in \tilde{K}_o$, $X \in S_\Delta$. Here $\mathrm{Ad}\,\tilde{k}$ ($\tilde{k} \in \tilde{K}_o$) acts on $\mathfrak{c}_{\mathfrak{m}}$ trivially. Then we have $\varphi(x) = \widetilde{\mathrm{Exp}}\,\mathrm{Ad}\,(\tilde{k}\varphi_W^{-1})\bar{\varphi}X$ and consequently $\bar{\varphi}X \in S_{\Delta'}$. Since $\mathrm{Exp}\,\varphi_W X = \mathrm{Exp}\,\bar{\varphi}X$, φ_W maps isomorphically $\{H \in \mathfrak{a}\,;\, \mathrm{Ad}\,(\exp 2X)H = H\}$ onto $\{H \in \mathfrak{a}\,;\, \mathrm{Ad}\,(\exp 2\bar{\varphi}X)H = H\}$. By an easy computation we see that $\{H \in \mathfrak{a}\,;\, \mathrm{Ad}\,(\exp 2X)H = H\} = \mathfrak{a} + \sum_{\gamma \in \Sigma + (\mathfrak{g},\mathfrak{k}) \cap \{\pi^\# - \Delta\}_Z} \mathfrak{m}_\gamma$ holds iff $X \in S_\Delta$ and $\{H \in \mathfrak{a}\,;\, \mathrm{Ad}\,(\exp 2Y)H = H\} = \mathfrak{a} + \sum_{\gamma \in \Sigma + (\mathfrak{g},\mathfrak{k}) \cap \{\pi^\# - \Delta'\}_Z} \mathfrak{m}_\gamma$ holds iff $Y \in S_{\Delta'}$. Note that right hand sides of these identities don't depend on the choice of $X \in S_\Delta$ and $Y \in S_{\Delta'}$ respectively. Thus we get $\bar{\varphi}(S_\Delta) \subset S_{\Delta'}$ (i.e., $\varphi\tilde{C}_\Delta \subset \tilde{C}_{\Delta'}$). Considering φ^{-1} we have $\varphi\tilde{C}_\Delta = \tilde{C}_{\Delta'}$. Now the lemma follows from Lemma 5. q.e.d.

For $\Delta \in \mathscr{D}$ we set $\tilde{W}_\Delta := \{\varphi\,;$ deck transformation of $p: \tilde{M} \to M$ such that $\varphi\tilde{C}_\Delta = \tilde{C}_\Delta\}$ and $\overline{W}_\Delta := \{\bar{\varphi} \in \tilde{W}_*(G,K)\,;\, \bar{\varphi}S_\Delta = S_\Delta\}$. Then clearly we get $p(\tilde{C}_\Delta) \cong \tilde{C}_\Delta / \tilde{W}_\Delta$.

Lemma 11. *$p(\tilde{C}_\Delta)$ is diffeomorphic to $(K/Z_\Delta \times S_\Delta)/\tilde{W}_\Delta$, where $\tilde{W}_\Delta$ acts on $K/Z_\Delta \times S_\Delta$ as in the same way as explained before the "remark" of this section.*

Proof. Recall that $\tilde{C}_\Delta \cong \tilde{K}_o/\tilde{Z}_\Delta \times S_\Delta$, where $\tilde{Z}_\Delta = \{\tilde{k} \in \tilde{K}_o\,;\, \widetilde{\mathrm{Exp}}\,Ad\,\tilde{k}\,X = \widetilde{\mathrm{Exp}}\,X$ for all $X \in S_\Delta\}$ (Lemma 4). We define a differentiable mapping $\Psi_\Delta: \tilde{K}_o/\tilde{Z}_\Delta \times S_\Delta \to (K/Z_\Delta \times S_\Delta)/\tilde{W}_\Delta$ by $\Psi_\Delta(\tilde{k}\tilde{Z}_\Delta, X) = p_\Delta(\pi(\tilde{k})Z_\Delta, X)$, where $p_\Delta: K/Z_\Delta \times S_\Delta \to (K/Z_\Delta \times S_\Delta)/\tilde{W}_\Delta$ is the canonical projection. Then Ψ_Δ is surjective and everywhere regular. In fact, first note that the covering projection $\pi: \tilde{G}_o \to G'$ induces a covering projection $\tilde{K}_o/\tilde{Z}_\Delta \to K'/Z'_\Delta$, where $Z'_\Delta := \{k \in K'\,;\, \mathrm{Exp\,Ad}\,k\,X = \mathrm{Exp}\,X$ for all $X \in S_\Delta\}$. Next because $K_o = C_K K'$ and $C_K \subset Z^o_\Delta := \{k \in K_o\,;\, \mathrm{Exp\,Ad}\,k\,X = \mathrm{Exp}\,X$ for all $X \in S_\Delta\}$, the inclusion $K' \subsetneq K_o$ induces a diffeomorphism $K'/Z'_\Delta \cong K_o/Z^o_\Delta$. Finally the inclusion $K_o \subsetneq K$ induces a diffeomorphism $K_o/Z^o_\Delta \cong K/Z_\Delta$ because of $K = K_o(K \cap A)$, $K \cap A \subset Z_\Delta$ (see the proof of Lemma 9). Thus the map $\tilde{K}_o/\tilde{Z}_\Delta \times S_\Delta \to K/Z_\Delta \times S_\Delta$ defined by $(\tilde{k}\tilde{Z}_\Delta, X) \to (\pi(\tilde{k})Z_\Delta, X)$ is everywhere regular and surjective. Now assume that $\Psi_\Delta(\tilde{k}Z_\Delta, X) = \Psi_\Delta(\tilde{k}'Z_\Delta, X')$. Then there exists a $\bar{\varphi} \in \overline{W}_\Delta$ such that $\pi(\tilde{k})Z_\Delta\varphi_W^{-1} = \pi(\tilde{k}')Z_\Delta$ and $X' = \bar{\varphi}X$. If we put $\varphi_W^{-1} = \mathrm{Ad}\,\tilde{h}$, $\tilde{h} \in \tilde{K}_o$, we get $\pi(\tilde{k}'^{-1}\tilde{k}\tilde{h}) \in Z_\Delta$, i.e., $\mathrm{Exp\,Ad}\,\pi(\tilde{k}'^{-1}\tilde{k}\tilde{h})X = \mathrm{Exp}\,X$ for all $X \in S_\Delta$. Thus there exists a $\bar{\psi} \in \tilde{W}_*(G,K)$ such that $\tilde{k}'^{-1}\tilde{k}\tilde{h}\tilde{h}_1 \in \tilde{Z}_\Delta$ and $\bar{\psi}X = X$, where we have put $\psi_W^{-1} = \mathrm{Ad}\,\tilde{h}_1$. Note that $\bar{\psi} \in \overline{W}_\Delta$. Since the fundamental group of riemannian symmetric space is commutative, we get

$$\begin{aligned}\psi\varphi(\tilde{k}\tilde{Z}_\Delta, X) &= \psi(\tilde{k}\tilde{h}\tilde{Z}_\Delta, \bar{\varphi}X) = (\tilde{k}\tilde{h}\tilde{h}_1\tilde{Z}_\Delta, \bar{\psi}\bar{\varphi}X)\\ &= (\tilde{k}'\tilde{Z}_\Delta, \bar{\varphi}\bar{\psi}X) = (\tilde{k}'\tilde{Z}_\Delta, X')\end{aligned}$$

and $\psi\varphi \in \tilde{W}_\Delta$. This completes the proof of the lemma. q.e.d.

Now we shall define an equivalence relation $\sim$ on $\mathscr{D}_o$: $\Delta_1 \sim \Delta_2$ iff there exists a deck transformation φ such that $\bar{\varphi}S_{\Delta_1} = S_{\Delta_2}$. We denote by $[\Delta]$ the equivalence class containing Δ. Next we define a partial order "$\subset$" on $\mathscr{D}_0/\sim$: $[\Delta_1] \subset [\Delta_2]$ iff there exists a deck transformation φ of p such that $\bar{\varphi}S_{\Delta_1} \subset S_{\Delta_2}$. Finally set $C_{[\Delta]} := P(\tilde{C}_\Delta)$ for $\Delta \in \mathscr{D}_0$. Then we get the following result due to M. Takeuchi.

Theorem 12 (Takeuchi [19]). *Let $M = G/K$ be a compact riemannian symmetric space. Then the whole conjugate locus M_s of M has the following structure: $M_s = \bigcup_{[\Delta] \in \mathscr{D}_0/\sim} C_{[\Delta]}$ which is the finite disjoint union such that*

(1) *$C_{[\Delta]}$ is an embedded submanifold of M which is diffeomorphic to $(K/Z_\Delta \times S_\Delta)/\overline{W}_\Delta$.*

(2) $\overline{C_{[\Delta]}} = \bigcup_{[\Delta'] \subset [\Delta]} C_{[\Delta']}$.

Proof. Since $M_s = p(\tilde{M}_s)$ and $\tilde{M}_s = \bigcup_{\Delta \in \mathscr{D}_0} \tilde{C}_\Delta$, we have $M_s = \bigcup_{[\Delta] \in \mathscr{D}_0/\sim} \tilde{C}_{[\Delta]}$. First assume that $C_{[\Delta]} \cap C_{[\Delta']} \neq \emptyset$, i.e., there exists a deck transformation φ such that $\varphi\tilde{C}_\Delta \cap \tilde{C}_{\Delta'} \neq \emptyset$. Then we get by Lemma 11, $\varphi\tilde{C}_\Delta = \tilde{C}_{\Delta'}$, i.e., $C_{[\Delta']} = C_{[\Delta]}$. Thus the above union is disjoint. Secondly we show that $C_{[\Delta]}$ is an embedded submanifold of M whose manifold topology coincides with the relative topology. This follows from Lemma 11 and the following. If $\{\operatorname{Exp} \operatorname{Ad} k_n X_n\}_{n=1}^\infty$ converges to $\operatorname{Exp} \operatorname{Ad} k\, X$, then $\{p_\Delta(k_n Z_\Delta, X_n)\}$ converges to $p_\Delta(kZ_\Delta, X)$. This is easily seen as in [17]. Finally let $x \in \overline{C_{[\Delta]}}$. Then there exists an $\tilde{x} \in p^{-1}(x) \cap \overline{C}_\Delta$. By Lemma 5 there exists a $\Delta' \subset \Delta$ such that $\tilde{x} \in C_{\Delta'}$ i.e., $x \in C_{[\Delta']}$, $[\Delta'] \subset [\Delta]$. q.e.d.

Now we shall consider the tangent cut locus $\tilde{C}$ of o in a compact riemannian symmetric space $M = G/K$. Note that M is not simply connected iff $F(\mathfrak{g}, \mathfrak{k}) \cap \tilde{C} \neq \emptyset$ by Crittenden's theorem. If $X \in F(\mathfrak{g}, \mathfrak{k})$ is a tangent cut point of o, there exists a regular element $Z \in \mathfrak{a}$, $Z \neq X$ such that $\|Z\| = \|X\|$ and $\operatorname{Exp} Z = \operatorname{Exp} X$ (Theorem 3). Then there exist $Y \in F(\mathfrak{g}, \mathfrak{k})$ and $\varphi_W \in W(\mathfrak{g}, \mathfrak{k})$ such that $\|X\| = \|Y\|$ and $\operatorname{Exp} \varphi_W Y = \operatorname{Exp} X$, i.e., $\varphi_W Y = X - A$ with $A \in \Gamma(G, K)\backslash\{0\}$. Now $\bar{\varphi} := t(A)\varphi_W$ belongs to $\tilde{W}_*(G, K)$ with $A = \bar{\varphi}(0)$ and X satisfies $\langle X, A^* \rangle = 1$ where we put $A^* = 2A/\langle A, A \rangle$. Conversely if $\langle X, A^* \rangle = 1$ holds for some $A = \bar{\varphi}(0)$, then there exists $Z = \varphi_W X \in \mathfrak{a}$, $Z \neq X$, $\|Z\| = \|X\|$ such that $\operatorname{Exp} Z = \operatorname{Exp} X$. Note that in this case we get $\widetilde{\operatorname{Exp}}\, X = \varphi \operatorname{Exp} \varphi_W \bar{\varphi}^{-1} X$ in the universal covering space. If we put $\Theta := \{\bar{\varphi}(0) ; \varphi \in \tilde{W}_*(G, K)\} \subset \overline{F(\mathfrak{g}, \mathfrak{k})} \cap \Gamma(G, K)$, then it is easy to see that $F(\mathfrak{g}, \mathfrak{k}) \cap \tilde{C} = \{X \in F(\mathfrak{g}, \mathfrak{k}) ; \langle X, A^* \rangle \leq 1$ holds for all $A \in \Theta$ and the equality $\langle X, A^* \rangle = 1$ holds for at least one $A \in \Theta\}$. For $\Delta \subset \pi^\#$, $\Phi \subset \Theta$ we define

$$S_{\Delta,\Phi} := \{H \in S_\Delta ; \langle H, A^* \rangle < 1 \quad \text{for } A \in \Phi \qquad = 1 \quad \text{for } A \in \Theta\backslash\Phi\} .$$

and put $\mathscr{E}: = \{(\Delta, \Phi); \Delta \subset \pi^{\#}, \Phi \subsetneqq \Theta, S_{\Delta,\Phi} \neq \emptyset\}$. Note that $(\Delta, \Phi) \in \mathscr{E}$ and $\Delta' \subset \Delta$, $\Phi' \subset \Phi$ then $(\Delta', \Phi') \in \mathscr{E}$. Then by definition $\tilde{C} \cap \overline{F(\mathfrak{g}, \mathfrak{k})} = \bigcup_{(\Delta,\Phi)\in\mathscr{E}} S_{\Delta,\Phi}$ gives a cell-decomposition of $\tilde{C} \cap \overline{F(\mathfrak{g}, \mathfrak{k})}$. Finally we put

$$\tilde{C}_{\Delta,\Phi}: = \operatorname{Exp} \operatorname{Ad} K_o S_{\Delta,\Phi} \qquad \text{in } M .$$

Then since $\Phi_\Delta: \tilde{K}_o/\tilde{Z}_\Delta \times S_\Delta \to \tilde{C}_\Delta$ is a diffeomorphism (Lemma 4) we have the following lemma.

Lemma 13. (1) *$\tilde{C}_{\Delta,\Phi}$ is an embedded submanifold of $\tilde{M}$ which is diffeomorphic to $\tilde{K}_o/\tilde{Z}_\Delta \times S_{\Delta,\Phi}$.*

(2) $\overline{\tilde{C}_{\Delta,\Phi}} = \bigcup_{\Delta'\subset\Delta,\Phi'\subset\Phi} \tilde{C}_{\Delta',\Phi'}$.

Lemma 14. *Let φ be a deck transformation of $p: \tilde{M} \to M$. Then we have the following:*

(1) *$\varphi\tilde{C}_{\Delta,\Phi} \cap \tilde{C}_{\Delta',\Phi'} \neq \emptyset$ iff $\varphi\tilde{C}_{\Delta,\Phi} = \tilde{C}_{\Delta',\Phi'}$.*

(2) *$\varphi\tilde{C}_{\Delta,\Phi} \cap \overline{\tilde{C}_{\Delta',\Phi'}} \neq \emptyset$ iff $\varphi\tilde{C}_{\Delta,\Phi} \subset \overline{\tilde{C}_{\Delta',\Phi'}}$.*

Proof. Assume that $\varphi\tilde{C}_{\Delta,\Phi} \cap \tilde{C}_{\Delta',\Phi'} \neq \emptyset$. Then there exist $\tilde{k}, \tilde{h} \in \tilde{K}_o$, $X \in S_{\Delta,\Phi}$, $Y \in S_{\Delta',\Phi'}$ such that $\varphi\, \widetilde{\operatorname{Exp}} \operatorname{Ad} \tilde{k} X = \widetilde{\operatorname{Exp}} \operatorname{Ad} \tilde{h}\, Y$, that is, $Y = \bar{\varphi}X$ and $\tilde{h}^{-1}\tilde{k}\tilde{h}_1 \in \tilde{Z}_\Delta$ where $\bar{\varphi} \in \tilde{W}_*(G, K)$ and $\varphi_W^{-1} = \operatorname{Ad} \tilde{h}_1$ ($\tilde{h}_1 \in \tilde{K}_o$). In particular we have $\bar{\varphi}S_{\Delta,\Phi} \cap S_{\Delta',\Phi'} \neq \emptyset$ and by Lemma 10, $\bar{\varphi}S_\Delta = S_{\Delta'}$. On the other hand if we put $I: = \{X \in F(\mathfrak{g}, \mathfrak{k}), \langle X, A^*\rangle < 1 \text{ for } A \in \Theta\}$, then we can easily see that I is a fundamental domain of $\tilde{W}_*(G, K)$ acting on $\overline{F(\mathfrak{g}, \mathfrak{k})}$, that is, $\overline{F(\mathfrak{g}, \mathfrak{k})} = \bigcup_{\bar{\varphi}\in\tilde{W}_*(G,K)} \bar{\varphi}\bar{I}$ and $\bar{\varphi}(I) \cap I \neq \emptyset$ implies that $\bar{\varphi}$ is the identity. $\bar{I}$ is a compact convex subset of $\overline{F(\mathfrak{g}, \mathfrak{k})}$ whose boundary ∂I consists of faces formed by $\{S_{\Delta,\Phi}\}_{(\Delta,\Phi)\subset\mathscr{E}}$ and $\{S_{\Delta,\Theta}\}_{\Delta\in\mathscr{D}}$. Since $\tilde{W}_*(G, K)$ acts on $\overline{F(\mathfrak{g}, \mathfrak{k})}$ as a congruence transformation group and permutes congruent convex subsets of $\{\varphi(\bar{I})\}_{\varphi\in\tilde{W}_*(G,K)}$, $\varphi(\partial I) \cap \partial I$ consists of some of above faces unless it is empty. Thus $\varphi S_{\Delta,\Phi} \cap S_{\Delta',\Phi'} \neq \emptyset$ implies $\varphi S_{\Delta,\Phi} = S_{\Delta',\Phi'}$. This proves (1) and (2) follows immediately from (1) and Lemma 13. q.e.d.

Now we define $\tilde{W}_{\Delta,\Phi}: = \{\varphi$: deck transformation of $p: \tilde{M} \to M$ such that $\varphi\tilde{C}_{\Delta,\Phi} = \tilde{C}_{\Delta,\Phi}\}$. Then clearly we get $p(\tilde{C}_{\Delta,\Phi}) = \tilde{C}_{\Delta,\Phi}/\tilde{W}_{\Delta,\Phi}$.

Lemma 15. *$p(\tilde{C}_{\Delta,\Phi})$ is diffeomorphic to $(K/Z_\Delta \times S_{\Delta,\Phi})/\tilde{W}_{\Delta,\Phi}$.*

Proof. Recall that $\tilde{C}_{\Delta,\Phi}$ is diffeomorphic to $\tilde{K}_o/\tilde{Z}_\Delta \times S_{\Delta,\Phi}$. We define a differentiable mapping $\Psi_{\Delta,\Phi}: \tilde{K}_o/\tilde{Z}_\Delta \times S_{\Delta,\Phi} \to (K/Z_\Delta \times S_{\Delta,\Phi})/\tilde{W}_{\Delta,\Phi}$ by $\Psi_{\Delta,\Phi}(\tilde{k}\tilde{Z}_\Delta, X): = p_{\Delta,\Phi}(\pi(\tilde{h})Z_\Delta, X)$, where $p_{\Delta,\Phi}: K/Z_\Delta \times S_{\Delta,\Phi} \to (K/Z_\Delta \times S_{\Delta,\Phi})/\tilde{W}_{\Delta,\Phi}$ denotes the canonical projection. In the above $\tilde{W}_{\Delta,\Phi}$ acts on $K/Z_\Delta \times S_{\Delta,\Phi}$ in the same manner as in Lemma 11. Then $\Psi_{\Delta,\Phi}$ is clearly well-defined everywhere regular, and surjective mapping. If $p_{\Delta,\Phi}(\pi(\tilde{k})Z_\Delta, X) = p_{\Delta,\Phi}(\pi(\tilde{k}')Z_\Delta, X')$ holds, then by the same argument as in Lemma 11, we can show that $\tilde{\psi}(\tilde{k}\tilde{Z}_\Delta, X) = (\tilde{k}', \tilde{Z}_\Delta, X')$ holds for some $\psi \in \tilde{W}_{\Delta,\Phi}$. This completes the proof of the lemma. q.e.d.

Now we define an equivalence relation $\sim$ on $\mathscr{E}$: $(\Delta, \Phi) \sim (\Delta', \Phi')$ iff there exists a $\bar{\varphi} \in \tilde{W}_*(G, K)$ such that $\bar{\varphi} S_{\Delta,\Phi} = S_{\Delta',\Phi'}$. Let $[\Delta, \Phi]$ denotes the equivalence class containing (Δ, Φ). Next we define a partial order "$\subset$" on $\mathscr{E}/\sim$ as follows: $[\Delta', \Phi'] \subset [\Delta, \Phi]$ iff there exists a $\bar{\varphi} \in \tilde{W}_*(G, K)$ such that $\bar{\varphi} S_{\Delta',\Phi'} \subset \overline{S_{\Delta,\Phi}}$. Finally we put $C_{[\Delta,\Phi]} := p(\tilde{C}_{\Delta,\Phi})$ which is clearly independent of the choice of the representative of $[\Delta, \Phi]$. Then we get

Theorem 16 (Takeuchi [19]). *Let $M = G/K$ be a compact riemannian symmetric space. Then the cut locus C of o in M has the following structure: $C = \bigcup_{(\Delta,\Phi)\mathscr{E}} C_{[\Delta,\Phi]}$ which is the disjoint union satisfying*

(1) *$C_{[\Delta,\Phi]}$ is a regular submanifold of M diffeomorphic to $K/Z_\Delta \times S_{\Delta,\Phi})/\tilde{W}_{\Delta,\Phi}$.*

(2) $\overline{C_{[\Delta,\Phi]}} = \bigcup_{[\Delta',\Phi']\subset[\Delta,\Phi]} C_{[\Delta',\Phi']}$.

The proof proceeds in the same way as that of Theorem 12. During the preparation of the present paper the author was partially supported by the "Sonderforshungbereich Theoretische Mathematik (SFB 40)" at the University of Bonn.

References

[1] Arnold, V. I.: *On a characteristic class entering in quantization condition,* Funct. Anal. Appl. **1** (1967), 1–13.

[2] Berger, M.: *On geodesics in riemannian geometry,* Tata Institute.

[3] Bishop, R. and Crittenden, R.: Geometry of Manifolds, Academic Press, New York, 1964.

[4] Buchner, M.: *Simplicial structure of the real analytic cut locus,* Proc. A. M. S. **64** (1977), 118–121.

[5] Cheeger, J. and Ebin, G.: Comparison theorems in Riemannian geometry, North Holland-American Elsevier, New York, 1975.

[6] Crittenden, R.: *Minimum and conjugate points in symmetric spaces,* Canad. J. Math., **14** (1962), 320–328.

[7] Duistermaat, J. J.: *On the Morse index in variational calculus,* Advances in Math. **21** (1976), 173–195.

[8] ——: *Fourier integral operators,* Courant Institute Lecture Notes, New York, 1973.

[9] Gromoll, D., Klingenberg, W. and Meyer, W.: *Riemannche Geometrie im Großen* (Lecture Notes in Math., 55) Springer Verlag, Berlin and New York, 1968.

[10] Helgason, S.: Differential Geometry and symmetric spaces., Academic Press, New York, 1962.

[11] Klingenberg, W.: *Der Indexsatz für geschlosene Geodätische,* Math., Z., **139** (1974), 231–256.

[12] Kobayashi, S.: *On conjugate and cut loci,* Studies in Global Geometry and Analysis, Math. Assoc. Amer., 1967, 96–122.

[13] Naitoh, H.: *On cut loci and first conjugate loci of the irreducible symmetric R-spaces and the irreducible compact hermitian symmetric spaces,* Hokkaido Math. J. **6** (1977), 230–242.

[14] Rauch, H. E.: *Geodesics and Jacobi equations on homogeneous manifolds,* Proc. U.S.-Japan Seminar in Diff. Geo. 1965, Nippon Hyoronsha, Tokyo, 1966.

[15] Sakai, T.: *On the geometry of manifolds of Lagrangean subspaces of a*

symplectic vecter space, to appear in J. Differential Geometry.
[16] ——: *On cut loci of compact symmetric spaces,* Hokkaido Math. J. **6** (1977), 136–161
[17] ——: *On the structure of cut loci in compact riemannian symmetric spaces,* to appear in Math. Ann.
[18] Takeuchi, M.: Modern theory of spherical functions (in Japanese), Iwanami, Tokyo, 1976.
[19] ——: *On conjugate loci and cut loci of compact symmetric spaces I, II,* Preprint.
[20] ——: *On the fundamental group and the group of isometries of a symmetric space*, J. Fac. Sci. Univ. Tokyo, **10** (1963).

DEPARTMENT OF MATHEMATICS
HOKKAIDO UNIVERSITY
SAPPORO, 060 JAPAN
AND
MATHEMATISCHES INSTITUT
DER UNIVERSITÄT BONN
WEGELERSTR. 10
5300 BONN, GERMANY

Minimal Submanifolds and Geodesics
Kaigai Publications, Tokyo, 1978, 209–215

ON A LOWER BOUND OF THE NUMBERS OF MINIMAL SURFACES

YOSHIHIRO SHIKATA

We first introduce a kind of Morse theory which connects a lower bound of numbers of (weak) minimal surfaces to homology group of a certain space.

Definition 1. A (topological) *variation problem* is a triple $V = (X, f : \varphi)$ of a locally connected topological space X, a (continuous) map f of X into itself and a real valued continuous function φ on X satisfying the following conditions:

1) $\varphi(f(x)) \leqq \varphi(x)$.
2) $\varphi(f(x)) = \varphi(x)$ if and only if $f(x) = x$.
3) The induced homomorphism f_* on $H_*(X; k)$ is the identity.
4) For any compact set A in X, any minimizing sequence in $\overline{\bigcup_n f^{(n)}(A)}$ has a converging subsequence.

Definition 2. For a variation problem $V = (X, f; \varphi)$, a fixed point of f is said to be a *critical point* of V and the set of all the critical points is denoted by $\gamma(V)$.

Proposition 1. *If there are given variation problems $V_1 = (X, f; \varphi)$, $V_2 = (X, g; \varphi)$, then $V_3 = (X, f \circ g; \varphi)$ also turns out to be a variation problem, if $\overline{\bigcup_n (f \circ g)^{(n)}(A)}$ is compact for any compact set A in X so that*

$$\gamma(V_3) = \gamma(V_1) \cap \gamma(V_2) .$$

It is known [S] that

Theorem 1. *Let $V = (X, f : \varphi)$ be a variation problem. Then*

$$^{\#}\{\gamma(V)\} \geqq \text{cup length of } H_*(X; k) .$$

It is obvious that

Proposition 2. *For a variation problem $V = (X, f; \varphi)$*

$$^{\#}\{\gamma(V)\} \geqq {}^{\#}\{\text{connected component of } X\} .$$

Let C be a Jordan curve in $\boldsymbol{R}^3$ whose projection into $\boldsymbol{R}^2$ has no self-intersection and bounds a simply connected domain D. Then we can construct a variation problem $V = (Y, f : \varphi)$ using the space X of Lipschiz surfaces bounded by C for the area integral φ.

Introduce the non parametric representation for a surface $\alpha \in X$ and define φ-topology in X by

$$d_\varphi(\alpha, \beta) = \max |\alpha(x) - \beta(x)| + |\varphi(\alpha) - \varphi(\beta)| .$$

Take subdivisions K_1, K_2 or $\boldsymbol{R}^2$ into squares dual to each other, and define surfaces $f_1(\alpha), f_2(\alpha)$ by

$$f_i(\alpha)|_{\Delta_i} = \begin{cases} \alpha|_{\Delta_i}, \textit{ if } \operatorname{dist}(\Delta_i, B') = 0 , \\ \textit{minimal surface } m(\alpha, \Delta_i) \textit{ bounded by} \\ \alpha|_{\partial\Delta_i}, \textit{ if } \operatorname{dist}(\Delta_i, B') > 0 \end{cases}$$

for each squares $\Delta_i \in K_i$.

As is proved by Radó and Neumann [R], $m(\alpha, \Delta_i)$ again satisfies the L-Lipschitz condition and we have thus a correspondence $f_1, f_2, f_1 \circ f_2$ of X into X.

Because of the approxiation property [R] of the minimal surfaces we see that $m(\alpha, \Delta_i)$ depends continuously on α in the uniform topology, and because of the uniqueness of $m(\alpha, \Delta_i)$, deduced from the convexity of Δ_i, we also have the continuity of the area integral φ of $m(\alpha, \Delta_i)$.

Hence we see that f_1, f_2 and therefore $f = f_1 \circ f_2$ are continuous on X.

The triple $(X, f; \varphi)$ thus obtained satisfies the conditions 1)–3) of Definition 1 and gives a variation problem. In fact, the uniqueness $m(\alpha, \Delta)$ yields

$$\varphi(\alpha|_\Delta) \leqq \varphi(m(\alpha, \Delta))$$
$$\varphi(\alpha|_\Delta) = \varphi(m(\alpha, \Delta)) \textit{ if and only if } a|_\Delta = m(\alpha, \Delta) ,$$

hence yields the properties 1), 2), on the other, taking a subdivision K_i^t which divides each square Δ of K_i into rectangles $\Delta(t), \Delta(1-t)$ of area $t\varphi(\Delta), (1-t)\varphi(\Delta)$, we can define a homotopy h_t^i by

$$h_t^i(\alpha) = \begin{cases} \textit{minimal surface bounded by } \alpha|_{\Delta(t)}, \textit{ on } \Delta(t) \\ \qquad \alpha|_{\Delta(1-t)} \textit{ on } \Delta(1-t) \end{cases}$$

which connects f_i to the identitity.

The compactness condition 4) can be deduced from the compactness in the uniform topology as follows: In any infinite sequence $\{\sigma_n = f^{(n)}(\beta_n)\}$ of $\overline{\bigcup f^{(n)}(A)}$, choose uniformly convergent subsequence $\{f^{(m-1)}(\beta_m)\}$, then $\{f^{(m)}(\beta_m)\}$ converges in X, because φf is continuous in the uniform topology.

Proposition 3. *For a compact set $Y \subset X$ in the uniform topology, the triple $V = (Y, f; \varphi)$ is a variation problem and the critical set $\gamma(V)$ consists of L-Lipschitz surfaces bounded by C which cover D simply and are (globally) minimum on*

$$[\bigcup \{\Delta_1/\text{dist}\,(\Delta_1, \partial B) > 0\}] \cup [\bigcup \{\Delta_2/\text{dist}\,(\Delta_2, \partial B) > 0\}] .$$

Since the subdivisions K_1, K_2 are arbitrary except that they are dual to each other, we may choose K_i so that the diameters of simplexes tend to zero which the distances of the simplexes from B' tend to zero in order to make $\gamma(V)$ consists only of the minimal surface.

One of the modifications of the above construction is to extended f over the closure $\bar{X}$ using that f is uniformly continuous on X, in this case the variation problem is given on $\{\alpha \in \bar{X}/\varphi(\alpha) \leq c\}$ and the geometric interpretation of the critical set become a little complicated.

Also a generalization of the construction to the case in which R^3 is an isometric convering space of a Riemannian mainfold is easy, and the modification of the variation problem to the case of higher topological type is straight-forward.

For these cases Proposition 2 applies rather than Theorem 1.

Another modification is obtained through reconstruction of a homotopy h_t between f and the identity.

Let $\Delta = [0, 1] \times [0, 1]$ be a square of R^2 and let $\{\Delta(m, n\colon k), m, n = 1, \cdots, 2^{k+1}\}$ be a subdivision of Δ by lines

$$y = x + i2^{-k} - 1$$
$$y = -x + j2^{-k} .$$

Then for $t = 2^{-k}$ define $h_t^1(\alpha)$ by

$$h_t^1(\Delta)|_{\Delta(m,n,k)} = \textit{minimal surface bounded by } \alpha|_{\partial\Delta(m,n,k)}$$

and for $2^{-k-1} \leqq t \leqq 2^{-k-1} + 2^{-k-2}$, by taking a subdivision $\{\Delta'\}$ of $\Delta(m, n, k)$ by lines

$$y = x + i2^{-k-1} - 1$$
$$y = -x + j2^{-k}$$
$$y = -x + 2^{-k} + 2t + 2^{-k-1}$$

the map $h_t^1(\alpha)$ is defined on a simplex Δ' of the subdivision by $h_t^1(\alpha)|_{\Delta'} =$ *minimal surface bounded by* $h_s^1(\alpha)|_{\partial\Delta'}$, for $s = 2^{-k-1}$, finally for $2^{-k-1} + 2^{-k-2} \leqq t \leqq 2^{-k}$, the map $h_t^1(\alpha)$ is defined also by

$$h_t^1(\alpha)|_{\Delta''} = \textit{minimal surface bounded by } h_s^1(\alpha)|_{\partial\Delta''} ,$$
$$\text{for } s = 2^{-k-1} + 2^{-k-2} ,$$

on a simplex Δ'' of the subdivision of $\Delta(m, n, k)$ by lines

$$y = x + i2^{-k} - 1$$
$$y = x + j2^{-k} + 2t$$
$$y = -x + 2^{-k} .$$

The homotopy h_t^1 thus obtained connects f_1 to the identity in the uniform topology in X. And a homotopy h_t^i similarly defined by use of subdivisions of K_i whose edges are in general position each other $i = 1, 2, 3$, connects f_2 to the identity in the uniform topology. The homotopy $h_t = h_t^1 \circ h_t^2 \circ h_t^3$ satisfies the following

Lemma 1. *For $\alpha \in X$, $t > 0$,*

(1) $\varphi(h_t(\alpha)) \leqq \varphi(\alpha)$,

(2) *if $t > 0$, $\varphi(h_t(\alpha)) = \varphi(\alpha)$ implies that α is minimal on* Int $K_1 \cap$ Int K_2.

Lemma 2. *For $\varepsilon > 0$, $\varphi h_t(\alpha)$ is continuous in the uniform topology on* $[\varepsilon, 1] \times X$.

Corollary 1. *If $\{\alpha_n\}$ converges to α in the uniform topology of X and if $\{t_n\}$ tends to $t > 0$, then there is a subsequence $\{h_{t_m}(\alpha_m)\}$ in $\{h_{t_n}(\alpha_n)\}$ which converges to $h_t(\alpha)$ in φ-topology of X.*

Let Y denote the space of the disc type surfaces in R^3 which bound given curve C and let J be a function of Y into $[0, 1]$ satisfying the following 1)–5).

1) $J(\alpha) > 0$ implies that α has a homomorphic projection into (x, y)-plane.

2) $J(\alpha) > 0$ implies also that (the non parametric representation of) α satisfies L-Lipschitz condition.

3) J is upper semi continuous in the uniform topology on $Y_0 = \{\alpha \in Y / J(\alpha) > 0\}$.

4) For $\varepsilon > 0$, $Jh_t(\alpha)$ is continuous in the uniform topology on $[\varepsilon, 1] \times Y_0$.

Define J-topology and (J, φ)-topology on Y using J, φ by

$$d_J(\alpha, \beta) = \max |\alpha - \beta| + |J(\alpha) - J(\beta)| .$$
$$d_{J,\varphi}(\alpha, \beta) = \max |\alpha - \beta| + |J(\alpha) - J(\beta)| + |\varphi(\alpha) - \varphi(\beta)| ,$$

5) There exists a homotopy g_t on Y_0 such that

$$J(g_t(\alpha)) \geqq J(\alpha) , \qquad g_0 = \text{id} ,$$

and g_1 is continuous from Y_0 of the uniform topology into Y_0 of (J, φ)-topoloty.

Proposition 4. *Denote by Y^ε the subspace*

$$\{\alpha \in Y_0 / J(h_{t_1} \cdot \cdots \cdot h_{t_n}(\alpha)) \geqq \varepsilon,\ t_i \in [0, 1]\} ,$$

and by $f(\alpha)$ the map $h_{J(\alpha)}(\alpha)$ of Y^ε into itself. Then the triple $(Y^\varepsilon, f: \varphi)$ turns out to be a variation problem, where the topology of Y^ε is (J, φ)-topology.

In fact, Lemmas 1, 2 and the condition 4) for J yield the continuity of f and the conditions 1), 2) of Definition 1, because $\alpha \in Y^\varepsilon$ implies $J(\alpha) \geq \varepsilon$, on the other, the condition 3) of Definition 1 is deduced from the condition 5) for J and the homotopy for the variation problem $(X, f: \varphi)$, making use of a homotopy $h_{tJ(\alpha)}(\alpha)$ on Y_0 of the uniform topology. Finally the compactness condition 4) of Definition 1 is deduced from the conditions 2)–4) of J as follows: take a minimizing sequence $\{\alpha_n = f^{(n)}(\beta_n) \in Y^\varepsilon\}$ of $\overline{\bigcup f^{(n)}(A)}$, since $J(f^{(n-1)}(\beta_n)) \geq \varepsilon$, there is a converging subsequence $\{f^{(m-1)}(\beta_m)\}$ in the uniform topology to $\beta \in Y_0$, so that $\{t_m = Jf^{(m-1)}(\beta_m)\}$ tends also to t, then the condition 3) implies that $J(\beta) \geq t$, therefore the condition 4) and Lemma 1 imply that $\{h_{t_m}(f^{(m-1)}\beta_m)\}$, $\{Jht_m(f^{(m-1)}\beta_m)\}$ tend to $h_t(\beta), Jh_t(\beta)$, respectively, that is, $\{\alpha_m\}$ converges to $h_t(\beta)$ with $Jh_t(\beta) \geq \varepsilon$, in (J, φ)-topology, thus the minimizing property yields that $h_t(\beta)$ is a fixed point of f, completing the proof.

Assertion. *There exists a function J satisfying* 1)–5).

In fact the function J is obtained from the Lipschitz constant of $\alpha \in Y$. Take a parametric representation $(\alpha^1(u, v), \alpha^2(u, v), \alpha^3(u, v))$ of α and define

$$L(\alpha) = \text{ess. sup}\left\{\frac{|\det(\alpha^3, \alpha^1)|}{|\det(\alpha^1, \alpha^2)|}\ \frac{|\det(\alpha^3, \alpha^2)|}{|\det(\alpha^1, \alpha^2)|}\right\},$$

where $\det(\alpha^i, \alpha^j)$ stands for

$$\det\begin{pmatrix} \dfrac{\partial\alpha^i}{\partial u} & \dfrac{\partial\alpha^j}{\partial u} \\ \dfrac{\partial\alpha^i}{\partial v} & \dfrac{\partial\alpha^j}{\partial v} \end{pmatrix}.$$

Then it is obvious that $L(\alpha)$ is independent of the choice of the parameter and is equivalent to (local expression of) Lipschitz constant if α is expressed in the non parametric form.

Moreover, if α is smooth, $L(\alpha) < \infty$ implies that α has a homeomorphic projection into (x, y)-plane. Thus, by taking a non decreasing continuous function ρ of $\boldsymbol{R}$ into $[0, 1]$, $J(\alpha)$ is defined to be $\rho(1/L(\alpha))$, for which the properties 1)–3) are obvious, since the (global) Lipschitz constant is lower semi continuous in the uniform topology. The homotopy required in 5) is given through the smoothing process $\sigma_t * \alpha$ of α by a mollifier $\sigma_t(x) = t^{-2}\sigma(xt^{-1})$ so as to approximate α in C' sense when t tends to zero. Finally the property 4) for J is deduced using the isothermal coordinates A, B of the 3 points condition for α, β as follows: Assume $h_t(\alpha)|_\Delta$, $h_t(\beta)|_\Delta$ both are minimal surfaces bounded by $\alpha|_{\partial\Delta}$, $\beta|_{\partial\Delta}$. Then with Poisson kernal P, it holds that

$$h_t(\alpha)|_\Delta \circ A = P^*(\alpha \circ A|_{\partial\Delta}) ,$$
$$h_t(\beta)|_\Delta \circ B = P^*(\beta \circ B|_{\partial\Delta}) ,$$

on the other, the 3 points condition implies that $\alpha \circ A$ tends to $\beta \circ B$ if α tends to β in the uniform topology, therefore $h_t(\alpha)|_\Delta \circ A$ tends to $h_t(\beta)_\Delta \circ B$ in C^∞ sense in the interior to Δ and yields that $J(h_t(\alpha)|_\Delta)$ tends to $J(h_t(\beta)|_\Delta)$ (for the boundary need several argument [N]), because the Lipschitz constant is independent of the coordinate system.

Let $P_i(i = 1, 2)$ be planes in R^3 transversal to each other and let h_t^i, J_i be the homotopy (nothing is said on the continuity) and the function defined for the plane P_i instead of (x, y)-plane in the preceding construction.

Define Z^ε by

$$Z^\varepsilon = \{\alpha \in Y / J_1(\alpha(t_1 s_1 \cdots t_j s_j)) \geq \varepsilon \text{ or } J_2(h_s^1(\alpha(t_1, s_1 \cdots t_j, t_j)) \geq \varepsilon \quad \text{for } s, s_i, t_i \in [0, 1]\} ,$$

where $\alpha(t_1, s_1, \cdots, t_j, s_j)$ stands for

$$h_{t_1}^2 \circ h_{s_1}^1 \circ \cdots \circ h_{t_j}^2 \circ h_{s_j}^1(\alpha) ,$$

and set

$$f^i(\alpha) = h^i_{J_i(\alpha)}(\alpha) , \qquad F(\alpha) = f^2(f^1(\alpha)) .$$

Then $\alpha \in Z^\varepsilon$ yields L-Lipschitz condition for α relative to at least one of the planes P_i and therefore the area integral φ is defined on Z^ε. It is obvious that

Lemma 3. *For $\alpha \in Z^\varepsilon$,*

(1) $\varphi(F(\alpha)) \leq \varphi(\alpha)$,

(2) $\varphi(F(\alpha)) = \varphi(\alpha)$ *implies* $F(\alpha) = \alpha$ *and that* α *is minimal* (*on a certain subset of* D).

Introduce $(\max (J_1, J_2), \varphi)$-topology into Z^ε using Frechét distance in place of the non parametric distance. It is not hard to see that if $J_i(\alpha) \geq \varepsilon$, then in a neighborhood of α in Z^ε, the topology is equivalent to that induced from the non parametric distance relative to P_i.

If at first $J_1(\alpha) \geq \varepsilon$, then it follows that $f^1(\alpha)$ is continuous in $(\max (J_1, J_2), \varphi)$-topology and is minimal except the 1-skelton of a subdivision.

Since the minimal surface over a convex domain is unique, f^2 alters $f^1(\alpha)$ only along the 1-skelton hence alters $J_i(f^1(\alpha))$ and $\varphi(f^1(\alpha))$ very small, provided $J_2(f^1(\alpha))$ is sufficiently small.

This indicates the continuity of F at α for which therefore the continuity of F on $J_1(\alpha) \geq \varepsilon$ follows easily from the property 4) of J and Lemma 2. On the other, if $J_2(f^1(\alpha)) \geq \varepsilon$, then the continuity of $F(\alpha) = f^2(f^1(\alpha))$ also follows directly from the property 4) of J and Lemma 2. Thus we get the continuity of F on Z^ε.

The compactness condition 4) and the homotopy condition 3) of Definition 1 for the triple $(Z^\varepsilon, F; \varphi)$ is deduced in a similar way as in Proposition 4, for the compactness proof we have only to be careful to choose from given sequence $\{\alpha_n\}$ in Z^ε a subsequence $\{\alpha_m\}$ such that $J_1(\alpha_m) \geq \varepsilon$ or a subsequence $\{\alpha_m\}$ such that $J_2(f^1(\alpha_1)) \geq \varepsilon$.

Hence we have

Theorem 2. *The triple $V = (Z^\varepsilon, F; \varphi)$ turns out to be a variatiation problem, whose critical points are minimal surfaces except a neighborhood of the boundary and satisfy that*

$$^\#\{\gamma(V)\} \geq \text{cup length of } H\ (Z^\varepsilon, k)\ .$$

Though Theorem 2 is stated for the construction with 2 planes P_1, P_2, it is possible to generalize the construction for many planes. For these and further generalization we hope to publish in near furture with a little better description and computation of homology groups.

References

[N] J. C. C. Nitsche, Vorlesungen uber Mimimalflachen, Springer, 1975.

[R] T. Radó, On the problem of Plateau, Chelsea, 1951.

[S] Y. Shikata and I. Mogi, *Some topological aspects of abstract variation theory*, Differential Geometry, in honor of K. Yano, Kinokuniya, Tokyo, 1972, 451–457.

Nagoya University
Nagoya, 464
Japan

Minimal Submanifolds and Geodesics
Kaigai Publications, Tokyo, 1978, 217–228

TOPOLOGY OF POSITIVELY CURVED MANIFOLDS WITH A CERTAIN DIAMETER

KATSUHIRO SHIOHAMA

It is interesting to study relations between curvature and topology of a connected and complete Riemannian manifold without boundary. A beautifull result due to Myers states that if the sectional curvature K of such an M satisfies $K \geqq \delta > 0$, then the diameter $d(M)$ of M is at most $\pi/\sqrt{\delta}$, and hence M is compact, and moreover the fundamental group $\pi_1(M)$ of M is finite. The so-called sphere theorem was then obtained by Berger [1] and Klingenberg [8] stated as follows. Let M be a connected, compact and simply connected Riemannian manifold of dimension $m \geqq 2$. 1 (*Berger*). Assume that m is even and $0 < \delta \leqq K \leqq 4\delta$. If $d(M) = \pi/2\sqrt{\delta}$, then M is isometric to one of the symmetric spaces of rank 1 of compact type. If $d(M) > \pi/2\sqrt{\delta}$, then M is homeomorphic to an m-sphere S^m. 2 (*Klingenberg*). If $0 < \delta \leqq K < 4\delta$, then M is homeomorphic to S^m. Recently the sphere theorem was generalized by Grove and the author in [7]. The generalization is stated as follows. If the sectional curvature K and the diameter $d(M)$ of a connected, compact Riemannian manifold M of dimension m satisfy $K \geqq \delta > 0$ and $d(M) > \pi/2\sqrt{\delta}$, then M is homeomorphic to S^m.

The purpose of the present paper is to investigate topology of a connected compact Riemannian m-manifold M whose sectional curvature and diameter satisfy

$$(*) \qquad K \geqq \delta > 0$$

and

$$(**) \qquad d(M) = \pi/2\sqrt{\delta}\,.$$

As is easily seen, there are several typical examples of such manifolds which are simply connected or not. A result obtained in [12] states that if $\delta \leqq K < 4\delta$ and $**$) are fulfilled for an M, then M is of constant curvature δ, and $\pi_1(M)$ has a fully reducible representation. Simply connected examples are seen in Berger's result as stated above.

From now on let M be a connected and compact Riemannian manifold of dimension $m \geqq 2$ without boundary and whose sectional curvature and diameter satisfy $*$) and $**$). Pick a pair of points $p, \bar{p} \in M$ with $d(p, \bar{p}) = d(M)$, where d is the distance function. Then the set A_p defined to be

$$A_p := \{x \in M;\ d(x, p) = d(M)\}$$

is a non-empty convex closed set (see Lemma 1.1 of [12]). This convex set gives a strong restriction to the topology of M. Recall that if M is not simply connected, then A_p has no boundary for any $p \in M$ with $d(p, \bar{p}) = d(M)$. Moreover the set B defined by

$$B := \bigcap \{A_q;\ q \in A_p\}$$

is again a non-empty closed convex set. Then the main result obtained here is

Main theorem. (1) *If both A_p and B have no boundary for some $p \in M$, then M is the union of two normal disk bundles over A_p and B joined along their common boundary.* (2) *If A_p has non-empty boundary for some p and if B has no boundary with* $\dim B > 0$, *then M has the same cohomology structure as one of the symmetric spaces of compact type of rank one.* (3) *If both A_p and B have non-empty boundaries, then M is homeomorphic to S^m.*

Remarks. 1. A contrapositive of Lemma 1.3 in [12] states that if A_p has nonempty boundary for some $p \in M$, then M is simply connected. Moreover a convex function is defined on A_p if it has non-empty boundary (see 8.10 Theoreme of [2]), and this function can be replaced by a strictly convex function (see Proposition 3 of [4]). This can be approximated by a smooth strictly convex function with a unique minimum (see Lemma 3 of [5]). Hence A_p has a neighborhood which is an embedded image of an m-disk.

2. As is done in [7], a function is defined to be the difference between the distance functions to A_p and B, which can be approximated by a smooth function whose gradient is non-vanishing outside the union of nieghborhoods of A_p and B. Thus (1) in the main theorem is a straightforward consequence of this.

3. (2) is shown by constructing a metric on M such that the tangent cut locus of some point in A_p is a sphere and its cut locus is B. Then (2) and (3) are direct consequences of Theorem 2.5 in [10]. Moreover it follows from Theorem 2.5 in [10] that m is equal to $(\lambda + 1)$ times positive integer and the dimension of B is equal to $m - (\lambda + 1)$, where $\lambda = 1, 3$, or 7.

In case where M is not simply connected, a slight extension of a theorem in [12] is obtained as follows. A minimizing geodesic $\gamma: [0, \pi/2\sqrt{\delta}] \to M$ joining p to $\bar{p}$ is said to be *isolated* if there exists an $a > 0$ such that any minimizing geodesic σ joining p to $\bar{p}$ makes an angle with γ at p which is not less than a. We then have the following

Theorem 1. *Assume that M is not simply connected. If there is an isolated minimizing geodesic joining p to $\bar{p}$ for some pair of points $p, \bar{p}$ at maximal distance and if* $\dim A_p \geqq 2$, *then M is isometric to a spherical space form whose fundamental group has a fully reducible representation.*

Note that if $\delta \leqq K < 4\delta$, then any minimizing geodesic with length $\pi/2\sqrt{\delta}$ is isolated because there is no pair of conjugate along it. Let G be the fundamental group of a spherical space form M of constant curvature δ. Then there exist fixed point free irreducible orthogonal representations $\sigma_1, \cdots, \sigma_r$ of G over $\boldsymbol{R}^{m+1}$ such that $M = S^m(\delta)/\{\sigma_1 \oplus \cdots \oplus \sigma_r\}(G)$ and $\pi_1(M) = G$. $\pi_1(M)$ is said to have a *fully reducible representation* if $r > 1$. It has been proved in [12] that the fundamental group of a spherical space form of constant curvature δ has a fully reducible representation if and only if its diameter is equal to $\pi/2\sqrt{\delta}$.

A basic tool for the proof of the above theorems is Theorem 2, the idea of which we have already used to prove a generalization of the sphere theorem in [7]. Let us define the function $f\colon M \to \boldsymbol{R}$ to be

$$f(x) := d(x, A_p) - d(x, B)\,, \qquad x \in M\,.$$

Theorem 2. *There exist open sets $V_2 \supset B$, $V_1 \supset A_p$ and a family of smooth functions $\{f_\rho\colon M \to \boldsymbol{R};\, 0 < \rho < \rho_0\}$ such that they satisfy*: (1) *both ∂V_1 and ∂V_2 are compact hypersurfaces and disjoint*, (2) *$f_\rho \to f$ uniformly on M as $\rho \to 0$*, (3) *$\|\nabla f_\rho\| \neq 0$ on $M - V_1 \cup V_2$*, (4) *∇f_ρ is transversal to the boundary hypersurfaces ∂V_1 and ∂V_2.*

The rest of this paper is organized as follows. We recall the properties of convex sets A_p and B in § 1, some of which are shown in [12]. We then give a proof of Theorem 1 assuming Theorem 2. Theorem 2 is proved in § 3. If both A_p and B have no boundary, then Theorem 2 is a direct consequence of what we have done in the proof of a generalization of sphere theorem. We shall show in § 2 the existence of a neighborhood of a closed convex set with non-empty boundary, which is the embbedded image of an m-disk. In § 4, we shall introduce a metric on M with respect to which there is a point on A_p (or B) with non-empty boundary at which the tangent cut locus is a sphere. Then the Main Theorem will be proved in § 4.

I would like to take this opportunity to express my deep gratitude to Professor T. Otsuki who in 1963 led me to the study of differential geometry and who has assisted and inspired me ever since. I owe it entirely to his thorough teaching and constant encouragement that I have become a differential geometer. Therefore it gives me very great pleasure to dedicate this paper to him on his 60th birthday.

1. Convex sets at maximal distance. As is shown in Lemma 1.1 of [12], every geodesic segment joining any points on A_p with length $<\pi/\sqrt{\delta}$ is contained in A_p, and hence it is convex. Theorem 1.6 of [3] states that A_p carries the structure of an embedded topological submanifold of M whose interior is smooth and totally geodesic. Recall that for every geodesic segment γ joining any points q and q' in A_p and for any minimizing geodesic σ joining q to p there exists a totally geodesic surface of constant curvature δ whose boundary consists of γ, δ and the minizing geodesic from q' to p whose tangent vector at q' is obtained by the parallel translation of $\dot{\sigma}(0)$ along γ. As a direct consequence of this fact, we have

Lemma 1.1. *If* $\dim M = 2$, *then* M *is either homeomorphic to* S^2 *or else isometric to the real projective space of constant curvature* δ.

The proof is obvious and hence omitted here.

For a pair of points $x, y \in M$ let $\Gamma(x, y)$ be the set of all minimizing geodesics from x to y. $\gamma \in \Gamma(x, y)$ is said to be *isolated by an angle* $a > 0$ if $\measuredangle(\dot{\gamma}(0), \dot{\sigma}(0)) \geqq a$ holds for any $\sigma \in \Gamma(x, y)$. We see that if $\Gamma(x, y)$ contains at most finitely many geodesics, then every $\sigma \in \Gamma(x, y)$ is isolated by the minimum angle among vectors tangent to minimizing geodesics. We say that $\gamma \in \Gamma(x, y)$ is *a-isolated* if it is isolated by a.

Proposition 1. *If* $\gamma \in \Gamma(p, q)$ *is an a-isolated geodesic for some* $q \in A_p$ *and if* $\dim A_p \geqq 2$, *then* A_p *has constant sectional curvature* δ.

Proof. It suffices to construct a local isometry between a small neighborhood of each point on A_p and the corresponding neighborhood in the k-sphere $S^k(\delta)$ of constant curvature δ, where $k = \dim A_p \geqq 2$. Let $r(M)$ be the convexity radius of M and let $B(x, r)$ be the open metric r-ball centered at x. It is possible to choose a number $\varepsilon \in (0, r(M))$ which has the following property: At each point $x \in A_p$, parallel translation along any geodesic triangle sketched in $A_p \cap B(x, \varepsilon)$ rotates vectors at most by an angle $a/2$. Recall that all the unit vectors tangent to minimizing geodesics from any $q' \in A_p$ to p are obtained by parallel translation along a geodesic segment joining q to q' in A_p. Hence at each point $q' \in A_p$ there is an a-isolated minimizing geodesic joining q' to p whose tangent vector at q' is transposed by parallel translation of $\dot{\gamma}(0)$ along a geodesic segment in A_p. Take a point $q' \in A_p$ and let $\gamma' \in \Gamma(q', q)$ be an a-siolated minimizing geodesic with $\gamma'(0) = q'$. From the choice of ε, it follows that each point $x \in B(q', \varepsilon) \cap A_p$ is joined to p by an a-isolated minimizing geodesic whose tangent vector is obtained by parallel translation of $\dot{\gamma}(0)$ along the unique minimizing geodesic from x to q'. Thus the unit vectors tangent to these a-isolated minimizing geodesics form a unit parallel normal field, say N, defined on $A_p \cap B(q', \varepsilon)$. For each $x \in A_p \cap B(q', \varepsilon)$ let $\gamma_x \colon [0, d(M)] \to M$ be the geodesic with $\dot{\gamma}_x(d(M)) = -N(x)$. Then $\gamma_x(0) = p$ and it is obvious that the set $\{\dot{\gamma}_x(0) \in M_p; x \in B(q', \varepsilon) \cap \mathring{A}_p\} =$

Σ is a non-empty convex open set in a great k-sphere of the unit hypersphere S_p centered at the origin of M_p. Note that the tangent space to this great k-sphere at $\dot{\gamma}_x(0)$ is obtained by parallel translation of the tangent space to A_p at $\gamma_x(d(M))$ along γ_x, and then by canonical translation from the origin of M_p to $\dot{\gamma}(0)$. Every great circle arc on $d(M)\cdot\Sigma$ is mapped by $\exp_p$ onto a geodesic segment in $\mathring{A}_p \cap B(q', \varepsilon)$. Hence $\exp_p | d(M)\cdot\Sigma$ is is regular. It is now clear that for a fixed point $p^* \in S^k(\delta)$, $\exp_{p^*} \circ (\exp_p | d(M)\cdot \sum)^{-1}$ is an isometry.

The following Lemma is used in the proof of Theorem 1 from Theorem 2, and which is essentially the same as the so-called "Berger's Lemma", see 6.2 Lemma in [2].

Lemma 1.2. *Let N be a complete Riemannian manifold and* $W \subset N$ *a submanifold. Assume that there are points* $w \in W$ *and* $n \in N - W$ *and an open set* $U \subset W$ *of w such that* $d(n, w) \geqq d(n, x)$ *for any* $x \in U$, *Then for any vector* $X \in W_w$, *there is a minimizing geodesic* $\gamma_X\colon [0, d(n, w)] \to N$ *with* $\dot{\gamma}_X(0) = w$, $\gamma_X(d(n, w)) = n$ *and* $\langle \dot{\gamma}_X(0), X\rangle \geqq 0$.

Proof of Theorem 1 *from Theorem* 2. Since M is not simply connected, both B and A_p have no boundary. It is clear that if either $\dim A_p = m - 1$ or else $\dim B = m - 1$, then M is isometric to the real projective space of constant curvature δ. We may assume without loss of generality that $\dim A_p \leqq m - 2$ and $\dim B \leqq m - 2$. Let $\tilde{M}$ be the universal Riemannian covering and $\pi\colon \tilde{M} \to M$ the covering projection. Set $\tilde{B}\colon = \pi^{-1}(B)$ and $\tilde{A}\colon = \pi^{-1}(A_p)$. Convexity of $\tilde{A}$ (and $\tilde{B}$ respectively) implies that the distance between any points on $\tilde{A}$ (and $\tilde{B}$ respectively) with respect to the induced Riemannian metric on $\tilde{A}$ (and $\tilde{B}$ respectively) is equal to that on $\tilde{M}$. Let $d_{\tilde{M}}$ be the distance function on $\tilde{M}$. By means of the maximal diameter theorem (see Satz 2, p. 213 of [6]) it suffices to show that $d_{\tilde{M}}(\tilde{A}) = \pi/\sqrt{\delta}$. This is achieved by showing that $\tilde{A}$ is simply connected. Indeed if $\tilde{A}$ is simply connected, then $\tilde{A}$ is isometric to $S^k(\delta)$. Suppose that $d_{\tilde{M}}(\tilde{A}) < \pi/\sqrt{\delta}$. Then there would exist from Lemma 1.2 at least two distinct minimizing geodesics on $\tilde{A}$ with length $= d_{\tilde{M}}(\tilde{A})$ joining two points at maximal distance on $\tilde{A}$. However this connot occur on $S^k(\delta)$, a contradiction.

The rest of the proof is to verify the simple connectness of $\tilde{A}$ which is derived below by the aid of Theorem 2.

Proposition 2. *Under the assumptions of Theorem* 1, $\tilde{A}$ *is simply connected.*

Proof. Let $\tilde{\gamma}$ be a geodesic loop in $\tilde{A}$ at a point $\tilde{x} \in \tilde{A}$. There is a homotopy $H\colon [0, 1] \times [0, 1] \to \tilde{M}$ between the point curve $\{\tilde{x}\}$ and $\tilde{\gamma}$. Let $U \supset \tilde{A}$ and $V \supset \tilde{B}$ be open sets such that $\pi(U) \supset A_p$ and $\pi(V) \supset B$ are obtained from Theorem 2. From $\tilde{A} \cap \tilde{B} = \varnothing$ it follows that $\dim \tilde{A} + \dim \tilde{B} \leqq m - 1$, and hence $m - \dim \tilde{B} \geqq \dim \tilde{A} + 1 \geqq 3$. Therefore we may consider that each component of $H([0,1],[0,1]) \cap \tilde{B}$ is a proper subset of B. Thus for each component there is a point q which is not contained in the component. Fix such a component. Then its intersection with $B(\tilde{q}, r(M))$ can be swept out continuously into the boundary. $\partial B(q, r(M)) \cup \partial V$. By iterating this process, we observe that the intersection of each component of new homotopy image with $\tilde{B}$ is contained in some $B(\tilde{q}', r(M))$, $\tilde{q}' \in \tilde{B}$. Cleraly a smooth frame field normal to $TB | B(\tilde{q}', r(M))$ is well-defined, along which the homotopy image is pushed out into ∂V. This is possible because $m - \dim \tilde{B} \geqq 3$. By means of Theorem 2 there is a smooth function, say, $\tilde{f}_\rho$, on $\tilde{M}$ whose gradient is nonvanishing on $\tilde{M} - U \cup V$ and transversal to the boundaries. Therefore the homotopy image can be pushed down along $-$gradient $\tilde{f}_\rho$ into ∂U. Since each point on $\tilde{U}$ has a unique minimizing geodesic perpendicular to $\tilde{A}$, this homotopy image can be pushed down along perpendiculars continuously into $\tilde{A}$. Hence $\tilde{A}$ is simply connected.

2. Convex functions on convex set of positive curvature. A function on a complete Riemannian manifold is said to be *convex* if it is restricted to any arc length parametrized geodesic γ, then

$$\tau \circ \gamma((1-\lambda)t_1 + \lambda t_2) \leqq (1-\lambda)\tau \circ \gamma(t_1) + \lambda \tau \circ \gamma(t_2)$$

holds for any $t_1, t_2 \in \boldsymbol{R}$ and any $\lambda \in [0,1]$. If the inequality holds for any γ, t_1, t_2 and λ, then τ is called *strictly convex*. A convex function is defined on every closed convex set C with non-empty boundary in a complete manifold of nonnegative sectional curvature as follows (see 8.10 Theorem in [2])

$$\tau(x) := -d(x, \partial C) , \qquad x \in C .$$

This convex function is replaced by a strictly convex function if the sectional curvature of the manifold is positive, (see Proposition 3 of [4]). Moreover Greene and Wu proved (see Lemma 3 in [5]) that a strictly convex function defined on a compact set C can be approximated by a family of smooth strictly convex functions

$$\{\tau_\varepsilon : U \to \boldsymbol{R} ; \varepsilon \in (0, \varepsilon_0)\} ,$$

where U is an open subset of C such that $\bar{U} \subset \mathring{C}$. If $C \subset M$ is a closed convex set with non-emtpy boundary, then $K > 0$ implies that the convex function $\tau : C \to \boldsymbol{R}$ defined above attains its minimum at a single point. For a closed convex set $C \subset M$, there is a metric r-ball $B(C, r)$ of C each point x on which has a unique minimizing geodesic $h_x : [0,1] \to M$ perpendicular to C with $h_x(0) \in C$, $h_x(1) = x$ and the length of $h_x = d(x, C)$. A possible radius r_0 of such a ball is given by $r_0 = \frac{1}{2}$ Min $\{r(M), \pi/\sqrt{\text{Max } K}\}$. In fact, let $x \in B(C, r_0)$ and h, h': $[0, 1] \to M$ be distinct minimizing geodesics with $h(0), h'(0) \in C$, $h(1) = h'(1) = x$ and their lengths are $d(x, C)$. The edge angles of the geodesic triangle $(x, h(0), h'(0))$ at $h(0), h'(0)$ are equal to the right angle. The circumference of this triangle is less than $2\pi/\sqrt{\text{Max } K}$ and each side of it is contained in the open $r(M)$-ball centered at the opposite edge. Hence Rauch's theorem (see 1.30 Cor. of Rauch I in [2] or Kor., p. 179 in [6]) is applied to get a contradiction.

Proposition 3. *Let $C \subset M$ be a closed convex set with non-empty boundary. Given $\eta > 0$, there is an embedding $E: \bar{V} \to M$ of a closed m-disk $\bar{V}$ into M which has the following properties.* (1) *$C \subset E(V)$ and $E(\partial V)$ is a compact hypersphere,* (2) *at each point $x \in E(\partial V)$ there is a unique geodesic h_x: $[0, 1] \to M$ perpendicular to C with $h_x(0) \in C$, $h_x(1) = x$ and the length of $h_x = d(x, C) \in [r_0/2, r_0]$* (3) *at each point $x \in E(\partial V)$ the angle between $\dot{h}_x(1)$ and the outer normal to $E(\partial V)$ is less than η.*

Proof. For any $a > 0$ the set $C^a := \{x \in C; \tau(x) \leqq -a\} = \{x \in C; d(x, \partial C) \geqq a\}$ is a convex set (see Theorem 1.9 of [3]), where τ is defined as above. Since τ is not locally constant, a strictly convex function $\tilde{\tau}: C \to R$ is obtained as follows (see Prop. 3 of [4]). Let χ be a smooth function such that $\chi(0) = 0$, $\chi' > 0$ and $\chi'' > 0$. Then $\tilde{\tau} = \chi \circ \tau$ is strictly convex. Every $\tilde{\tau}$-level set is a τ-level set. Choose a small number $a_0 > 0$ such that if $a \in (0, a_0)$ then $\bar{B}(C, r_0/2) \subset B(C^a, r_0)$. There is an open set $U \subset C$ of C^a with $\bar{U} \subset \mathring{C}$. For each $a \in (0, a_0)$ there is a family of smooth strictly convex functions $\{\tau_\varepsilon: U \to \boldsymbol{R}; \varepsilon \in (0, \varepsilon_0)\}$ such that $\tau_\varepsilon \to \tilde{\tau}$ uniformly on C^a as $\varepsilon \to 0$ (see Lemma 3 of [5]). We can choose ε_0 and $a_1 \in (0, a_0)$ so that if $a \in (0, a_1)$ and $\varepsilon \in (0, \varepsilon_0)$ then (1) $\tau_\varepsilon^{-1}(-a) \subset U - C^{a_0}$, (2) $\bar{B}(C, r_0/2) \subset C(W, r_0)$, where $W = \{x \in U; \tau_\varepsilon(x) = -a\}$ and (3) for each $x \in \bar{B}(W, r_0) - B(C, r_0/2)$, let k_x: $[0, 1] \to M$ be a unique minimizing geodesic perpendicular to W with $k_x(1) = x$, $k_x(0) \in W$. Then the angle between $\dot{k}_x(1)$ and $\dot{h}_x(1)$ is less than $\eta/3$.

Fix such an $\varepsilon \in (0, \varepsilon_0)$. $\tau_\varepsilon|W$ takes its minimum at a single point. In fact if y and y' are points in W at which τ_ε takes minimum. Convexity of W implies the existence of a geodesic segment in W joining y to y', on which τ_ε must be constant. However this is impossible for a strictly convex function. Let $q_0 \in W$ be the point at which τ_ε takes minimum. As is shown in the proof of Theorem 3 in [5], there is a diffeomorphism $j: \bar{D}^k \to W$ such that $j(0, \cdots, 0) = q_0$ and every straight line segment in $\bar{D}^k$ passing through the origin is mapped by j onto an integral curve of $\tau_\varepsilon/\|\nabla\tau_\varepsilon\|^2$.

We shall construct a map between a neighborhood of $\bar{D}^k$ in $\boldsymbol{R}^m$ and $B(C, r_0/2)$, by use of which the desired embedding will be obtained. Let $n_{k+1}, \cdots, n_m$ be an orthonormal basis for the normal space $N_{q_0}W \subset M_{q_0}$ of W at q_0. Translate them parallely along every integral curve of $\nabla\tau_\varepsilon/\|\nabla\tau_\varepsilon\|^2$

to obtain normal fields $N_{k+1}, \cdots, N_m$ defined over W. At each point they form an orthonormal basis for $N_q W$ and at each point $y \in \partial W, N_{k+1}, \cdots, N_m$ and $\nabla\tau_\varepsilon/\|\nabla\tau_\varepsilon\|$ form an orthonormal basis for the normal space to ∂W at y. Define $E: \boldsymbol{R} \to M$ as follows. For each $x \in R^m$ let us decompose $x = x' + x''$ by $x' = (x_1, \cdots, x_k)$ and $x'' = (x_{k+1}, \cdots, x_m)$. Set

$$E(x) := \begin{cases} \exp_{j(x')} \sum_{i=k+1}^{m} x_i N_i \circ j(x') , & \text{for } \|x'\| < 1 \\ \exp_{j(x'/\|x'\|)} \left[\sum_{i=k+1}^{m} x_i N_i \circ j\left(\frac{x'}{\|x'\|}\right) \right. \\ \qquad \left. + (\|x'\| - 1) \frac{\nabla\tau_\varepsilon \circ j(x'/\|x''\|)}{\|\nabla\tau_\varepsilon \circ j(x'/\|x''\|)\|} \right] , & \text{for } \|x'\| \geqq 1 . \end{cases}$$

Clearly E is continuous and if it is restricted to the closed r_0-ball $\bar{B}(\bar{D}^k, r_0)$ of $\bar{D}^k$ then it is 1–1 and regular. Moreover for any $x \in B(\bar{D}^k, r_0)$ the curve $t \to E(t(x - x') + x')$ is the unique minimizing geodesic from $E(x)$ to W which is perpendicular to W at $j(x')$ if $\|x'\| < 1$, and similarly if $\|x'\| \geqq 1$, then the curve $t \to E(((1 - 1/\|x'\|)x' + x'')t + x'/\|x'\|)$ has the same property. It is possible to choose an open set $V \subset \boldsymbol{R}^m$ of $\bar{D}^k$ which is diffeomorphic to D^m and has the following properties: (1) $E(\partial V) \subset E(B(\bar{D}^k, r_0)) - B(C, r_0/2)$ (2) if q is any point on $E(\partial V)$, then the outer normal to $E(\partial V)$ at q has an angle with $\dot{k}_q(1)$ at most $\eta/3$. Thus we observe that $E(\bar{V})$ has the desired properties. Indeed we only show that the angle between $\dot{h}_q(1)$ and the outer normal to $E(\partial V)$ at q is less than η. This is clear because of $\measuredangle(\dot{h}_q(1), \dot{k}_q(1)) \leqq \eta/3$.

Remark. In case where $\partial C = \emptyset$, we set $V := B(C, r_0/2)$. Then it satisfies the desired properties in Proposition 3. However in this case V is not necessarily an m-disk.

3. A smooth appoximation of f. Recall that $f: M \to \boldsymbol{R}$ is defined to be

$$f(x) := d(x, A_p) - d(x, B) , \qquad x \in M .$$

f is continuous on M and smooth outside the set $\mathscr{C} := \mathscr{C}(A_p) \cup \mathscr{C}(B)$ of measure zero, where $\mathscr{C}(A_p)$ and $\mathscr{C}(B)$ are the cut locus of A_p and B respectively. Note that $B \subset \mathscr{C}(A_p)$ and also $A_p \subset \mathscr{C}(B)$. For each point $x \in M - B(B, r_0/2) \cup B(A_p, r_0/2)$ let $\alpha: [0, a] \to M$ and $\beta: [0, b] \to M$ be minimizing geodesics from x to A_p and B respectively such that $\alpha(0) = \beta(0) = x$, $\alpha(a) \in A_p$, $\beta(b) \in B$ and $a = d(x, A_p)$, $b = d(x, B)$. Then the triangle comparison theorem by Toponogov implies that

$$\measuredangle(\dot{\alpha}(0), \dot{\beta}(0)) = \cos^{-1}\{-\cot a\sqrt{\delta} \cot b\sqrt{\delta}\} \geqq \pi/2 .$$

More precisely we have the following

Lemma 3.1. *Given* $r \in (0, \pi/4\sqrt{\delta})$, *there exists a positive number* ξ *such that if* $x \in M - B(A_p, r) \cup B(B, r)$ *is any point and if* $\alpha: [0, a] \to M$ *and* $\beta: [0, b] \to M$ *are any minimizing geodesics as above, then*

$$\measuredangle(\dot{\alpha}(0), \dot{\beta}(0)) \geqq \pi/2 + \xi .$$

Proof. Suppose that there is no such $\xi > 0$. Then there are families $\{\alpha_i: [0, a_i] \to M]$, $\{\beta_i: [0, b_i] \to M\}$ of shortest geodesics and a sequence of points $\{x_i\}$ in $M - B(A_p, r) \cup B(B, r)$ such that $\alpha_i(0) = \beta_i(0) = x_i$, $\alpha_i(a_i) \in A_p$, $\beta_i(b_i) \in B$ and $a_i = d(x_i, A_p)$, $b_i = d(x_i, B)$ and they satisfy.

$$\lim_{i\to\infty} \measuredangle(\dot{\alpha}_i(0), \dot{\beta}_i(0)) = \pi/2 .$$

Compactness of $M - B(A_p, r) \cup B(B, r)$ implies the existence of converging subsequences of $\{x_i\}, \{\alpha_i\}, \{\beta_i\}$. Passing to the limit, we find geodesics $\alpha: [0, a] \to M$, $\beta: [0, b] \to M$ and a point $x = \alpha(0) = \beta(0) \in M - B(A_p, r) \cup B(B, r)$ such that $\dot{\alpha}(0) = \lim_{i\to\infty} \dot{\alpha}_i(0)$, $\dot{\beta}(0) = \lim_{i\to\infty} \dot{\beta}_i(0)$, $x = \lim_{i\to\infty} x_i$ and moreover $\measuredangle(\dot{\alpha}(0), \dot{\beta}(0)) = \pi/2$. By means of Toponogov's triangle comparison theorem we get

$$\cos a\sqrt{\delta} \cdot \cos b\sqrt{\delta} + \sin a\sqrt{\delta} \cdot \sin b\sqrt{\delta} \cdot \cos\theta = 0 ,$$

where $\theta \in (0, \pi/2]$ is the angle of the triangle sketched on $S^2(\delta)$ with edge lengths, $\pi/(2\sqrt{\delta})$, a and b opposite $\pi/(2\sqrt{\delta})$, and equality holoding if and only if $a = d(M)$ or $b = d(M)$. But $a = d(M)$ implies $x \in B$, and from $b = d(M)$ we get $x \in A_p$, a contradiction.

By a standard convolution process we have a family $\{f_\rho: M \to \boldsymbol{R}; 0 < \rho < \rho_0\}$ of smooth approximations of f as follows.

$$f_\rho(x) := \int_{B(x,\rho)} f(\cdot)\Phi_\rho(x, \cdot)d\mu_x , \qquad x \in M ,$$

where $\rho \in (0, r(M))$ and $d\mu_x$ is the Lebesgue measure on $B(x, \rho)$ induced from M_x through the exponential map at x, and Φ_ρ is the kernel function which is smooth and

$$\int_{B(X,\rho)} \Phi_\rho(X, \ )d\mu_x = 1$$

(see [7]). Theorem 2.2 of [7] states that given $\varepsilon > 0$, there is a ρ_0 such that if $\rho \in (0, \rho_0)$, then

$$\left\| \nabla f_\rho(x) - \int_{B(X,\rho)} \Phi_\rho(X, \cdot)P_x(\nabla f(\cdot))d\mu_x \right\| < \varepsilon ,$$

for all $x \in M$. In order to find a positive lower bound for the norm of the integrated vectors, we are lead to investigate the local behavior of ∇f near the cut locus. For this prupose we define, in the same way as is done in [7], unit vector fields $X, \bar{X}$ on $M - B(A_p, r) \cup B(B, r)$. For each point $x \in M - B(A_p, r) \cup B(B, r)$ let $\alpha: [0, b] \to M$ be minimizing geodesics as above. Denoting by $\mathscr{B}_s(u)$ the closed s-ball on the unit hypersphere $S_x \subset M_x$ centered at $u \in S_x$, we have

$$-\dot{\alpha}(0) \in \bigcap_{\beta} \mathscr{B}_{\pi/2-\xi}(\dot{\beta}(0)) \,, \qquad -\dot{\beta}(0) \in \bigcap_{\alpha} \mathscr{B}_{\pi/2-\xi}(\dot{\alpha}(0)) \,.$$

Let X_x (respectively $\bar{X}_x$) be the center of the smallest closed ball in S_x which contains the closed covex set $\bigcap_\beta \mathscr{B}_{\pi/2-\xi}(\dot{\beta}(0)) \subset S_x$ (respectively, $\bigcap_\alpha \mathscr{B}_{\pi/2-\xi}(\dot{\alpha}(0)) \subset S_x$). Clearly $X - \bar{X} = \nabla f$ on $M - \mathscr{C}$. Theorem 1.7 of [7] implies that for any $x \in M - B(A_p, r) \cup B(B, r)$ and for any $\xi' \in (0, \xi)$, there is a $\rho(\xi', x) \in (0, r(M))$ so that (1) parallel translation along any geodesic triangle in $B(x, \rho(\xi', x))$ rotates vectors at most by $\xi'/3$, and (2) if $y \in B(x, \rho(\xi', x)) - B(A_p, r) \cup B(B, r) \cup \mathscr{C}$, then

$$\sphericalangle (X_x, P_x(\nabla f(y))) < \pi/2 - \xi + \xi' \,,$$

and similarly

$$\sphericalangle (-X_x, P_x(\nabla f(y))) < \pi/2 - \xi + \xi' \,,$$

where $P_x: TM \,|\, B(x, r(M)) \to M_x$ is by definition the parallel translation along minimizing radial geodesics to x. We also note that the function

$$x \longrightarrow \rho(\xi', x)$$

is not continuous but upper semi continuous (see [7]).

Proof of Theorem 2. Let ξ be the constant obtained by Lemma 3.1 for $r = r_0/6$. For $\eta = \xi/6$, let $V_1 \supset A_p$ and $V_2 \supset B$ be open sets obtained by Proposition 3 (if the convex set has non-empty boundary) or by the remark after the proof of Proposition 3 (if the convex set has no boundary). Then (1) and (2) are clear. To prove the rest, note first that for any $x \in M - V_1 \cup V_2$ we have $B(x, r_0/3) \subset M - B(A_p, r_0/6) \cup B(B, r_0/6)$. Let ξ': $\xi/3$. Then there is an open cover $\{B(x, \rho(\xi', x)); x \in M - V_1 \cup V_2\}$ for the compact set $M - V_1 \cup V_2$. Let $2\lambda_0$ be the Lebesgue number for this cover. For any $x \in M - V_1 \cup V_2$ there is an $x_0 \in M - V_1 \cup V_2$ $x_0 \in M - V_1 \cup V_2$ such that $B(x, \lambda_0) \subset B(x_0, \rho(\xi', x_0))$. Note also that $\dot{h}_x(1) = -\bar{X}_x$ if $x \in B(A_p, r_0)$ and $\dot{h}_x(1) = -X_x$ if $x \in B(B, r_0)$. Hence we can choose a $\lambda \in (0, \mathrm{Min}\,\{r_0/3, \lambda_0\})$ so that if $y \in \partial V_1 \cup \partial V_2$ then $\sphericalangle (\dot{h}_y(1), P_y(\dot{h}_z(1)) < \xi/6$ for any $z \in B(B, \lambda)$. Then computations show that if $x \in M - V_1 \cup V_2$ and if $y \in B(x, \lambda) - \mathscr{C}$,

$$\sphericalangle (P_x(X_{x_0}), P_x(\nabla f(y))) < \pi/2 - \xi + \xi' + \xi'/3 ,$$

and similarly

$$\sphericalangle (P_x(-\bar{X}_{x_0}), P_x(\nabla f(y))) < \pi/2 - \xi + \xi' + \xi'/3 .$$

Since $\|\nabla f(y)\| > \sqrt{2}$, we have

$$\left\langle P_x(X_{x_0}), \int_{B(x,\lambda)} \Phi_\lambda(x, y) P_x(\nabla f(y)) d\mu_x \right\rangle > \sqrt{2} \sin \frac{5\xi}{9} .$$

And the same inequality holds for $P_x(-\bar{X}_{x_0})$. To observe the transversality we take $x \in \partial V_1$ and $y \in B(x, \lambda) - \mathscr{C}$. Then the angle between the outer normal to ∂V_1 at x and $P_x(\nabla f(y))$ is less than $\eta + \sphericalangle(\dot{h}_x(1), P_x(\nabla f(y))) \leqq \eta + \sphericalangle(\dot{h}_x(1), P_x(\dot{h}_{x_0}(1))) + \sphericalangle(P_x(\dot{h}_{x_0}(1)), P_x(\nabla f(y))) < \pi/2 - 2\xi/9$. Thus we can choose ε in Theorem 2.2 of [7] so small that (3) and (4) are satisfied for f_ρ.

As a consequence of theorems 1 and 2 we have

Theorem 3. *Assume that M is simply connected. If there is an isolated minimizing geodesic in $\Gamma(p, \bar{p})$ for some pair of points at maximal distance and if* $\dim A_p$ *and* $\dim B$ *are greater than* 1, *then M is homeomorphic to S^m.*

Proof. $\dim A_p \geqq 2$ implies that it has a non-empty boundary. In fact, suppose that $\partial A_p = \varnothing$. Then it follows from Propositions 1 and 2 that A_p is of constant curvature and $d(A_p) = \pi/\sqrt{\delta}$, contradicting $**$). Hence $V_1 \supset A_p$ can be taken to be an embedded m-disk. Since the same is true for B, we ovserve that M is exhibited as the union of two embedded m-disks and a cylinder joined along their common boundary. Thus the proof is complete.

4. Proof of Main Theorem. In case where A_p and B have no boundary, we take $V_1 \supset A_p$ and $V_2 \supset B$ to be the metric $(r_0/2)$-balls around A_p and B respectively. From Theorem 2 there is a smooth function on M whose gradient is non-vanishing on $M - V_1 \cup V_2$ and transversal to the boundaries of V_1 and V_2. Thus M is the union of normal disk bundles over A_p and B joined along their common boundaries. Note that if $\dim A_p = \dim B = 0$, then M is homeomorphic to S^m. To prove (2), we shall construct a new metric on M with respect to which the tangent cut locus at $\bar{p}$ (if the boundary of B is empty and $\dim B > 0$) or p (if the boundary of A_p is empty and $\dim A_p > 0$) is a sphere. Assuming the existence of such a metric, we derive the conclusion from Theorem 2.5 of [10]. Without loss of generality we may assume that $\partial A_p \neq \varnothing$. Then Proposition 3 ensures an embedded m-disk V_1 around A_p. Taking $V_2 \supset B$ as above, we know that there is a smooth function which satisfies the properties stated in Theorem 2. Thus $M - \bar{V}_2$ is an embedded m-disk. By means of Proposition C in [14] there is a new metric on M which agrees with the original one on a neighborhood of $\bar{V}_2$ and moreover there is a point $\bar{p} \in A_p$ with respect to which $\exp_{\bar{p}}$ restricted to the unit ball around the origin of $M_{\bar{p}}$ is a diffeomorphism onto $M - \bar{V}_2$. Therefore every geodesic emanating from $\bar{p}$ meets B orthogonally at the same length, and it does not intersect with any other geodesic segment emanating from $\bar{p}$ up to B. This fact means that the tangent cut locus at $\bar{p}$ is a sphere. Thus the proof of (2) is a straightforward consequence of [10]. Finally (3) is obtained as a special case of the above.

References

[1] Berger, M.: *Les variétés riemanniennes* (1/4)*-pincées,* Ann. Scuola Norm. Sup. Pisa, Ser. III, **14** (1960), 161–170.

[2] Cheeger, J. and Ebin, D.: Comparison theorems in riemannian geometry, North-Holland Mathematical Library, **9** (1975).

[3] Cheeger, J. and Gromoll, D.: *On the structure of complete manifolds of nonnegative curvature,* Ann. of Math., **96** (1972), 413–443.

[4] Greene, R. and Wu, H.: *Integrals of subharmonic functions on manifolds of nonnegative curvature,* Inventiones Math., **27** (1974), 265–298.

[5] Greene, R. and Wu, H.: C^∞ *convex functions and manifolds of positive curvature,* Acta Math., **137** (1976), 209–245.

[6] Gromoll, D., Klingenberg, W. and Meyer, W.: Riemannsche Geometrie im Grossen, Lecture notes in Mathematics 55, Springer 1968, Berlin-Heidelberg-New York.

[7] Grove, K. and Shiohama, K.: *A generalized sphere theorem,* Ann. of Math., **106** (1977), 201–211.

[8] Klingenberg, W.: *Über riemannsche Mannigfaltigkeiten mit positiver Krümmumg,* Comment Math. Helv., **35** (1961), 47–54.

[9] Myers, S. B.: *Riemannian manifolds in the large,* Duke Math. J., **1** (1935), 39–49.

[10] Nakagawa, H. and Shiohama, K.: *Geodesic and curvature structures characterizing projective spaces,* Differential Geometry, in honor of K. Yano, Kinokuniya, Tokyo, 1972, 305–315.

[11] Rauch, H.: *A contribution to differential geometry in the large,* Ann. of Math., **54** (1951), 38–55.

[12] Sakai, T. and Shiohama, K.: *On the structure of positively curved manifolds with certain diameter,* Math. Z., **127** (1972), 75–82.

[13] Shiohama, K.: *The diameter of* δ*-pinched manifolds,* J. Differential Geometry, **5** (1971), 61–74.

[14] Weinstein, A.: *The cut locus and conjugate locus of a Riemannian manifold,* Ann. of Math., **87** (1968), 29–41.

University of Tsukuba
Ibaraki, 300-31
Japan

Minimal Submanifolds and Geodesics
Kaigai Publications, Tokyo, 1978, 229–247

SURFACES MINIMIZING THE INTEGRALS OF CRYSTALLINE INTEGRANDS

JEAN E. TAYLOR[1]

Minimal surfaces are surfaces which are stationary for the area integrand. The study of surfaces which minimize the integrals of other integrands is interesting both in its own right and in the light it can shed on area minimizing surfaces. One class of integrands of codimension one which is particularly noteworthy in both respects is the class of crystalline integrands. A theory of the structure of surfaces which minimize the integrals of these integrands is being developed, and for such integrands in $\boldsymbol{R}^3$, an algorithm for the explicit construction of at least one local minimum now exists for a dense set of boundaries which lie on convex sets.

These crystalline integrands are inspired by the surface tension functions of solids. Surface tension is commonly thought of as a fluid phenomenon, but in fact there is a notion of surface tension for the interface between *any* two substances, or between two phases of the same substance (such as a solid and its melt. The surface tension is surface energy per unit surface area, and it arises from the fact that the atoms (or molecules, or ions) of a given substance which lie on the interface have a different environment from those that lie in the interior of the substance.

In this article we will deal with "surface tension functions" which are an outgrowth of the surface tensions of solids having their atoms arranged in some regular way. If one fixes the orientation of the lattice in $\boldsymbol{R}^3$, then the environment of an atom on a planar interface between something else and such a regular structure can be different for different plane directions. Thus the surface tension between one substance and another can be a function

$$F : G_o(3, 2) \longrightarrow \boldsymbol{R}^+$$

where $G_o(3, 2)$ is the Grassmannian of oriented 2-planes through the origin in $\boldsymbol{R}^3$ and $\boldsymbol{R}^+$ is the set of positive real numbers. For the interface between, say, soapy water and air, the function F would be identically a constant; but for that between, say, a single crystal of ordinary salt

[1] This work was partially supported by NSF Grant MCS76-06424 and by a Sloan Fellowship.

(NaCl) and air, F would depend quite strongly on direction. If one wishes to allow the regular structure to vary over relatively long distances, one can even obtain a function

$$F: \boldsymbol{R}^3 \times G_o(3, 2) \longrightarrow \boldsymbol{R}^+ .$$

(This notion of surface tension for solid bodies is not at all esoteric, by the way: it is commonly used by metallurgists and others who deal with the surface properties of materials.)

Functions from $G_o(3, 2)$ to $\boldsymbol{R}^+$ are already well known in the calculus of variations as treated in the context of geometric measure theory. They are called 2-dimensional constant coefficient oriented integrands on $\boldsymbol{R}^3$, since they can be integrated over any surface S which has an oriented tangent plane $Tan(S, x)$ at at least almost every point: the integral of F over S is then

$$\int_S F(Tan(S, x)) d\mathscr{H}^2 x .$$

(Here $\mathscr{H}^2$ is Hausdorff 2-dimensional area, which agrees with any other reasonable notion of area on smooth 2-dimensional submanifolds of $\boldsymbol{R}^3$ and additionally gives a precise meaning to the area of surfaces with singularities.) The typical problem is to minimize the integral of an integrand, subject to various possible side conditions such as a surface having a given boundary (in any one of a variety of senses) and/or enclosing one or more given volumes. For integrands coming from surface tension, this amounts to finding the surface of least energy satisfying the side conditions (we are, of course, neglecting here a large number of other possible contributions to energy, such as energy from curvature—including edges and corners—but the above is still usually a reasonable model). Since physical systems are in equilibrium if and only if they are at a (local) minimum of total energy, it is of great importance to known such things as the existence, uniqueness, regularity, singularity structure, and construction of shapes minimizing surface energy.

Mathematically, these questions have been addressed in the past primarily for the area integrand $F \equiv 1$ (minimizing the integral of this integrand produces minimal surfaces), and more generally for smooth elliptic integrands [A1]. But integrands which are surface tensions of solids with regular lattice structures are not elliptic, as will be indicated in Sections 2 and 3.

A class of non-elliptic integrands which includes such surface tension functions is isolated below and called crystalline. My object is to consider all the above problems—existence, uniqueness, singularity structure, and construction—for surfaces having a prescribed boundary and minimizing the integral of a crystalline integrand on $\boldsymbol{R}^3$. It will be illustrated how this knowledge can be used to give more information on surfaces minimizing the integrals of any integrand on $\boldsymbol{R}^3$. Many of the results of this article already extend to higher dimensions; extension of the others is an active area of research.

This article is in five sections. The first gives a classical construction, the Wulff construction, for a given integrand F and a proof that the result of the construction is the shape which has uniquely the least surface integral for the volume it contains. This shape is called the *crystal* of the integrand F, and F's whose crystals are *polyhedral* are called *crystalline*. The second section gives examples of integrands and their crystals.

The third section illustrates the problems in solving the prescribed boundary problem for crystalline integrands; the fourth gives ways to avoid the problems and the resulting theorems. In particular, by using convex crystalline integrands and a volume-maximizing condition, solutions are shown to have their tangent planes restricted to a specific finite set provided a condition is imposed on the boundary (boundaries satisfying this condition are, however, C^0-dense in the space of all boundaries). Finally, in Section 5, a procedure for the explicit construction of $\boldsymbol{F}$-minimal surfaces satisfying an additional boundary condition is outlined (without proof). By the results of Section 2, this procedure produces explicit approximations to a surface minimizing the integral of any given integrand.

Before we begin, however, a little terminology is in order. (For the terminology of geometric measure theory in general, see [F1] or [F2].) The class of surfaces to consider in minimizing surface energy should involve oriented surfaces of finite area forming at least part of the boundary of a region and having a well-defined oriented unit normal, at least $\mathscr{H}^2$ almost everywhere. Such surfaces, if they also have nice boundaries, are represented naturally by elements of the class $I_2(\boldsymbol{R}^3)$ of integral currents of dimension 2 on $\boldsymbol{R}^3$. For the purely mathematical questions, also, integral currents are natural domain of the problem. A 2-dimensional integral current S considered in this article has in particlar the following ingredients: (1) a measure $\|S\|$ on $\boldsymbol{R}^3$, which is simply a positive-integer-valued function m (the multiplicity) times $\mathscr{H}^2$ restricted to a set, $\operatorname{spt} S$, which can be arbitrarily closely approximated (in $\mathscr{H}^2$ measure) by a 2-dimensional C^1 manifold, (2) a $G_o(3,2)$-valued function $\vec{S}$ defined for $\|S\|$ almost all x; $\vec{S}(x)$ is just the oriented tangent plane to that underlying set at x, and (3) an oriented boundary, ∂S, which has properties like (1) and (2) but with 1 replacing 2 throughout. Thus the integral $\boldsymbol{F}(S)$ of $\boldsymbol{F}$ over the integral current S is

$$F(S) = \int F(\vec{S}(x))d\,\|S\|\,x\,.$$

The mass of S, $M(S)$, is defined to be $\|S\|\,(R^3)$ (which is just $\int_{\mathrm{spt}\,S} m(x)d\mathscr{H}^2x$). Since the integral currents considered in this article correspond to "(γ, δ) restricted sets" [A2], one need not be concerned here with the more pathological currents.

We will also consider top dimensional integral currents (elements of $I_3(R^3)$) in the case where volume constraints are used. An open set with positive orientation, finite volume, and piecewise C^1 boundary which also has finite area is represented by such a current, as is any element in the closure of the space of such sets under the distance N defined by

$$N(S, T) = M(S - T) + M(\partial S - \partial T)\,.$$

In fact, using the strong approximation theorem [F1 ; 4.2.20], one can show that this closure includes all integral currents with positive orientation and multiplicity function $+1$ everywhere.

A very important property of integral currents is compactness: every sequence of integral currents whose supports lie in a bounded region, whose masses are bounded, and whose boundaries have bounded masses, has a subsequence which converges (in the "flat" topology—see [F1] or [F2]) to an integral current satisfying the same bounds [F2 ; 4.2.17].

Another notion we will use occasionally is that of a varifold. A varifold V, in the context of this article, also has a measure $\|V\|$ which is $\mathscr{H}^2$ restricted to a relatively nice set and a "tangent plane" function defined at $\|V\|$ almost all points; the difference is that the "tangent plane" can be *independent* of the actual tangent plane of the surface and it can be a probability distribution of planes, rather than a single plane. A simple example arises in Section 3. If the "tangent plane" distribution *does* agree with the actual tangent plane to the underlying surface at $\|V\|$ almost all points, then we say V is the varifold naturally associated to the surface. See [Ad] for further details.

1. The Wulff construction and the crystal of an integrand

We call

$$F : G_o(n + 1, n) \longrightarrow R^+$$

a (constant coefficient, codimension 1, oriented) integrand on $\boldsymbol{R}^{n+1}$. Although integrands can be quite wild, to each integrand F there is associated a convex oriented geometric object, its *crystal* (defined below), which seems to contain all the essential information about F. In particular, any two integrands having the same crystal also have the same geometrical solutions to the problem of minimizing—subject to having a given boundary—the integral of the integrand (Theorem 4.1). And in a sense, the wilder the integrand is, the simpler its crystal becomes, with many of the more intricate integrands having polyhedral crystals. In this case, which is the main one considered in this article, the crystal alone provides many barriers (see Section 4); these are then the major tools used in proving the regularity of the solutions and in constructing some of them. There is even no need to have the integrand be bounded; extended integrands

$$F: G_o(n+1, n) \longrightarrow \{t : 0 \leq t \leq \infty\}$$

have crystals, and an example of such an extended integrand is given in Section 2. The basic property of the crystal of an integrand is that it is the shape having the least surface integral for the volume it contains (Theorem 2.1), and hence if F is the surface tension function of a physical material, the crystal of F should be the equilibrium shape of a lump of that material. (since such problems as mass transport and non-equilibrium growth in fact occur for any sizable lump, the crystal is sometimes called the infinitesimal equilibrium shape.)

The crystal of an (extended) integrand F is defined to be the result of the Wulff construction:

Wulff's construction. *Let F be an integrand or extended integrand on $\boldsymbol{R}^{n+1}$ and let F^* be the dual function defined on the unit sphere $\partial B^{n+1}(0,1)$ by $F^*(v) = F(*v)$ for each v in $\partial B^{n+1}(0,1)$; here $*v$ denotes the dual of v and hence can be regarded as an element of $G_o(n+1, n)$.*

Plot F^* radially: for each v in $\partial B^{n+1}(0,1)$, go out a distance $F^*(v)$ in the direction v (see Figure 1). *Now at each point $F^*(v)v$ of this polar plot, discard the oriented half space whose boundary contains that point and has tangent plane $-(*v)$* (see Figure 2). *The result of the Wulff construction is the remaining set in $\boldsymbol{R}^{n+1}$, with positive orientation, and is denoted W_F (or W, if F is clear from context).*

In terms of a formula,

$$W_F = \sum_{v \in \partial B^{n+1}(0,1)} \{x \in \boldsymbol{R}^{n+1} : \langle x, v \rangle \leq F^*(v)\}$$

with positive orientation.

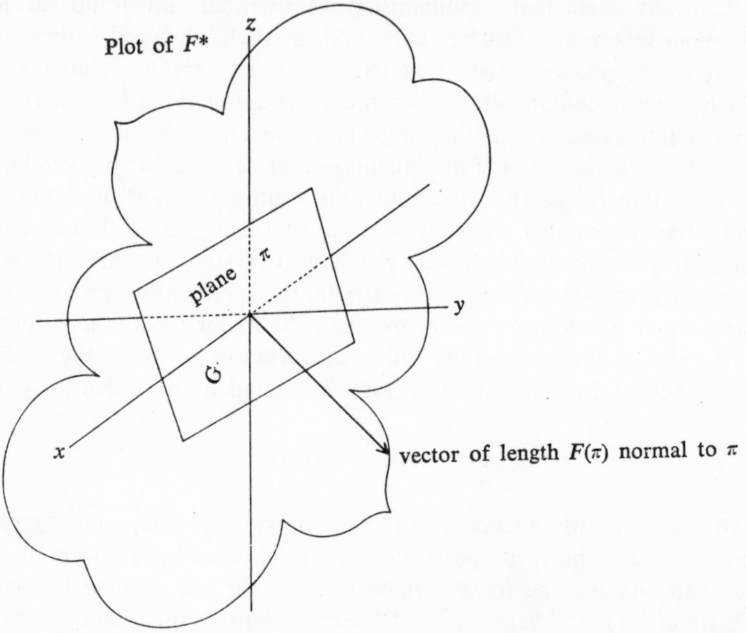

Fig. 1

An alternate description of the Wulff construction developed in [Fu] is as follows: with $\langle\cdot,\cdot\rangle$ the standard inner product on $\boldsymbol{R}^3$ and $|\cdot|$ the standard norm, we define the operator

$$W: \text{integrands} \longrightarrow \text{integrands}$$

$$W(F)(\pi) = \inf\{F(^*w)\langle{}^*\pi, w\rangle^{-1}: w \in \boldsymbol{R}^3,\ |w| = 1,\ \langle{}^*\pi, w\rangle > 0\}$$

$$\text{for any integrand } F \text{ and each } \pi \in G_o(n+1, n)\ .$$

Then the crystal of F, W_F, is the open set with positive orientation whose boundary is the radial plot of the function $(W(F))^*$.

A property of integrands and their crystals which is obvious from the first definition of W_F is that the crystal of any integrand is convex (and hence $W_F \in I_{n+1}(R^{n+1})$). To see other properties, we follow [Fu] and define three more operators, I, A, and C, from integrands to integrands as follows: for every integrand F and every $\pi \in G_o(n+1, n)$, we let

$$I(F)(\pi) = 1/F(\pi)$$

(I is the multiplicative inverse operator),

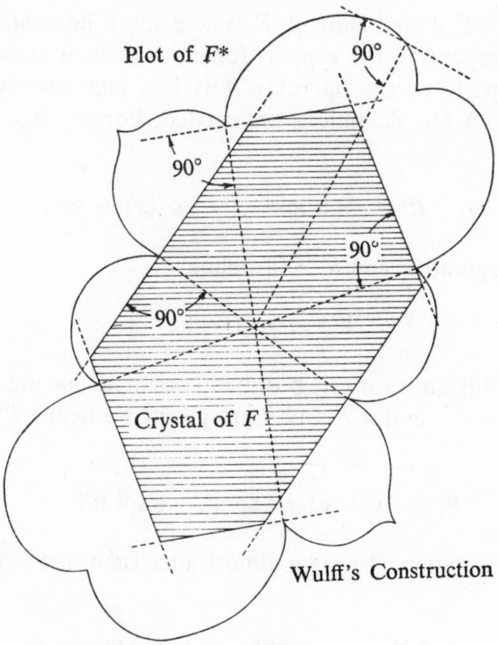

Fig. 2

$$A(F)(\pi) = \sup \{F(^*w)\langle ^*\pi, w\rangle : w \in \mathbf{R}^3, |w| = 1\}$$

(so $A = I \circ W \circ I$ and $W = I \circ A \circ I$); we let $C(F)$ be the integrand such that the radial plot of $(C(F))^*$ is the boundary of the convex hull of the radial plot of F^*.

One checks that $C = W \circ A$, and for a given integrand F, if H is any integrand which satisfies $W(H) = C(F)$, then $H \geq A(F)$. From this, it follows that for any convex set K containing the origin in its interior, there is an integrand F such that the crystal of F is K; one such F is just $A(k)$, where k is the integrand such that the radial plot of k^* is the boundary of K.

One property that some integrands have is convexity: an integrand F is *convex* (equivalently, semielliptic) if and only if $C \circ I(F) = I(F)$. Note that the integrand $A(k)$ having the given convex set K as its crystal is in fact a convex integrand, and hence that for any integrand F, there is a unique convex integrand (namely, $A \circ W(F)$) having the same crystal; this is also the smallest integrand having that crystal. The consequences of convexity will be discussed in the next section.

(In the terminology of [R], $(A \circ W(F))^*$, extended to a function on all of $\mathbf{R}^{n+1}$ by positive homogeneity, is the *support function* of W_F; note that $A \circ W(F) = F$ if and only if F is a convex integrand, so convex integrands correspond to the support functions of their crystals.)

Another property of the operator A is less immediately obvious but quite important. A few definitions are needed. For any $h > 0$, define the homothety

$$\mu_h : \mathbf{R}^{n+1} \longrightarrow \mathbf{R}^{n+1}, \qquad \mu_h(x) = hx .$$

For a given integrand F and $h > 0$, define

$$W_F^h = \mu_{h\#}(W_F) .$$

If $P \in \mathbf{I}_{n+1}(\mathbf{R}^{n+1})$ is an open set, positively oriented, having ∂P piecewise C^1 and $\mathbf{M}(\partial P) < \infty$, and if F and h are given, we define $P^h \in \mathbf{I}_{n+1}(\mathbf{R}^{n+1})$ by

$$P^h = \{x + y : x \in \operatorname{spt} P,\ y \in \operatorname{spt} W_F^h\}$$

with positive orientation. Then for almost all x such that $\overrightarrow{\partial P}(x)$ exists, we have

$$\lim_{h \to 0} h^{-1} \operatorname{dist}(x, \operatorname{spt} \partial P^h) = A \circ W(F)(\overrightarrow{\partial P}(x)) .$$

A fundamental fact about the crystal of an integrand is the following classical theorem. It was first stated without proof by G. Wulff [W], and successively more complete proofs were given by [Ln], [Le], [D], and [H], among others. A complete proof of the unoriented case in general dimensions appears in [T1] and [T2].

Theorem 1.1 (Wulff's Theorem). *Suppose F is an (extended) integrand. Then $\mathbf{F}(\partial W_F) < \mathbf{F}(\partial T)$ for every $T \in \mathbf{I}_{n+1}(\mathbf{R}^{n+1})$ such that $\mathbf{M}(T) = \mathbf{M}(W_F)$, T is positively oriented $\|T\|$ almost everywhere, and T is not W_F or a translation of W_F.*

Proof. We abbreviate W_F by W. Let P in $\mathbf{I}_{n+1}(\mathbf{R}^{n+1})$ be an open set, positively oriented, with piecewise C^1 boundary. Then, in the terminology defined above,

$$\begin{aligned} \mathbf{F}(\partial P) &= \int F(\overrightarrow{\partial P}(x))\, d\|\partial P\|\, x \\ &\geq \int A \circ W(F)(\overrightarrow{\partial P}(x))\, d\|\partial P\|\, x \\ &= \lim_{h \to 0} (\mathbf{M}(P^h) - \mathbf{M}(P))/h . \end{aligned}$$

By the Brunn-Minkowski theorem [F2, 3.2.41],

$$M(P^h) \geq (M(P)^{1/(n+1)} + M(W^h)^{1/(n+1)})^{n+1};$$

since $M(W^h) = h^{n+1}M(W)$ and $M(W) = M(P)$, we get

$$\lim_{h\to 0} (M(P^h) - M(P))/h \geq (n + 1)M(W) .$$

If $P = W$, then all the inequalities above are *equalities*. Therefore, running back up through the argument with W replacing P, we get $(n + 1)M(W) = F(\partial W)$, and hence $F(\partial W) \leq F(\partial P)$.

If T is as in the statement of the theorem and has multiplicity 1 $\|T\|$ almost everywhere, then T is in the N-closure of currents such as P (see the terminology in the introduction) and if $\{P_i, i = 1, 2, \cdots\}$ is a sequence of such currents converging to some T as in the theorem then $F(P_i)$ converges to $F(T)$. The condition of multiplicity 1 can be removed since the masses of currents and their boundaries scale under μ_h by different powers of h.

Uniqueness of the solution, compared to currents P with piece-wise C^1 boundary as above, follows from the fact that at least one of the inequalities is strict if P is not W or a translation of it. Uniqueness in the larger class follows from [T2] and is sufficiently involved that it will not be reproduced here. ∎

Remark. W is actually the unique solution even in a varifold sense to this problem of minimizing the surface integral subject to enclosing a given volume; see [T2].

If the crystal of F is not centrally symmetric (i.e. if F is not equivalent to an integrand on the unoriented Grassmannian $G(n + 1, n)$), then the *inversion* of the crystal is also important; this integral current is defined by taking the central inversion of the crystal and giving it a negative orientation.

Corollary. *The inversion of W_F minimizes F among all integral currents T with $M(T) = M(W_F)$ and negative orientation.*

2. Examples of Wulff construction

(1) Trivial example: $F \equiv 1$. Then the crystal of F is the unit sphere. This F is called the area integrand, since $F(S)$ is the n-dimensional area of S (counting multiplicity) for any $S \in I_n(R^{n+1})$. (Actually, the example is not so trivial, in that the Brunn-Minkowski inequality, which is the heart of the proof of the Wulff construction, is a standard means of proving that the sphere has least area for the volume it surrounds.)

(2) Typical example: Material with a lattice structure. Computing the energy holding a substance together by using only nearest neighbor

bonds, and then obtaining a surface energy function by seeing what bonds are broken, one can arrive at an approximate surface tension function; there are also techniques for direct measurement of the surface tension function (e.g. [AC]). Figure 3 (taken from [H]) shows a typical function; the crystals of these functions are polyhedral.

(3) Unusual example: Catching fish using a sailboat. One can use the Wulff construction to decide the best path to sail when trawling for fish with a sailboat. The speed a particular sailboat can sail is a function of the direction it is sailing, given a wind of constant velocity and otherwise constant conditions. The velocity profile of a typical sailboat is is given in Figure 4. Let $F: G_o(2,1) \to \boldsymbol{R}^+ \cup \{\infty\}$ be given by $F(\pi) =$ the time to sail one mile in direction π for each π in $G_o(2,1)$ (note that $F = 1/v$ if v is the velocity function). The polar plot of F^* and the result of the Wulff construction are given in Figure 5 (note that the coordinates appear to be rotated because one plots F^*, not F). The crystal of F is

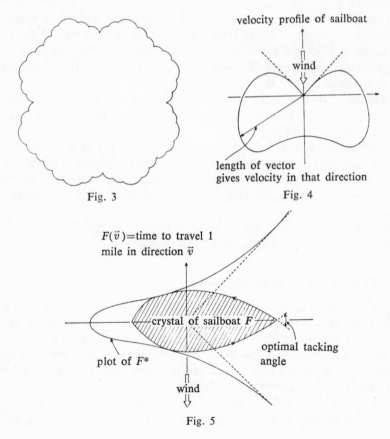

Fig. 3

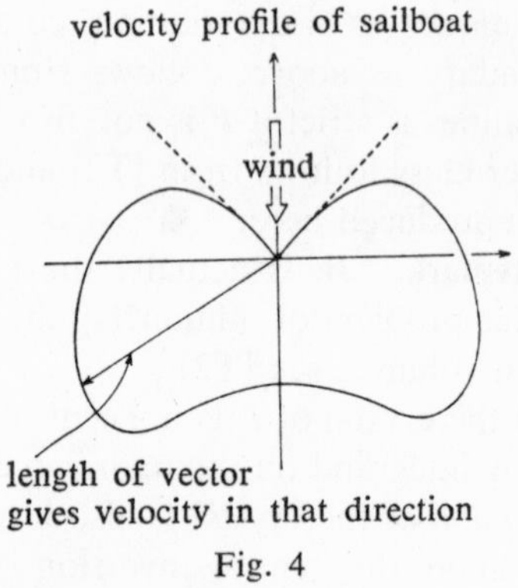

Fig. 4

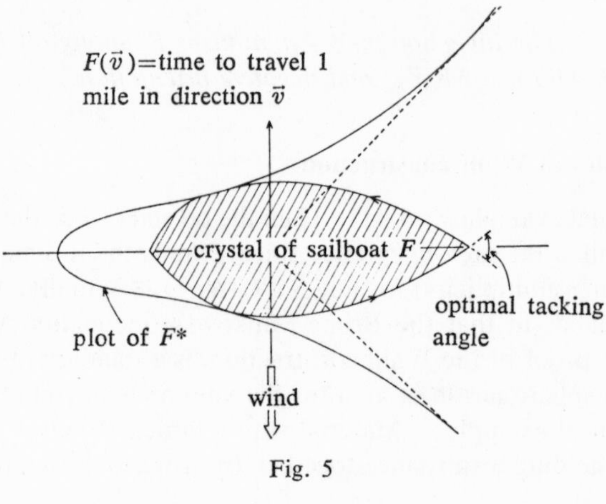

Fig. 5

that vaguely fish-shaped object and is the region of given area such that the total time to sail around it is a minimum; equivalently, it is the shape of the largest area that can be surrounded in a fixed amount of time. Therefore if the fish are uniformly distributed, the most fish are surrounded in the given time by following this path. Note that the best tack and jibe angles appear automatically. The direction to sail is counterclockwise since the region surrounded is to have positive orientation; if you preferred to sail clockwise, you would use the inversion of the crystal as the best path.

3. Crystalline integrands and the prescribed boundary problem: problems

An integrand F is defined to be *crystalline* if and only if its crystal is polyhedral; thus in the examples of the previous section, surface tension functions for physical crystals are crystalline, but the time function for the sailboat is not. We study crystalline integrands for several reasons: those with appropriate symmetries provide a mathematical model for question involving surface tensions of solids; the finiteness of the number of plane directions in the boundary of the crystal leads, under certain circumstances, to a similar finiteness and computability for solutions to the prescribed boundary problem; and the interplay between elliptic integrands and crystalline integrands may yield new information on surfaces minimizing the integrals of elliptic integrands. In this section we consider that interplay and the extent to which the crystal of an integrand determines the solutions to the prescribed boundary problem for that integrand.

Crystalline integrands can clearly be either convex or not. An integrand F is said to be *strictly crystalline* if F is crystalline and

$$F(\pi) > A \cdot W(F)(\pi)$$

whenever $\pi \in G_0(3, 2)$ is *not* a tangent plane of the crystal of F. An illustration of a convex integrand and a strictly crystalline integrand having the same crystal is given in Figure 6. Elliptic integrands, on the other hand, are those which have *uniformly* convex crystals (the area integrand, for instance, is elliptic). Thus the set of elliptic integrands is disjoint from the set of crystalline integrands.

Theorem 3.1. *If F is an elliptic integrand, then in the C^0 topology F is in the closure of the class of crystalline integrands. If G is convex and crystalline, then in the C^0 topology G is in the closure of the class of elliptic integrands.*

Proof. Approximate the crystal of F by convex polyhedral bodies; approximate the crystal of G by uniformly convex bodies. ■

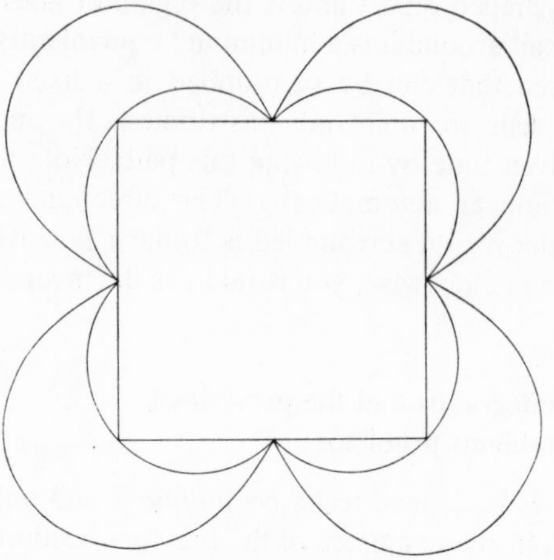

Fig. 6

This theorem implies in particular that any construction procedure for surfaces minimizing the integrals of crystalline integrands will produce explicit approximations to surfaces minimizing the integrals of elliptic integrands. Since such a construction procedure seems possible (see the last section), this is another major reason for studying crystalline integrands.

Convexity of an integrand is important because it implies the lower semicontinuity of its integrals: if S_i, $i = 1, 2, \cdots$, is a sequence of integral currents converging to S, then $\lim F(S_i) \geq F(S)$. Thus we have:

Theorem 3.2. *If F is convex and $B \in I_{n-1}(\mathbf{R}^{n+1})$ with $\partial B = 0$, then there exists S in $I_n(\mathbf{R}^{n+1})$ with $\partial S = B$ such that $F(S) \leq F(S')$ for any S' in $I_n(\mathbf{R}^{n+1})$ with $\partial S' = B$.*

Proof. The direct method in the calculus of variations—taking the limit of a (sub)sequence of currents whose integrals approach the infimum—produces a solution. ■

One example of non-lower-semicontinuity is provided by the sailboat. Suppose you want to go directly up-wind. If you just point the boat that way, it will take infinite time, since the speed you cay sail in that direction is zero (in truth, it is negative). So you make one or more tacks. Neglecting the time it takes to tack, as we did implicitly before, if you take any finite number of tacks, and always sail along one of the sides of the best tack angle in between tacks, then all paths leaving from the starting point and arriving at the finishing point require the same length of time, which is also the least possible time (Theorem 4.1). But the limit (in the usual, namely integral current, sense) of a sequence of such paths can be a path heading straight up-wind, which is clearly not a path of the same or smaller time. (See Figure 7.) (One path which does have the same time, however, is the (rotated) graph of a Cantor type function.) To make the limit also be a solution, one would have to go to varifolds, rather than integral currents; with varifolds, the limiting "tangent plane" distribution of this sequence at each point of the limit line would be half π_1 and half π_2, where π_1 and π_2 form the best tack angle.

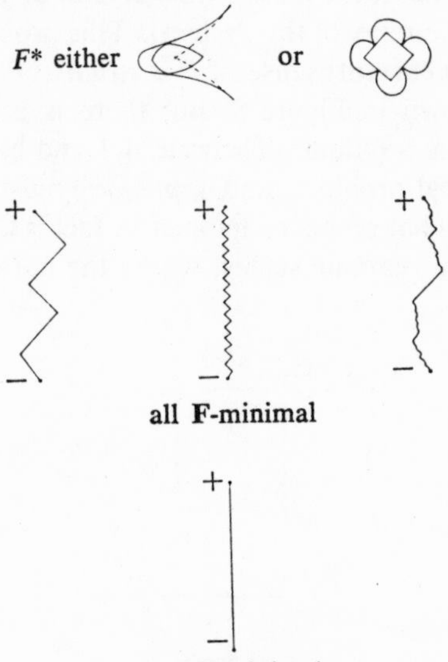

Fig. 7

The reason this lack of lowersemicontinuity is a problem is that there is a standard "direct method" for finding solutions to variational problems: Take a sequence of surfaces $\{S_i\}$ for which $\{F(S_i)\}$ converges to the infimum of $F(S)$ over all reasonable comparison surface S. Associate a measure integral current or the natural varifold) to the surfaces $\{S_i\}$; if one can show that the surfaces can be assumed to lie in a bounded region of space, then there is a convergent subsequence and a limit measure V (integral current or varifold) for this sequence of measures. One then tries to show that there is a "nice" surface S such that V is the measure naturally associated to S and such that $F(S)$ is the minimum. In the absence of lowersemicontinuity, this last step cannot necessarily be done; in the above case, it depends on the particular minimizing sequence chosen.

In R^3, one cannot avoid the problem by choosing other minimizing sequences. Consider the problem of taking a single crystal of a substance, such as table salt, whose surface tension function F has a cube as its crystal and fixing a rectangular boundary B in a plane π parallel to one edge of the cube and at a 45° angle to the other two edges (this could be accomplished, say, by painting the whole crystal with an insoluble paint andn the slicing off a chunk parallel to an edge of the crystal as above), then trying to find the interface of least energy having this boundary (say by putting the painted, sliced crystal in a saturated solution of the salt so that either depositing more mass on the crystal or dissolving mass does not change the total energy of the system). This problem has no solution in the usual (integral current) sense, if F is *strictly* crystalline: a minimizing sequence is shown in Figure 8, but there is a solution only in the varifold sense, and it is unique (Theorem 4.1 and barriers). Since this is a reasonable physical problem, and *something* must happen, our model must break down; what seems to happen in fact is that everything works fine until one reaches atomic scales, where the notion of tangent planes becomes untenable.

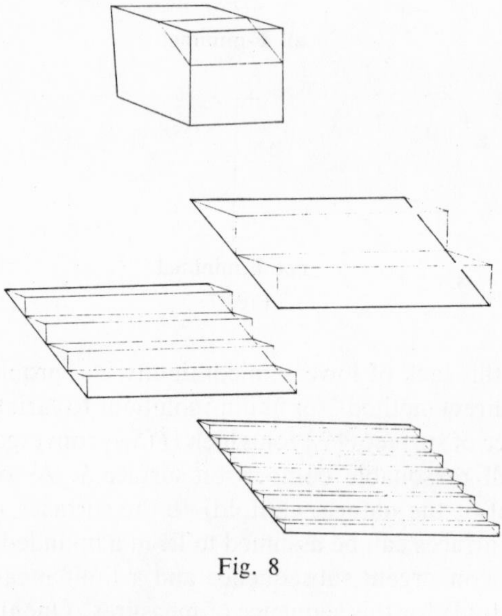

Fig. 8

4. Crystalline integrands and the prescribed boundary problem: solutions

It is thus useful to consider only convex integrands if possible. By Section 1, to every integrand there corresponds a convex crystalline integrand. The following theorem says we lose nothing by using the convex integrand instead of the original one.

Theorem 4.1. *Suppose G is any crystalline integrand having the same crystal as the convex crystalline integrand F. Then if $S' \in I_2(\boldsymbol{R}^3)$ minimizes* *the integral of G (among all integral currents with the same boundary), it minimizes the integral of F. Conversely, if* $S \in I_2(\boldsymbol{R}^3)$ *then there is a varifold V with* $\operatorname{spt}\|V\| = \operatorname{spt}\|S\|$ *which minimizes the integral of G in the family of all varifolds which are varifold limits of currents with boundary* ∂S.

Situations like that in the example of Section 3, where the graphs of even Cantor-type functions can be F-minimal for a given boundary, illustrate that surfaces minimizing the integrals of nonelliptic integrands can be quite irregular without some further selection. The situation becomes considerably simplified if, in the class of *all* integral currents minimizing the integral of F, subject to having a given boundary, one looks only at those satisfying a local volume maximizing principle. One such condition is that of obeying local crystal corner barriers:

> If at some point x in $\operatorname{spt} S$ there exists a neighborhood N of x such that a tangent cone to a corner of the crystal can be "pushed up" through S restricted to N without moving Bndry ($N \cap \operatorname{spt} S$) then S does not obey local crystal corner barriers. If for every point x in S there is no such neighborhood, then S does obey local crystal corner barriers.

Theorem 4.2. *Let B in* $I_1(\boldsymbol{R}^3)$ *have* $\partial B = 0$ *and let F be a convex crystalline integrand. Then there exists S in* $I_2(\boldsymbol{R}^3)$ *such that* $\partial S = B$, *S is* $\boldsymbol{F}$*-minimal among all integral currents with boundary B, and S obeys local crystal corner barriers. Furthermore, the collection of possible tangent planes to any such S form a one-dimensional subfamily of* $G_0(3, 2)$.

One can reduce the set of possible tangent planes of an $\boldsymbol{F}$-minimal surface to a finite set by putting conditions on the support of the boundary. Note that the set of boundaries satisfying the theorem is C^0-dense in the space of all boundaries.

Theorem 4.3. *Suppose F is convex and crystalline. If* $B \in I_1(\boldsymbol{R}^3)$ *has* $\partial B = 0$ *and* $\operatorname{spt} B$ *piecewise linear, with each linear segment parallel to an edge of the crystal, and if* $S \in I_2(\boldsymbol{R}^3)$ *has* $\partial S = B$, *is (locally)* $\boldsymbol{F}$*-minimal, and obeys local crystal corner barriers, then* $\|S\|$ *almost all points in* $\operatorname{spt} S$ *have a tangent plane in the finite set P consisting of those planes which are spanned by lines parallel to edges of the crystal.*

One woule hope that this finiteness would imply that some finite explicit procedure for the computation of $\boldsymbol{F}$-minimal surfaces under these conditions would be possible. This is in fact the case and such a procedure is outlined in the next section. Before we leave this section, however, let us investigate the conditions under which the set P of planes in the theorem above is just the set of directions of the faces of the crystal itself, and hence under which the convexity condition on F can be removed without having to go to varifold solutions.

Proposition 4.4. *Suppose that G is crystalline and F is the convex integrand with the same crystal as F. Suppose that every face of this crystal has an even number of edges and that for every face, opposite edges are always parallel. Then*:

(1) *The plane determined by lines through the origin parallel to any two edges of the crystal is parallel to two faces of the crystal* (*of opposite orientation*).

(2) *Under the conditions of the preceding theorem, S is in fact* **G**-*minimal.*

Theorem 4.5. *If F is crystalline, all* **F**-*minimal cones can be determined by using tangent cones to the crystal of F as barriers.*

For proofs of these results, see [TJ3].

5. Crystalline integrands and the prescribed boundary problem: construction

The finite number of possible tangent planes gives hope that ***F***-minimal surfaces which locally maximize volume may be computable under appropriate boundary conditions. An explicit finite procedure for producing at least one integral current which at least locally minimizes ***F*** now exists if the support of the boundary *B* lies on the boundary of a convex set (and is piecewise linear with each piece being parallel to an edge of the crystal).

This procedure has three stages, which can be summarized as follows:

(1) Cut in, with cones derived from the crystal, at the part of the boundary of the *negatively* oriented convex set which has boundary *B*. A surface which obeys crystal corner barriers is produced.

(2) Remove non-minimal intersections of planes by inserting segments of the crystal appropriately, and (if necessary) separate some plane segments into two or more segments—with appropriate segments of the crystal as patches between them—according to a fairly simple criterion.

(3) Vary the position of each plane segment whose position is not fixed from both sides by local crystal barriers (this includes varying planes off the boundary line segments, with appropriate patching from the crystal) to find the best overall position. (This amounts to solving a particular system of linear equalities and inequalities.)

The resulting surface satisfies five local properties which can then be shown to imply that the surface is locally ***F***-minimal. A complete description of the procedure and the proof that it produces a surface which is locally ***F***-minimal will be published later.

Another ***F***-minimal surface can be obtained by working with the other half of the boundary of the convex set; it will produce a "locally volume minimizing" surface of the wrong orientation which can be converted to a locally volume maximizing surface of the right orientation simply by changing orientation and then applying local corner barriers. If this second surface agrees with the surface obtained by the first procedure, then in fact *all* $\boldsymbol{F}$-minimal surfaces can be determined from it.

An undergraduate student at Rutgers University is currently attempting to put this procedure on a computer. Some examples of $\boldsymbol{F}$-minimal surfaces obtained by the procedure for various F's are shown in Figures 9 through 10.

Several comments should be made about the significance of this procedure. First of all, it explicitly produces exact solutions to a large class of problems which are reasonable mathematical models for physical phenomena involving surface tension. (Note that the finiteness of the number of plane segments means that atomic dimensions can easily be avoided, making surface tension a meaningful concept.) Secondly, since any integrand in $\boldsymbol{R}^3$ (in particular, any elliptic integrand) can be approximated by crystalline integrands (Theorem 3.1), the procedure produces a sequence of explicit approximations to surfaces minimizing the integral of the original integrand. (Approximations for area minimizing surfaces with boundaries on convex sets when there is known to be a unique

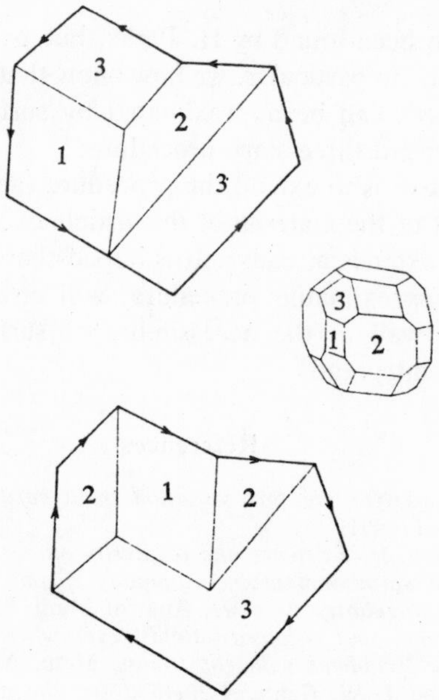

Fig. 9

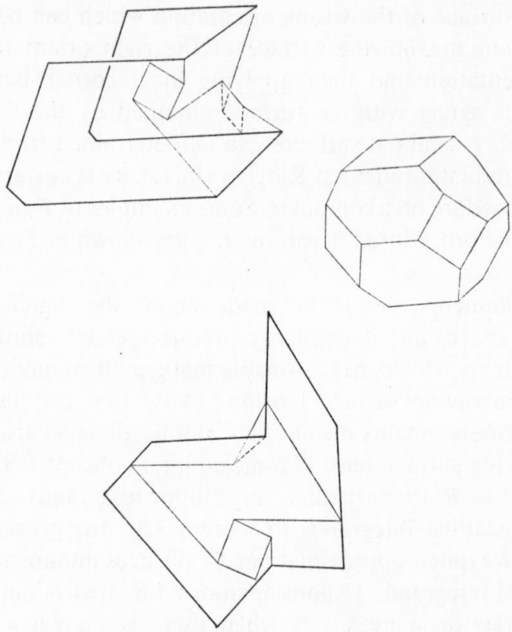

Fig. 10

solution have also been found by H. Parks, but by an entirely different method [P].) Thus, in particular, we now know that extreme locally area-minimizing surfaces can be approximated by surfaces obtained in this rather straightforward three-stage procedure.

A major objective is to extend the procedure (and its proof) to higher dimensions (most of the material of this article except for the construction fairly easily extends already). It is hoped that the construction, together with the approximating procedure, will give new information on the structure as well as the accessibility of surfaces minimizing any codimension one integrand.

References

[Ad] W. K. Allard, *On the first variation of a varifold,* Ann. of Math. **95** (1972), 417–491.

[A1] F. J. Almgren, Jr. *Existence and regularity almost everywhere of solutions to elliptic variational problems among surfaces of varying topological type and singularity structure,* Ann. of Math. **87**, 321–391 (1968).

[A2] ——, *Existence and regularity almost everywhere of solutions to elliptic variational problems with constraints,* Mem. A. M. S. 4, Number 165.

[AC] E. Arbel and J. W. Cahn, *A method for the absolute measurement of anisotropic solid-melt surface free energies,* preprint.

[C] J. W. Cahn, personal communication.

[D] A. Dinghas, *Über einen geometrischen Satz von Wulff für die Gleichgewichtsform von Kristallen,* Z. Krist., **105** (1944), 304–314.

[F1] H. Federer, *Geometric Measure Theory*, Springer, New York, 1969.

[F2] H. Federer, Colloquium lectures on Geometric Measure Theory, Summer Meeting of the A. M. S., August 15–18, 1977, Seattle, Washington, Bull. Amer. Math. Soc. **84** (1978), 291–338.

[Fu] J. H. G. Fu, *A mathematical model for crystal growth and related problems,* Junior paper, Princeton University.

[H] C. Herring, *Some theorems on the free energy of crystal surfaces,* Phys. Rev., **82** (1951), 87–93.

[Le] M. V. Laue, *Der Wulffsche Satz für die Gleichgewichtsform von Kristallen,* Z. Krist., **105** (1943), 124–133.

[Ln] H. Liebmann, *Der Curie-Wulff'sche Satz über Combinationsformen von Krystallen,* Z. Krist., **53** (1913), 171–177.

[P] H. R. Parks, *Explicit determination of area minimizing hypersurfaces,* Duke Math. J., **44** (1977), 519–534.

[R] R. T. Rockafellar, *Convex Analysis,* Princeton University Press, Princeton, New Jersey, 1970.

[T1] J. E. Taylor, *Existence and structure of solutions to a class of nonelliptic variational problems,* Symposia Mathematica, XIV (1974), 499–508.

[T2] ——, *Unique structure of solutions to a class of nonelliptic variational problems,* Proc. Symp. P. Math. XXVII (1974), 481–489.

[T3] J. E. Taylor, *Crystalline variational problems,* Bull. Amer. Math. Soc., **84** (1978), 568–588.

[W] G. Wulff, *Zur Frage der Geschwindigkeit des Wachsthums und der Auflösung der Krystallflachen,* Z. Krist., **34** (1901), 449–530.

RUTGERS UNIVERSITY
NEW BRUNSWICK, NJ 08903
AND
INSTITUTE FOR ADVANCED STUDY
PRINCETON, NJ 08540
U.S.A.

Minimal Submanifolds and Geodesics
Kaigai Publications, Tokyo, 1978, 249–254

THE EQUATION OF A VIBRATING GENERAL MEMBRANE

TAKUICHI HASEGAWA

Introduction

The equation of a vibrating string in the Euclidean two-plane E^2 or that of a vibrating membrane in E^3 is classical. We can, however, vibrate other membranes of general kind, such as, geometrically, stable domains on a minimal surface in E^3. In this note, expressing a small motion normal to the equilibrium state by a family of normal vector fields parametrized by time, we shall derive the equation of a vibrating general membrane (minimal submanifold) of arbitrary dimension and codimension in a curved space. In this situation it turns out that the classical Laplacian is replaced by the Jacobi differential operator, which has been appearing in the second variation formula of minimal submanifolds. I wish to express my gratitude to Professor T. Otsuki who listened critically to my first draught of this note. His suggestions made me use directly the second variation formula, from which I could easily remove the restriction of the codimension-one-ness of the minimal submanifold.

Let N be an n-dimensional Riemannian manifold, M an m-dimensional minimal submanifold of N, and D a compact connected domain of M with piecewise C^∞ boundary. In this situation we introduce the physical notions, terminologies and principles in its suitably generalized meaning in the following way. First we regard D as a strained homogeneous membrane with a constant tension μ_0 and a mass density ρ_0 in the space N. Secondly we make a small deformation of D in the normal direction with its boundary fixed. And we suppose that the mechanical properties of the system are determined by two quantities, the potential and kinetic energy. The potential energy is considered to be proportional to the change of area. Its proportionality factor is the tension, which is assumed to be preserved under such a small deformation. Thus the state that D is minimal in the sense of geometry is equivalent to the state that D is in an equilibrium position in the physical sense. Deformed nearby surfaces are expressed, through the normal exponential mapping, as cross sections of the normal bundle of D, i.e. as normal vector fields on D. Therefore any small motion of D normal to it can be expressed by a family of normal vector fields on D parametrized by the time t such that the direction of vectors at each point is constant throughout the time.

Theorem. *Suppose that $V(x, t)$ is a C^2-family of normal vector fields which corresponds to such a motion. Then it obeys the following differential equation (a generalization of the equation of a vibrating membrane):*

$$\rho_0 V_{tt} + \mu_0 JV = 0 .$$

Here $V_{tt} := \partial^2 V / \partial t^2$, and $J := -\nabla^2 + \bar{R} - \tilde{A}$ is a strongly elliptic, self-adjoint second order differential operator acting on normal vector fields on D, cosisting of the Laplace operator ∇^2, symmetric transformations $\bar{R}$ and $\tilde{A}$ of the normal bundle $TM^{\perp}|_D$. J has been used in the second variation formula of minimal submanifolds in [2], [3] and is to be called the *Jacobi differential operator* on the ground that solutions of $JU = 0$ are just Jacobi fields.

If $\rho_0/\mu_0 = 1$ and V can be expressed as $V(x, t) = \alpha(t)U(x)$, the product of a factor $\alpha(t)$ depending only on the time and a factor $U(x)$, the equation is divided into two parts:

$$JU = \lambda U \text{ and } U(x) = 0 , \quad x \in \partial D : \alpha''(t) = -\lambda\alpha(t) , \ \lambda = \text{a constant.}$$

The former can be considered as a generalization of the Helmholtz equation. For positive eigenvalues λ, α becomes periodic functions of time, in which case periodic motions i.e. vibrations occur. But J may have negative eigenvalues, which are at most finite in number, and in that case aperiodic motions occur instead of vibrations.

Proof of Theorem. Using a partition of unity, the problem can be considered to be a local one, so we may assume that D is covered by one coordinate neighborhood $(x^1, \cdots, x^m) : D \cong \mathcal{O} \subset \boldsymbol{R}^m$. We can set $V(x, t) = f(x, t)Y(x)$, where $\langle Y(x), Y(x)\rangle = 1$, $\langle \ , \ \rangle$ being the inner product of vectors. The potential energy P is given by

$$P = \mu_0(\text{area } (D_t) - \text{area } (D))$$

where $D_t = \{\exp_x^{\perp} V(x, t) \,|\, x \in D\}$ is the state at the time t of the motion, and $\exp^{\perp}$ is the normal exponential mapping of M. Then the question is how to approximate P. In this regard we follow the same principle as has been taken in the classical cases. Namely the principle goes that, motions being so small, higher powers of $f, f_1, \cdots$ and f_m may be neglected compared with lower powers. Here $f_i := \partial f / \partial x^i$.

Let $(Y, S_{m+2}, \cdots, S_n)$ be a frame of $TM^{\perp}$ over D, and $(x^1, \cdots, x^m, y^{m+1}, \cdots, y^n)$ the coordinate of the point $\exp(y^{m+1}Y(x) + y^{m+2}S_{m+2}(x) + \cdots + y^n S_n(x))$, $(x = (x^1, \cdots, x^m))$. We define a map $v_t : D \to N$, by

$v_t(x) = (x^1, \cdots, x^m, f(x, t), 0, \cdots, 0) = \exp_x^\perp V(x, t)$ which we simply write $v_t(x) = (x, f)$. Then

$$v_{t*}\frac{\partial}{\partial x^i} = \frac{\partial}{\partial x^i} + f_i \frac{\partial}{\partial y^{m+1}} \, .$$

We set $\tilde{g}_{ij}(x, t) := \langle v_{t*}(\partial/\partial x^i), v_{t*}(\partial/\partial x^j)\rangle$, $1 \leqslant i, j \leqslant m$, and $(\tilde{g}^{jj}) :=$ the inverse matrix of $(\tilde{g}_{ij})$, $\tilde{g}_{ij}(x, 0) := g_{ij}(x)$, $\tilde{g}^{ij}(x, 0) := g^{ij}(x)$. Then

$$\tilde{g}_{ij}(x, f) = g_{ij}(x, f) + f_i g_{m+1,j}(x, f) + f_j g_{m+1,i}(x, f) + f_i f_j g_{m+1,m+1}(x, f)$$

where

$$g_{m+1,j}(x, f) := \left\langle \left(\frac{\partial}{\partial y^{m+1}}\right)_{v_t(x)}, \left(\frac{\partial}{\partial x^j}\right)_{v_t(x)} \right\rangle ,$$

$$g_{ij}(x, f) := \left\langle \left(\frac{\partial}{\partial x^i}\right)_{v_t(x)}, \left(\frac{\partial}{\partial x^j}\right)_{v_t(x)} \right\rangle \text{ etc.}$$

$$\text{area } (D_t) = \int_{\mathcal{O}} \sqrt{\det (\tilde{g}_{ij}(x, f))} dx^1 \cdots dx^m$$

$$\text{area } (D) = \int_{\mathcal{O}} \sqrt{\det (g_{ij}(x))} dx^1 \cdots dx^m \, .$$

We set $G(f, f_1, \cdots, f_m) := \sqrt{\det (\tilde{g}_{ij})}$. Regarding G as a function of independent variables $f, f_1, \cdots$ and f_m, we seek the principal part of the Taylor expansion of $G(f, f_1, \cdots, f_m) - G(0, \cdots, 0)$ centered at $(0, \cdots, 0)$. Its integration over $\mathcal{O}$ will give the desired approximation of the potential energy.

$$\frac{\partial G}{\partial f} = \frac{1}{2}\sqrt{\det (\tilde{g}_{ij})}\left\{\left(\frac{\partial g_{ij}}{\partial y^{m+1}} + f_i \frac{\partial g_{m+1,j}}{\partial y^{m+1}} + f_j \frac{\partial g_{m+1,i}}{\partial y^{m+1}} + f_i f_j \frac{\partial g_{m+1,m+1}}{\partial y^{m+1}}\right) \cdot \tilde{g}^{ij}\right\} ,$$

$$\frac{\partial G}{\partial f_k} = \frac{1}{2}\sqrt{\det (\tilde{g}_{ij})}\{(2g_{m+1,j} + 2f_j g_{m+1,m+1}) \cdot \tilde{g}^{kj}\} ,$$

$$\frac{\partial G}{\partial f}(0, \cdots, 0) = -G(0) \cdot \left\langle g^{ij} B\left(\frac{\partial}{\partial x^i}, \frac{\partial}{\partial x^j}\right), Y\right\rangle = 0 ,$$

where B is the second fundamental form of M, $g^{ij}(x)B(\partial/\partial x^i, \partial/\partial x^j)$ its mean curvature, which is zero from the assumption. $G(0) := G(0, \cdots, 0)$.

$$\frac{\partial G}{\partial f_k}(0, \cdots, 0) = 0 \quad \text{because} \quad g_{m+1,j}(x, 0) = \left\langle Y(x), \left(\frac{\partial}{\partial x^j}\right)_x \right\rangle = 0 \, .$$

Here we have used the Einstein convention for $1 \leqslant i, j, k \leqslant m$. Thus the part of first order derivatives vanishes. For the calculation of the part of second order derivatives, we make use of the second variation formula of the area integral in the theory of minimal submanifolds. We set $H(\lambda) := G(\lambda f, \lambda f_1, \cdots, \lambda f_m)$, $-\varepsilon_0 \leqslant \lambda \leqslant \varepsilon_0$. Then

$$H'(\lambda) = \frac{\partial G}{\partial f}(\lambda f, \cdots, \lambda f_m) f + \frac{\partial G}{\partial f_k}(\lambda f, \cdots, \lambda f_m) f_k ,$$

$$H''(0) = \frac{\partial^2 G}{\partial f^2}(0) f^2 + \frac{\partial^2 G}{\partial f_k \partial f}(0) f_k f + \frac{\partial^2 G}{\partial f_i \partial f_j}(0) \cdot f_i f_j .$$

Thus $\frac{1}{2}H''(0)$ gives the part of second order derivatives of the Taylor expansion. On the other hand if we set

$$\tilde{D}_\lambda(x) := \exp_x^\perp \lambda V(x, t) , \qquad \int_{\mathcal{O}} H(\lambda) dx^1 \cdots dx^m$$

gives the area $\mathscr{A}(\lambda)$ of the surface $\tilde{D}_\lambda$. Therefore, from the second variation formula ([3]),

$$\mathscr{A}''(0) = \int_{\mathcal{O}} H''(0) dx^1 \cdots dx^m$$

$$= \int_D (\langle \nabla V, \nabla V \rangle + \langle \bar{R}(V) - \tilde{A}(V), V \rangle) \cdot v_g .$$

Thus the first approximation of the potential energy as well as its geometric interpretation is given by

$$P = \frac{1}{2} \mu_0 \cdot \int_D (\langle \nabla V, \nabla V \rangle + \langle \bar{R}(V) - \tilde{A}(V), V \rangle) \cdot v_g .$$

Namely the potential energy of the state is approximately equal to $\frac{1}{2}\mu_0$ multiplied by the second variation of the area integral in such a direction (normal vector field) that corresponds to the state through the normal exponential mapping.

Here we explain notations used above.

v_g: the canonical measure of the Riemannian metric g.

∇: the canonical covariant derivative of the normal bundle $TM^\perp$.

A: the second fundamental form of M as a cross section of Hom $(TM^\perp, SM)$, where SM is the bundle of symmetric transformations of TM. A and B above are related by $\langle A^w(x), y \rangle = \langle B(x, y), w \rangle$ for $x, y \in TM_p$, $w \in TM_p^\perp$.

$\tilde{A} := {}^tA \circ A$, a cross section of Hom $(TM^\perp, TM^\perp)$, where ${}^tA$ denotes the transpose of A.

$\bar{R}$: a cross section of Hom $(TM^{\perp}, TM^{\perp})$, defined by

$$\bar{R}(v) := \sum_{i=1}^{m} (\bar{R}_{e_i,v}(e_i))^{\perp}$$

where the latter $\bar{R}$ is the curvature tensor of N, $v \in TM_p^{\perp}$, (e_i) a frame of TM_p, and $\perp$ denotes the normal part of vectors.

On the other hand the kinetic energy K is approximately given by

$$K = \frac{1}{2}\int_D \rho_0 \cdot \langle V_t, V_t \rangle \cdot v_g = \frac{1}{2}\rho_0 \cdot \int_D f_t^2 \cdot v_g$$

where $V_t = \partial V/\partial t$, $f_t = \partial f/\partial t$.

Then, from Hamilton's principle, the actual motion V makes the value of the integral T defined by

$$\begin{aligned} T(V) :&= \int_{t_0}^{t_1} (K - P)dt \\ &= \frac{1}{2}\int_{t_0}^{t_1}\int_D (\rho_0\langle V_t, V_t\rangle - \mu_0\langle \nabla V, \nabla V\rangle - \mu_0\langle \bar{R}(V) \\ &\qquad - \tilde{A}(V), V\rangle)\cdot dt \otimes v_g \end{aligned}$$

stationary with respect to all neighboring virtual motions which lead from the initial to the final position of the system in the same interval $[t_0, t_1]$ of time. Thus, for an arbitrary C^∞ family W of normal vector fields on D such that $W(x, t_0) = W(x, t_1) = 0$ $(x \in D)$, and $W(x, t) = 0$ $(x \in \partial D,$ $t \in [t_0, t_1])$

$$\begin{aligned} 0 &= \lim_{\varepsilon \to 0} \frac{1}{\varepsilon}(T(V + \varepsilon W) - T(V)) \\ &= \int_{t_0}^{t_1}\int_D \langle -\rho_0 V_{tt} + \mu_0(\nabla^2 V - \bar{R}(V) + \tilde{A}(V)), W\rangle \cdot dt \otimes v_g\,. \end{aligned}$$

Therefore, V being a C^2-function of x and t,

$$\rho_0 V_{tt} - \mu_0(\nabla^2 V - \bar{R}(V) + \tilde{A}(V)) = 0 \quad \text{i.e.} \quad \rho_0 V_{tt} + \mu_0 JV = 0$$

Q.E.D.

References

[1] R. Courant and D. Hilbert, Methods of Mathematical Physics, vol. 1, Interscience, New York, 1953.

[2] R. Herman, *The Second Variation for Minimal Submanifolds,* J. Math. Mech. **16** (1966), 473–491.

[3] J. Simons, *Minimal varieties in riemannian manifolds*, Ann. of Math. (2) **88** (1968), 62–105.

TOKYO INSTITUTE OF TECHNOLOGY
TOKYO, 152
JAPAN

Minimal Submanifolds and Geodesics
Kaigai Publications, Tokyo, 1978, 255–271

ON ISOTROPIC IMMERSIONS OF SPACE FORMS INTO A SPACE FORM

HISAO NAKAGAWA & TAKEHIRO ITOH

Introduction

A Riemannian manifold of constant curvature is called a *space form*. The minimal immersions of spheres into a sphere have been completely determined by do Carmo and Wallach [1]. Let $H^{r,n}$ be the space of all spherical harmonic polynomials of degree r on an n-dimensional sphere S^n, where $\dim H^{r,n} = (n + 2r - 1)(n + r - 2)!/r!\,(n - 1)! = : N(r) + 1$. For an orthonormal basis $\{f_1, f_2, \cdots, f_{N(r)+1}\}$ of $H^{r,n}$, we define an immersion ι_r of S^n into an $(N(r) + 1)$-dimensional Euclidean space $R^{N(r)+1}$ by $\iota_r(x) = (f_1(x), f_2(x), \cdots, f_{N(r)+1}(x))/(N(r) + 1)^{1/2}$. Then the image by ι_r is contained in the unit sphere $S^{N(r)}(1)$ in $R^{N(r)+1}$, and by means of a theorem of Takahashi [6], it is seen that ι_r is a minimal isometric immersion of S^n into $S^{N(r)}(1)$ and $\iota_r(S^n)$ is not contained in the great sphere of $S^{N(r)}(1)$. The so-called Veronese manifold can be considered as the case of the degree 2. With regard to the degree of the immersions in the sense of Wallach [7], they also showed that the degree of ι_r is equal to r and if $r \leqq 3$, then ι_r is rigid.

On the other hand, O'Neill [4] defined a notion of isotropic immersions. In terms of isotropic immersions, the local version of Veronese manifolds has been treated by the second named author and K. Ogiue [3]. The purpose of this paper is to study the local version of the situation corresponding to the immersion ι_3 of S^n into $S^{N(3)}(1)$.

We would like to take this opportunity to express our deep gratitude to Professor T. Otsuki who has given us helpful suggestions and advice constantly. We feel it our great honor to dedicate this paper to him on his sixtieth birthday.

1. Preliminaries

First of all we begin with the self-contained discussion about Riemannian submanifolds immersed isometrically into a space form to fix our notation and state several properties for later use. We denote by $M^m(c)$ an m-dimensional connected space form of constant curvature c. Let M be an $n(\geqq 2)$-dimensional connected and orientable Riemannian manifold and ι an isometric immersion of M into $\tilde{M} = M^{n+p}(\tilde{c})$ with Riemannian metric $\tilde{g}$. When the argument is local, M need not be distinguished from $\iota(M)$, and to simplify the discussion, we may identify a point x in M with the point $\iota(x)$, and a tangent vector X in a tangent space $T_x(M)$ at x with the tangent vector $\iota_*(X)$ in the subspace $\iota_*(T_x(M))$ in $T_{\iota(x)}(\tilde{M})$ via the differential ι_* of ι.

Now, we choose an orthonormal local frame field $\{e_1, \cdots, e_n, e_{n+1}, \cdots, e_{n+p}\}$ in $\tilde{M}$ in such a way that, restricted to M, the vectors $e_1, \cdots, e_n$ are tangent to M and hence $e_{n+1}, \cdots, e_{n+p}$ are normal to M. With respect to this field of frames of $\tilde{M}$, let $\{\tilde{\omega}_1, \cdots, \tilde{\omega}_n, \tilde{\omega}_{n+1}, \cdots, \tilde{\omega}_{n+p}\}$ be the dual field. Here and in the sequel we use the following convention on the range of indices:

$$\begin{aligned} A, B, C, \cdots &= 1, \cdots, n, n+1, \cdots, n+p\ , \\ i, j, k, \cdots &= 1, \cdots, n\ , \\ \alpha, \beta, \gamma, \cdots &= n+1, \cdots, n+p\ . \end{aligned}$$

Then, associated to the frame field $\{e_1, \cdots, e_n, e_{n+1}, \cdots, e_{n+p}\}$, there exist real valued differential forms $\tilde{\omega}_{AB}$ on $\tilde{M}$, which are usually called *connection forms* on $\tilde{M}$, so that they satisfy the following structure equations of $\tilde{M}$:

$$d\tilde{\omega}_A + \textstyle\sum_B \tilde{\omega}_{AB} \wedge \tilde{\omega}_B = 0\ , \qquad \tilde{\omega}_{AB} + \tilde{\omega}_{BA} = 0\ , \tag{1.1}$$

$$\begin{aligned} d\tilde{\omega}_{AB} + \textstyle\sum_C \tilde{\omega}_{AC} \wedge \tilde{\omega}_{CB} &= \tilde{\Omega}_{AB}\ , \\ \tilde{\Omega}_{AB} &= -\tfrac{1}{2} \textstyle\sum_{C,D} \tilde{R}_{ABCD} \tilde{\omega}_C \wedge \tilde{\omega}_D\ , \end{aligned} \tag{1.2}$$

where $\tilde{\Omega}_{AB}$ (resp. $\tilde{R}_{ABCD}$) denotes the curvature form (resp. the curvature tensor) on $\tilde{M}$. Since $\tilde{M}$ is a space form of constant curvature $\tilde{c}$, we have

$$\tilde{R}_{ABCD} = \tilde{c}(\delta_{AD}\delta_{BC} - \delta_{AC}\delta_{BD})\ . \tag{1.3}$$

Restricting the forms $\tilde{\omega}_A$ and $\tilde{\omega}_{AB}$ to M, we denote them by ω_A and ω_{AB} without tilde, respectively. Then we have

$$\omega_\alpha = 0\ . \tag{1.4}$$

The metric on M induced from the Riemannian metric $\tilde{g}$ in the ambient space $\tilde{M}$ under the immersion ι is given by $g = 2 \sum_i \omega_i \omega_i$. Then $\{e_1, \cdots, e_n\}$ is an orthonormal local frame field with respect to the induced metric, and $\{\omega_1, \cdots, \omega_n\}$ is the dual field, which consists of real valued, linearly independent 1-forms on M. They are called *canonical forms* on M. It follows from (1.4) and Cartan's lemma that we obtain

(1.5) $$\omega_{\alpha i} = \sum_j h^\alpha_{ij}\omega_j \,, \qquad h^\alpha_{ij} = h^\alpha_{ji} \,.$$

The quadratic form $\sum_{ij} h^\alpha_{ij}\omega_i\omega_j$ is called the *second fundamental form* of the immersion ι on M in the direction of e_α. The second fundamental form σ of M can be written as

(1.6) $$\sigma(X, Y) = \sum_{\alpha,i,j} h^\alpha_{ij}\omega_i(X)\omega_j(Y)e_\alpha$$

for any tangent vectors X and Y on M.

By the structure equations (1.1) and (1.2), we have the following structure equations on the submanifold M:

(1.7) $$d\omega_i + \sum_j \omega_{ij} \wedge \omega_j = 0 \,, \qquad \omega_{ij} + \omega_{ji} = 0 \,,$$

(1.8) $$d\omega_{ij} + \sum_k \omega_{ik} \wedge \omega_{kj} = \Omega_{ij} \,,$$
$$\Omega_{ij} = -\tfrac{1}{2} \sum_{k,l} R_{ijkl}\omega_k \wedge \omega_l \,,$$

where ω_{ij} (resp. Ω_{ij}) denotes the connection form (resp. the curvature form) on the submanifold. Moreover, we have

(1.9) $$d\omega_{\alpha\beta} + \sum_\gamma \omega_{\alpha\gamma} \wedge \omega_{\gamma\beta} = \Omega_{\alpha\beta} \,,$$
$$\Omega_{\alpha\beta} = -\tfrac{1}{2} \sum_{k,l} R_{\alpha\beta kl}\, \omega_k \wedge \omega_l \,,$$

where $\{\omega_{\alpha\beta}\}$ defines the connection form induced in the normal bundle $N(M)$ of M and $\Omega_{\alpha\beta}$ is called the *normal curvature form* of M. From (1.2), (1.3) and (1.8) we have the equation of Gauss

(1.10) $$R_{ijkl} = \tilde{c}(\delta_{il}\delta_{jk} - \delta_{ik}\delta_{jl}) + \sum_\alpha (h^\alpha_{il}h^\alpha_{jk} - h^\alpha_{ik}h^\alpha_{jl}) \,,$$

and also the equation of Ricci

(1.11) $$R_{\alpha\beta kl} = \sum_i (h^\alpha_{il}h^\beta_{ik} - h^\alpha_{ik}h^\beta_{il}) \,.$$

The Ricci form and the scalar curvature R on M can be expressed as follows:

(1.12) $$\sum_{ij} R_{ij}\omega_i \otimes \omega_j \,, \qquad R_{ij} = R_{ji} = \sum_k R_{kijk} \,,$$

(1.13) $$R_{ij} = (n-1)\tilde{c}\delta_{ij} + \sum_{\alpha,k} (h^\alpha_{kk}h^\alpha_{ij} - h^\alpha_{ik}h^\alpha_{kj}) \,,$$

(1.14) $$R = n(n-1)\tilde{c} + \sum_{\alpha,k,i} (h^\alpha_{kk}h^\alpha_{ii} - h^\alpha_{ik}h^\alpha_{ik}) \,.$$

For the second fundamental form σ of the immersion of M, $\|\sigma\|$ denotes its length, namely

(1.15) $$\|\sigma\| = \{\sum_{\alpha,i,j} h^\alpha_{ij}h^\alpha_{ij}\}^{1/2} \,.$$

The vector $\sigma(e_i, e_i)$ is called a *normal curvature vector* in the direction of a unit vector e_i, and the *mean curvature vector* $\mathfrak{h}$ of M is given by $\sum_i \sigma(e_i, e_i)/n = \sum_{\alpha,i} h^\alpha_{ii} e_\alpha/n$. Its length H is the *mean curvature* of σ, so it satisfies

$$n^2H^2 = \sum_{\alpha,i,j} h^\alpha_{ii} h^\alpha_{jj} \,. \tag{1.16}$$

If every normal curvature vector has the same length for any unit tangent vector X at x, then the immersion σ is said to be *isotropic at* x. If σ is isotropic at any point on M, that is, the length of a normal curvature vector depends only on the point, then the immersion is called *isotropic*. In particular, if the length is equal to λ, the immersion is called *λ-isotropic*, and λ is a function on M, the square of which is differentiable on M. The immersion ι is λ-isotropic at a point x if and only if the second fundamental form σ satisfies

$$\sum_\alpha (h^\alpha_{ij} h^\alpha_{kl} + h^\alpha_{jk} h^\alpha_{il} + h^\alpha_{ki} h^\alpha_{jl}) = \lambda^2(\delta_{ij}\delta_{kl} + \delta_{jk}\delta_{il} + \delta_{ki}\delta_{jl}) \,. \tag{1.17}$$

The condition is equivalent to

$$\langle \sigma(X, X), \sigma(X, Y) \rangle = 0 \tag{1.18}$$

for any orthonormal vectors X and Y at x.

Now, we denote by $\tilde{\nabla}$ the operator of covariant differentiation in $\tilde{M}$ with respect to the Riemannian metric $\tilde{g}$ and by ∇ the operator of covariant differentiation in M with respect to the induced metric g. For any tangent vectors $X = \sum_i x^i e_i$ and $Y = \sum_i y^i e_i$ we have

$$\nabla_X Y = \sum_k \{dy^k(X) - \sum_i y^i \omega_{ik}(X)\} e_k \,, \tag{1.19}$$

$$\tilde{\nabla}_X Y = \nabla_X Y + \sigma(X, Y) = \nabla_X Y + \sum_{\alpha,i,j} h^\alpha_{ij} x^i y^j e_\alpha \,, \tag{1.20}$$

$$\tilde{\nabla}_X e_\alpha = -\sum_k \omega_{\alpha k}(X) e_k - \sum_\beta \omega_{\alpha\beta}(X) e_\beta \,. \tag{1.21}$$

The connection on the tangent bundle $T(M)$ is the unique Riemannian connection induced by the induced metric g, and the connection $\nabla^\perp$ in the normal bundle $N(M)$ is defined as follows: for any normal vector field $\tilde{Z}$ of M and any tangent vector X of M, we put

$$\nabla^\perp_X \tilde{Z} = (\tilde{\nabla}_X \tilde{Z})^N \,, \tag{1.22}$$

where $(\ )^N$ denotes the projection into $N(M)$. In particular, we have

$$\nabla^\perp_X e_\alpha = -\sum_\beta \omega_{\alpha\beta}(X) e_\beta \,. \tag{1.23}$$

The smooth section ξ of the normal bundle is said to be *parallel*, if it satisfies $\nabla^{\perp}_X \xi = 0$ for any tangent vector field X on M. Then the length of a parallel section ξ is constant and $\tilde{\nabla}_X \xi$ is tangent to M, so we can define a symmetric tensor field T of type $(1, 1)$ on M by $\tilde{\nabla}_X \xi = -TX$. The cross section is said to be *isoperimetric*, if $\operatorname{Tr} T$ is non-zero constant.

Now we define the covariant derivative h^α_{ijk} of h^α_{ij} by

$$(1.24) \qquad \textstyle\sum_k h^\alpha_{ijk}\omega_k = dh^\alpha_{ij} - \sum_l h^\alpha_{lj}\omega_{li} - \sum_l h^\alpha_{il}\omega_{lj} - \sum_\beta h^\beta_{ij}\omega_{\beta\alpha} .$$

We then have

$$(1.25) \qquad h^\alpha_{ijk} - h^\alpha_{ikj} = 0 .$$

If $h^\alpha_{ijk} = 0$ for all indices i, j, k and α, then the second fundamental form of M is said to be *parallel*. We denote by ∇' the operator of covariant differentiation on the Whitney sum $T(M) \oplus N(M)$. Then we have

$$(\nabla'\sigma)(X, Y, Z) = \textstyle\sum_{\alpha,i,j,k} h^\alpha_{ijk}\omega_i(X)\omega_j(Y)\omega_k(Z)e_\alpha .$$

Next, we suppose that $h^\alpha_{i_1\cdots i_m}$ are defined for some $m \geqq 2$. Inductively we can define the covariant derivative $h^\alpha_{i_1\cdots i_m j}$ of $h^\alpha_{i_1\cdots i_m}$ by

$$(1.26) \qquad \begin{aligned} \textstyle\sum_j h^\alpha_{i_1\cdots i_m j}\omega_j = dh^\alpha_{i_1\cdots i_m} &- \sum_{r=1}^m \textstyle\sum_j h^\alpha_{i_1\cdots i_{r-1} j i_{r+1}\cdots i_m}\omega_{j i_r} \\ &- \textstyle\sum_\beta h^\beta_{i_1\cdots i_m}\omega_{\beta\alpha} . \end{aligned}$$

From the above equation we have the following Ricci formula;

$$\begin{aligned} h^\alpha_{i_1\cdots i_m jk} - h^\alpha_{i_1\cdots i_m kj} = &\sum_{i=1}^m \textstyle\sum_l h^\alpha_{i_1\cdots i_{r-1} l i_{r+1}\cdots i_m} R_{i_r ljk} \\ &- \textstyle\sum_\beta h^\beta_{i_1\cdots i_m} R_{\beta\alpha jk} , \end{aligned}$$

which is reduced to

$$(1.27) \qquad \begin{aligned} h^\alpha_{i_1\cdots i_m jk} - h^\alpha_{i_1\cdots i_m kj} = \tilde{c} &\sum_{r=1}^m (h^\alpha_{i_1\cdots i_{r-1} j i_{r+1}\cdots i_m}\delta_{i_r k} \\ &\qquad - h^\alpha_{i_1\cdots i_{r-1} k i_{r+1}\cdots i_m}\delta_{i_r j}) \\ + \sum_{r=1}^m \textstyle\sum_{\beta,l} & h^\alpha_{i_1\cdots i_{r-1} l i_{r+1}\cdots i_m}(h^\beta_{i_r k}h^\beta_{lj} - h^\beta_{i_r j}h^\beta_{lk}) \\ + \textstyle\sum_{\beta,l} & h^\beta_{i_1\cdots i_m}(h^\alpha_{lk}h^\beta_{lj} - h^\alpha_{lj}h^\beta_{lk}) . \end{aligned}$$

2. Submanifolds in a space form

Let M be an $n(\geqq 3)$-dimensional Riemannian submanifold immersed in an $(n + p)$-dimensional space form $\tilde{M} = M^{n+p}(\tilde{c})$ of constant curvature $\tilde{c}$. Let S be the square of the length of the second fundamental form σ.

Then, using the notation in § 1, we can write the invariant S on M as follows:

$$S = \|\sigma\|^2 = \textstyle\sum_{\alpha,i,j} h_{ij}^\alpha h_{ij}^\alpha \ . \tag{2.1}$$

Furthermore we can consider the following invariant on M:

$$L_N = \textstyle\sum_{\alpha,\beta,i,j,k,l} h_{ij}^\alpha h_{kl}^\alpha h_{ij}^\beta h_{kl}^\beta \ . \tag{2.2}$$

Now, we consider the following tensor on M:

$$H_{ijkl} = \textstyle\sum_\alpha h_{ij}^\alpha h_{kl}^\alpha - \lambda_1 \delta_{ij}\delta_{kl} - \lambda_2(\delta_{ik}\delta_{jl} + \delta_{il}\delta_{jk}) \ , \tag{2.3}$$

where

$$\lambda_1 = \{n^2(n+1)H^2 - 2S\}/n(n+2)(n-1)$$

and

$$\lambda_2 = (S - nH^2)/(n+2)(n-1) \ .$$

Computing the square of its length, we have

$$\textstyle\sum_{i,j,k,l} H_{ijkl}H_{ijkl} = L_N + \dfrac{4nH^2S - 2S^2 - n^3(n+1)H^4}{(n+2)(n-1)} \ . \tag{2.4}$$

As is easily seen by the definition of the tensor H_{ijkl}, its vanishing means that the immersion is λ-isotropic and M is of constant curvature. Thus we find

Lemma 2.1. *If the following inequality*

$$L_N \leqq \frac{n^3(n+1)H^4 - 4nH^2S + 2S^2}{(n+2)(n-1)}$$

holds on M, then the immersion is λ-isotropic and M is of constant curvature, where $\lambda^2 = (n^2H^2 + 2S)/n(n+2)$.

We define a positive semi-definite symmetric matrix A of order p at each point on M by $A = (A_\beta^\alpha)$, $A_\beta^\alpha := \sum_{i,j} h_{ij}^\alpha h_{ij}^\beta$. A corresponds to a symmetric linear transformation on the tangent space which is positive semi-definite. It is trivial that $A = 0$ at a point if and only if the point is a geodesic one. Furthermore, suppose that the rank of A is equal to 1. Then we can choose an orthonormal frame $\{e_1, \cdots, e_{n+p}\}$ at the point such that $h_{ij}^{n+1} = H\delta_{ij}$ and $h_{ij}^\alpha = 0$ for any i, j and $\alpha > n+1$, which means that the point is umbilical. As for this matrix, we have

Lemma 2.2. *If the matrix A has exactly two distinct eigenvalues $\mu(\neq 0)$ and 0 at each point on M, then the following assertions are valid:*

(1) *The rank of A is not greater than $n(n+1)/2$ or $n(n+1)/2-1$, according as $H \neq 0$ or $H = 0$.*

(2) *If the rank of A attains the maximal value $n(n+1)/2$ or $n(n+1)/2-1$ at a certain point, then M is isotropic at that point.*

Proof. Suppose that the multiplicity of non-zero eigenvalue μ is q. It follows from (2.4) that we have

$$qn^3(n+1)H^4 - 4qnH^2S + \{2q-(n+2)(n-1)\}S^2 \leqq 0\,,$$

because the definition of A implies $S = \operatorname{Tr} A$ and $L_N = \operatorname{Tr} A^2$. The left hand side of the above inequality is regarded as a quadratic polynomial in S. If the coefficient of S^2 is positive, then the discriminant must be non-negative, which implies the assertion (1).

In particular, the rank of A attains the maximum at a point if and only if the equality holds in the above inequality at that point, from which we can apply Lemma 2.1 to this case. Q.E.D.

We give a simple application of Lemma 2.1. Lemma 2.2 will play a little more important rôle in the forthcoming paper, in which we shall study a certain isotropic immersion of M into $\tilde{M} = M^{n+p}(\tilde{c})$ with degree 3. It follows from the relation between S and L_N that we have $S^2 \geqq L_N$, where the equality holds if and only if A is a zero matrix or it has exactly two distinct eigenvalues μ and 0, and the multiplicity of μ is equal to 1. This is equivalent to

$$\{n^2(n+1)H^2 + (n^2+n-4)S\}(nH^2 - S) \leqq 0\,.$$

Since H^2 and S are non-negative, from the above inequality we have $nH^2 - S \leqq 0$. Then, as a direct consequence of Lemma 2.1, Gauss' equation and the property of A, we obtain

Proposition 2.3. *Let M be an n-dimensional Riemannian submanifold immersed in an $(n+p)$-dimensional space form of constant curvature $\tilde{c}$. If the inequality*

$$n(n-1)(H^2+\tilde{c}) \leqq R \quad \text{or} \quad (n-1)S \leqq R - n(n-1)\tilde{c}$$

holds on M, then M is totally geodesic or totally umbilical.

3. Parallel sections

From now on let $M = M^n(c)$ be an n-dimensional submanifold of constant curvature c which is immersed in an $(n+p)$-dimensional space form $\tilde{M} = M^{n+p}(\tilde{c})$ of constant curvature $\tilde{c}$. We assume that the immersion is λ-isotropic with constant λ. Furthermore we may suppose that λ is positive, because if $\lambda = 0$, M is totally geodesic. In this section, we derive several invariants from the above situation and show the existence of a parallel section in the normal bundle of M. First of all, Gauss' equation gives

$$(3.1)\quad \sum_\alpha h^\alpha_{il}h^\alpha_{jk} - \sum_\alpha h^\alpha_{ik}h^\alpha_{jl} = C(\delta_{il}\delta_{jk} - \delta_{ik}\delta_{jl})\ , \qquad C = c - \tilde{c}\ .$$

Since the immersion is λ-isotropic and M is of constant curvature c, from (1.17) and (3.1) we have

$$(3.2)\quad \sum_\alpha h^\alpha_{ij}h^\alpha_{kl} = \tfrac{1}{3}\{(\lambda^2 + 2C)\delta_{ij}\delta_{kl} + (\lambda^2 - C)(\delta_{ik}\delta_{jl} + \delta_{il}\delta_{jk})\}\ .$$

Since λ is constant, differentiating (3.2) covariantly, we have

$$\sum_\alpha h^\alpha_{ij}h^\alpha_{klm} + \sum_\alpha h^\alpha_{ijm}h^\alpha_{kl} = 0\ .$$

Thus we get

$$\begin{aligned}\sum_\alpha h^\alpha_{ij}h^\alpha_{klm} &= -\sum_\alpha h^\alpha_{ijm}h^\alpha_{kl} = \sum_\alpha h^\alpha_{im}h^\alpha_{jkl} = -\sum_\alpha h^\alpha_{ikm}h^\alpha_{jl}\\ &= \sum_\alpha h^\alpha_{km}h^\alpha_{ijl} = -\sum_\alpha h^\alpha_{klm}h^\alpha_{ij}\ ,\end{aligned}$$

because h^α_{ijk} is symmetric with respect to all indices i, j and k. This implies

$$(3.3)\qquad \sum_\alpha h^\alpha_{ij}h^\alpha_{klm} = 0 \qquad \text{for any } i, j, k, l \text{ and } m\ .$$

On the other hand, it is easily seen that H and S satisfy

$$(3.4)\qquad S = \|\sigma\|^2 = n^2H^2 - n(n-1)C\ ,$$

$$(3.5)\qquad 3nH^2 = (n+2)\lambda^2 + 2(n-1)C\ ,$$

and moreover we have

$$(3.6)\qquad \sum_{\alpha,k} h^\alpha_{ij}h^\alpha_{kk} = nH^2\delta_{ij}\ ,$$

which implies that the immersion is pseudo-umbilical. Since such minimal submanifolds in $\tilde{M}$ have been determined in [1, 7], we may suppose that H is a positive constant. Then, we can choose an orthonormal local frame field $\{e_{n+1}, \cdots, e_{n+p}\}$ such that $e_{n+1} = \mathfrak{h}/H$, where $\mathfrak{h}$ is the mean curvature vector, so we get

$$(3.7)\qquad h^{n+1}_{ij} = H\delta_{ij}\ , \quad \sum_k h^\alpha_{kk} = 0 \qquad \text{for } \alpha > n+1\ .$$

It follows from (3.3) and (3.7) that we have

$$(3.8)\quad h^{n+1}_{ijk} = 0, \text{ i.e., } \sum_{\alpha>n+1} h^\alpha_{ijk}h^\alpha_{lm} = 0 \qquad \text{for any } i, j, k, l \text{ and } m\ .$$

From (1.24) and (3.7) we have $\sum_k h^{n+1}_{ijk}\omega_k = -\sum_{\beta>n+1} h^\beta_{ij}\omega_{\beta n+1}$ for any i and j, which together with (3.8) implies that

$$\sum_{\beta>n+1} h^{\beta}_{ij}\omega_{\beta n+1} = 0 .$$

Multiplying this by $h^{\alpha}_{ij}(\alpha > n+1)$ and summing up over indices i and j, we have $\sum_{\beta>n+1} A^{\alpha}_{\beta}\omega_{\beta n+1} = 0$ for $\alpha > n+1$. We have from (1.24) and (3.7)

$$nH\omega_{\beta n+1} = \sum_{j,m} h^{\beta}_{mmj}\omega_j \qquad \text{for } \beta > n+1 . \tag{3.9}$$

Making use of these equations, we can prove the following

Lemma 3.1. *If M is not minimal, then the mean curvature vector $\mathfrak{h}$ is parallel in the normal bundle $N(M)$ and it is isoperimetric.*

Proof. In order to prove this lemma, by (1.23) and the definition of parallel sections we have only to show $\omega_{\beta n+1} = 0$ or $\sum_m h^{\beta}_{mmj} = 0$ for $\beta > n+1$, because H is non-zero constant.

We can now choose an orthonormal frame $\{\hat{e}_{n+2}, \cdots, \hat{e}_{n+p}\}$ such that $\{e_{n+1}, \hat{e}_{n+2}, \cdots, \hat{e}_{n+p}\}$ is an orthonormal frame satisfying (3.7) and $\hat{A}^{\alpha}_{\beta} = 0$ if $\alpha \neq \beta$, $\alpha, \beta > n+1$. Then, a matrix $\hat{A}$ defined by $\hat{A} = (\hat{A}^{\alpha}_{\beta})$ is symmetric, positive semi-definite and of order $p-1$ and it has two eigenvalues $2(\lambda^2 - C)/3$ and 0. Suppose that there is an umbilical point on M. Then $\hat{A}$ becomes a zero matrix at the point, so M is totally umbilical, because $\lambda^2 - C$ is constant. Hence the assertion is proved. Thus we may suppose that M has no umbilical points. In this case, A has exactly two distinct eigenvalues and the multiplicity of the non-zero eigenvalue is $(n+2)(n-1)/2$. We have also

$$\hat{A}^{\alpha}_{\alpha}\hat{\omega}_{\alpha n+1} = 0 \qquad \text{for } \alpha > n+1 .$$

We easily see that if $\hat{\omega}_{\alpha n+1} = 0$ for $\alpha > n+1$, then $\omega_{\alpha n+1} = 0$ for $\alpha > n+1$. Hence we may prove that $\hat{\omega}_{\alpha n+1} = 0$ for any $\alpha > n+1$. Then we can consider the following two cases: $\hat{A}^{\alpha}_{\alpha} = 0$, i.e., $h^{\alpha}_{ij} = 0$ for any indices i and j or $\hat{\omega}_{\alpha n+1} = 0$, i.e., $\sum_m h^{\alpha}_{mmj} = 0$ for any index j.

Let I be the set of all indices $\alpha > n+1$. Furthermore we consider the subsets I_1 and I_2 defined by $I_1 = \{\alpha \in I : h^{\alpha}_{ij} = 0 \text{ for } i \text{ and } j\}$ and $I_2 = \{\alpha \in I : \sum_m h^{\alpha}_{mmj} = 0 \text{ for } j\}$, and we put $I_3 := I_1 - I_1 \cap I_2$ and $I_4 := I_2 - I_1 \cap I_2$.

We consider here a fixed number α in I_3. Then equations $h^{\alpha}_{ij} = 0$ and $\sum_m h^{\alpha}_{mmk} \neq 0$ for any indices i, j and k hold at a point x_0 of M. Consequently, there exists a neighborhood $U(x_0, \alpha)$ of x_0, in which we have $dh^{\alpha}_{ij} = 0$ for i and j. Because $d\omega_{\alpha i} = d(\sum_j h^{\alpha}_{ij}\omega_j) = 0$, we get

$$H\omega_{n+1\alpha} \wedge \omega_i + \sum_{\beta\in I_4} \sum_k h^{\beta}_{ik}\omega_{\beta\alpha} \wedge \omega_k = 0 ,$$

which, together with (3.9), implies

$$\sum_{j,m} h^\alpha_{mmj}\omega_j \wedge \omega_i - n \sum_{\beta\in I_4} \sum_k h^\beta_{ik}\omega_{\beta\alpha} \wedge \omega_k = 0 \quad \text{for } \alpha \in I_3 . \tag{3.10}$$

On the other hand, we develop a discussion for a fixed number $\beta \in I_4$ similar to that proceeded above. Then we can choose a neighborhood $V(x_0, \beta)$ of x_0, in which we have $h^\beta_{ij} \neq 0$ and $\sum_m h^\beta_{mmk} = 0$ for any i, j and k. Since $d(\sum_m h^\alpha_{mmk}) = 0$ at x_0, from (1.26) for $m = 3$, we have

$$\sum_{\alpha\in I_3} \sum_m h^\alpha_{mmi}\, \omega_{\beta\alpha} = \sum_{m,j} h^\beta_{mmij}\omega_j \qquad \text{for } \beta \in I_4 .$$

Multiplying (3.9) by $\sum_n h^\alpha_{nnl}$ and summing up over indices α, by the above equation we have

$$\sum_{\alpha\in I_3} \sum_{m,p} h^\alpha_{mmj} h^\alpha_{ppl} + \sum_{\beta\in I_4} \sum_{i,m} h^\beta_{ij} h^\beta_{mmli} = 0 \qquad \text{for } j \text{ and } l .$$

Taking account of (3.3) and the above equation, we can show that

$$\sum_m h^{n+1}_{mmij} = 0 \qquad \text{for } i \text{ and } j \text{ at } x_0 .$$

(1.26) for $m = 3$ and this equation mean that $\sum_m h^\alpha_{mmj}$ vanishes for $\alpha \in I_3$, which contradicts to the assumption of I_3.

Thus the subset I_3 must be empty, and the assertion is proved. Q.E.D.

Since the mean curvature vector $\mathfrak{h}$ is parallel with respect to the normal connection and the mean curvature H is non-zero constant on M, it can be shown by means of a theorem of Smyth [5] that M must be minimally immersed into a totally umbilical hypersurface $\bar{M} = M^{n+p-1}(\bar{c})$ of $\tilde{M}$ with constant curvature $\bar{c} = \tilde{c} + H^2$ which is orthogonal to the mean curvature vector $\mathfrak{h}$, if c is positive. The positive semi-definiteness of the matrix $\hat{A}$ and the relation (3.5) between the mean curvature and the isotropy constant imply that $H^2 + \tilde{c} \geqq c$. This means that the hypersurface $\bar{M}$ is congruent to a (small) hypersphere in a Euclidean space, a sphere or a hyperbolic space. When we regard the immersion ι of M into $\tilde{M}$ as that of M into $\bar{M}$, we can easily verify that the immersion of a space form M into a space form $\bar{M}$ is also isotropic, whose isotropy constant is equal to $\lambda^2 - H^2 = 2(n-1)(\lambda^2 - C)/3n$. Thus, in conclusion, we can mention that it suffices to study a λ-isotropic and minimal immersion of $M^n(c)$ into $M^{n+p-1}(\bar{c})$.

Next, we give several invariants for the next section. The Ricci formula (1.27) implies

$$h^\alpha_{ijkl} - h^\alpha_{ijlk} = -\sum_m h^\alpha_{mi} R_{mjkl} - \sum_m h^\alpha_{mj} R_{mikl} - \sum_\beta h^\beta_{ij} R_{\beta\alpha kl} .$$

Using (3.2), from the above equality we have

$$(3.11)\quad \begin{aligned} &h^\alpha_{ijkl} - h^\alpha_{ijlk} \\ &= \tfrac{1}{3}(\tilde{c} - 4c + \lambda^2)(h^\alpha_{il}\delta_{jk} - h^\alpha_{ik}\delta_{jl} + h^\alpha_{jl}\delta_{ik} - h^\alpha_{jk}\delta_{il})\,, \end{aligned}$$

which implies

$$\textstyle\sum_k h^\alpha_{ijkk} = \sum_k h^\alpha_{kkij} - \tfrac{1}{3}(4c - \tilde{c} - \lambda^2)(\sum_k h^\alpha_{kk}\delta_{ij} - nh^\alpha_{ij})\,.$$

Hence, setting $S_2 = \sum_{\alpha>n+1}\sum_{i,j} h^\alpha_{ij}h^\alpha_{ij}$ and $S_3 = \sum_{\alpha>n+1}\sum_{i,j,k} h^\alpha_{ijk}h^\alpha_{ijk} = \|\nabla'\sigma\|^2$, we have

$$(3.12)\quad \begin{aligned} S_3 &= -\frac{n}{3}(4c - \tilde{c} - \lambda^2)S_2 \\ &= \frac{n^2}{n+2}\Big(H^2 + \tilde{c} - \frac{2(n+1)}{n}c\Big)S_2\,, \end{aligned}$$

because S_2 is constant. Hence, if $S_3 = 0$, then $\nabla'\sigma = 0$, that is, the second fundamental form is parallel.

By a discussion similar to that developed above, we have the next invariant S_4 defined by $S_4 = \sum_{\alpha>n+1}\sum_{i,j,k,l} h^\alpha_{ijkl}h^\alpha_{ijkl} = \|\nabla'^2\sigma\|^2$. Using the Ricci formula (1.27) and calculating directly, though a little complicated, we can express $h^\alpha_{ijklm} - h^\alpha_{mlijk}$ as a polynomial in h^α_{ijk} and the Kronecker delta, from which we have

$$\frac{1}{2}\Delta S_3 = S_4 + n\Big\{\frac{3(n+1)}{n}c - \tilde{c} - H^2\Big\}S_3\,.$$

Since S_3 is constant, we see

$$(3.13)\quad S_4 = n\Big(H^2 + \tilde{c} - \frac{3(n+1)}{n}c\Big)S_3\,.$$

4. The decomposition of the normal space

In this section, we consider the decomposition of the normal space of the immersion ι of $M = M^n(c)$ into $\tilde{M} = M^{n+p}(\tilde{c})$. At any point x in M we have the normal space $N_x(M) = (\iota_*(T_x(M)))^N$, where $(\ \ )^N$ is the operator taking the orthogonal complement in the tangent space $T_{\iota(x)}(\tilde{M})$ of the ambient space. The second fundamental form σ at x is a bilinear map of $T_x(M) \times T_x(M)$ into $N_x(M)$ defined by $(\tilde{\nabla}_X Y)^N$, where X and Y are local vector fields on a neighborhood of x in M. For convenience' sake, σ_x is regarded as a linear map of a symmetric square $S^2(M_x)$ of $T_x(M)$ into the normal space $N_x(M)$. We put $N^1_x(M) = \sigma_x(S^2(M_x))$, which is called the *first normal space* of ι at x. This means that $N^1_x(M)$ is the linear subspace of $N_x(M)$ spanned by vectors h_{ij} with components $(h_{ij}^{n+1}, \cdots, h_{ij}^{n+p})$ for any indices i and j, so we see that the dimension of $N_x^1(M)$ is not greater than $n(n+1)/2$. If ι is minimal, then $\dim N_x^1(M) \leqq (n+2)(n-1)/2$. By means of (3.2), we have

$$\sum_\beta A_\beta^\alpha A_\gamma^\beta = \tfrac{2}{3}(\lambda^2 - C)A_\gamma^\alpha + \tfrac{1}{3}(\lambda^2 + 2C) \operatorname{Tr} H^\alpha \operatorname{Tr} H^\gamma ,$$

where H^α is a symmetric matrix of order n defined by $H^\alpha = (h_{ij}^\alpha)$ for any fixed index α. Thus the linear transformation A of $N_x^1(M)$ has at most three distinct eigenvalues 0, $2(\lambda^2 - C)/3$ and nH^2. In particular, the multiplicity of the last eigenvalue is 1, because we may regard $X = (\operatorname{Tr} H^\alpha)$ as a vector in a p-dimensional vector space and it follows from (3.6) that $AX = nH^2X$. Let r be the multiplicity of the eigenvalue $2(\lambda^2 - C)/3$. Then we have $\operatorname{Tr} A = n\{2r(H^2 - C) + (n+2)H^2\}/(n+2)$. On the other hand, $\operatorname{Tr} A$ is easily calculated by (3.2) and (3.4), and we have $\operatorname{Tr} A = n\{(nH^2 + 2C) + n(n+1)(H^2 - C)\}/(n+2)$. These relations imply $r = (n+2)(n-1)/2$. Thus we find

Lemma 4.1. *The dimension of the first normal space $N_x(M)$ is $(n+2)(n-1)/2$ or $n(n+1)/2$, according as $H = 0$ or $H \neq 0$.*

Now, we put $O_x^1(M) = \iota_*(T_x(M)) \oplus N_x^1(M)$ for any point x of M, which is called the *first osculating space* of ι at x. We define a trilinear map τ of $T_x(M) \times T_x(M) \times T_x(M)$ into $(O_x^1(M))^N$ by

$$\tau_x(X, Y, Z) = (\tilde{\nabla}_X(\sigma(Y, Z)))^{N_1}$$

for any tangent vectors X, Y and Z, where $(\ \ )^{N_1}$ denotes the orthogonal projection into $(O_x^1(M))^N$. Then τ is well-defined and is symmetric, because M is of constant curvature, so it induces a linear map $\tau_x : S^3(T_x(M)) \to (O_x^1(M))^N$ of the symmetric third power of $T_x(M)$ into the orthogonal complement of the first osculating space. The linear space $N_x^2(M) = \tau_x(S^3(T_x(M)))$ is called the *second normal space* of ι at x and the linear space $O_x^2(M) = O_x^1(M) \oplus N_x^2(M)$ is called the *second osculating space* of ι at x. Then $N_x^2(M)$ is the linear subspace of $(O_x^1(M))^N$ spanned by vectors h_{ijk} with components $(h_{ijk}^{n+1}, \cdots, h_{ijk}^{n+p})$ for any indices i, j and k, because of (3.3). So we see that the dimension of $N_x^2(M)$ is not greater than $n(n+1)(n+2)/6$. In particular, if ι is minimal, then $\dim N_x^2(M) \leqq n(n+4)(n-1)/6$.

We here assume that the mean curvature H satisfies

$$0 < H^2 \leqq \frac{3(n+2)}{n} c - \tilde{c} . \tag{4.1}$$

Then we can choose a local orthonormal frame field $\{e_{n+1}, \cdots, e_{n+p}\}$

satisfying (3.7). Making use of equalities obtained in § 3, we see that if $S_2 = 0$, then $h^\alpha_{ij} = 0$ for i, j and $\alpha > n + 1$, so hence M is totally umbilical. Accordingly we may suppose that S_2 is a positive constant. We consider the following tensor:

$$H^\alpha_{ijkl} = h^\alpha_{ijkl} + \frac{S_3}{n(n+4)S_2}\{(n+2)(h^\alpha_{ij}\delta_{kl} + h^\alpha_{jk}\delta_{il} + h^\alpha_{ki}\delta_{jl}) - 2(h^\alpha_{kl}\delta_{ij} + h^\alpha_{il}\delta_{jk} + h^\alpha_{jl}\delta_{ki})\} \tag{4.2}$$

for $\alpha > n + 1$. Then, using (3.13) and (4.2), we have

$$\begin{aligned} \textstyle\sum_{\alpha>n+1} \sum_{i,j,k,l} H^\alpha_{ijkl} H^\alpha_{ijkl} &= S_4 - \frac{3(n+2)S_3^2}{n(n+4)S_2} \\ &= \frac{n(n+1)}{n+4}\left\{H^2 + \tilde{c} - \frac{3(n+2)}{n}c\right\}S_3 \,. \end{aligned} \tag{4.3}$$

By (4.3) and a theorem in [3], we can easily prove the following

Lemma 4.2. *If H satisfies* (4.1), *then we have the following*:

(1) $H^2 = \dfrac{k(n+k-1)}{n}c - \tilde{c}\,, \qquad k = 1, 2, 3\,.$

(2) *If $k = 1$, then ι is totally umbilical, the degree of ι is equal to* 2 *and the dimension of the first normal space is equal to* 1.

(3) *If $k = 2$, then ι is parallel, the degree of ι is equal to* 2 *and the dimension of the first normal space is $n(n+1)/2$. In particular, if ι is full, then $p = n(n+1)/2$.*

Now, we consider the case $k = 3$. From (1.16) and (3.7) we have

$$\textstyle\sum_{\alpha>n+1} (h^\alpha_{ij}h^\alpha_{kl} + h^\alpha_{jk}h^\alpha_{il} + h^\alpha_{ki}h^\alpha_{jl}) = (\lambda^2 - H^2)(\delta_{ij}\delta_{kl} + \delta_{jk}\delta_{il} + \delta_{ik}\delta_{jl})\,.$$

It follows from (3.5), (3.12), (4.2) and the assumption of H that we have

$$h^\alpha_{ijkl} = \frac{c}{n+2}\{2(\delta_{ij}h^\alpha_{kl} + \delta_{jk}h^\alpha_{il} + \delta_{ik}h^\alpha_{jl}) - (n+2)(h^\alpha_{ij}\delta_{kl} + h^\alpha_{jk}\delta_{il} + h^\alpha_{ik}\delta_{jl})\} \tag{4.5}$$

for $\alpha > n + 1$. Since $h^{n+1}_{ijk} = 0$ and $h^{n+1}_{ijkl} = 0$ for i, j, k, and l, we have

$$-\textstyle\sum_{\alpha>n+1} h^\alpha_{ijk}h^\alpha_{lmn} = \dfrac{c}{n+2}\sum_\alpha \{2(\delta_{ij}h^\alpha_{kl} + \delta_{jk}h^\alpha_{il} + \delta_{ik}h^\alpha_{jl})h^\alpha_{mn} - (n+2)(h^\alpha_{ij}\delta_{kl} + h^\alpha_{jk}\delta_{il} + h^\alpha_{ik}\delta_{jl})h^\alpha_{mn}\}\,. \tag{4.6}$$

On the other hand, (3.2) is reduced to

$$(4.7)\quad \sum_{\alpha>n+1} h^\alpha_{ij}h^\alpha_{kl} = \frac{2(n+3)}{n(n+2)}c\{-2\delta_{ij}\delta_{kl} + n(\delta_{ik}\delta_{jl} + \delta_{il}\delta_{jk})\}\,.$$

Substituting (4.7) into (4.6), we have

$$\begin{aligned}
&\sum_{\alpha>n+1} h^\alpha_{ijk}h^\alpha_{lmn}\\
&\quad = -\frac{2(n+3)}{(n+2)^2}[-2(\delta_{ij}\delta_{kl} + \delta_{jk}\delta_{il} + \delta_{ki}\delta_{jl})\delta_{mn}\\
(4.8)\quad &\quad + 2\delta_{ij}(\delta_{km}\delta_{ln} + \delta_{kn}\delta_{lm}) + \delta_{jk}(\delta_{im}\delta_{ln} + \delta_{in}\delta_{lm})\\
&\quad + \delta_{ki}(\delta_{jm}\delta_{ln} + \delta_{jn}\delta_{lm})\\
&\quad - (n+2)\{\delta_{kl}(\delta_{im}\delta_{jn} + \delta_{in}\delta_{jm}) + \delta_{il}(\delta_{jm}\delta_{kn} + \delta_{jn}\delta_{km})\\
&\quad + \delta_{jl}(\delta_{km}\delta_{in} + \delta_{kn}\delta_{im})\}]\,.
\end{aligned}$$

We consider now the dimension of the second normal space $N^2_x(M)$ at any point x of M. The tangent space $T_x(M)$ at x to M is the n-dimensional vector space over $\boldsymbol{R}$ and the tangent space $T_{\iota(x)}(\tilde{M})$ at $\iota(x)$ to $\tilde{M}$ is the $m(=n+p)$-dimensional vector space over $\boldsymbol{R}$. Furthermore the normal space $N_x(M)$ at x to M is the $(m-n)$-dimensional vector space, and the second fundamental form σ is the symmetric bilinear map from $T_x(M) \times T_x(M)$ into $N_x(M)$ which satisfies the property (1.17), because the immersion is λ-isotropic. Since the immersion is pseudo-umbilical and not minimal, we get the symmetric bilinear map $\tilde{\sigma}$ of $T_x(M) \times T_x(M)$ into the normal space $\tilde{N}_x(M)$ of M, which is the orthogonal complement of $\mathfrak{h}$ in $N_x(M)$ and of $(p-1)$-dimensional, as follows;

$$\tilde{\sigma}(X, Y) = \sigma(X, Y) - \langle X, Y\rangle\mathfrak{h} \qquad \text{for } X \text{ and } Y \in T_x(M)\,.$$

Then, we easily see from (4.7) that $\tilde{\sigma}$ satisfies

$$\begin{aligned}
&\sum_{i=1}^n \tilde{\sigma}(e_i, e_i) = 0\,,\\
(4.9)\quad &\langle\tilde{\sigma}(X, Y), \tilde{\sigma}(W, Z)\rangle = \frac{2(n+3)}{n(n+2)}c\{-2\langle X, Y\rangle\langle W, Z\rangle\\
&\qquad + n(\langle X, W\rangle\langle Y, Z\rangle + \langle X, Z\rangle\langle Y, W\rangle)\}
\end{aligned}$$

for any vectors X, Y, W and Z of $T_x(M)$, where $\langle\ ,\ \rangle$ is the inner product. The third fundamental form τ is the symmetric trilinear map of $T_x(M) \times T_x(M) \times T_x(M)$ into $(O^1_x(M))^N$, and $\langle\tau(e_i, e_j, e_k), \tau(e_l, e_m, e_n)\rangle$ is equal to the right hand side of (4.8). This means $\dim N^2_x(M) = {}_{n+2}C_3 - n$. Thus we have

Lemma 4.3. *The dimension of the second normal space is* $n(n+4)(n-1)/6$.

Let $F(M)$ and $F(\tilde{M})$ be the orthonormal frame bundles over M and $\tilde{M}$.

Let $\{\tilde{\omega}_A\}$ and $\{\tilde{\omega}_{AB}\}$ be the canonical 1-forms and the connection 1-forms on $F(\tilde{M})$, and let $F(\tilde{M}, M)$ be the set of frames $\{x, e_1, \cdots, e_n, e_{n+1}, \cdots, e_{n+p}\}$ such that $\{x, e_1, \cdots, e_n\} \in F(M)$ and $\{\iota(x), \iota_*(e_1), \cdots, \iota_*(e_n), e_{n+1}, \cdots, e_{n+p}\} \in F(\tilde{M})$. Then $F(\tilde{M}, M)$ becomes naturally a differentiable manifold and $\pi: F(\tilde{M}, M) \to M$ is considered as a principal bundle over M with the group $GL(n, \boldsymbol{R}) \times GL(p, \boldsymbol{R})$ in a natural way. Taking account of these frame bundles, we can prove the following

Lemma 4.4. *If H satisfies the equality of* (4.1) *and if ι is full, then* $p = n(n^2 + 6n - 1)/2$.

Proof. Using the matrix $A = (A^\alpha_\beta)$ we define a map f_2 of $N_x(M) \times N_x(M)$ into the reals $\boldsymbol{R}$ by $f_2(X, Y) = \sum_{\alpha,\beta} A^\alpha_\beta x^\alpha y^\beta$, where $X = \sum_\alpha x^\alpha e_\alpha$ and $Y = \sum_\alpha y^\alpha e_\alpha$. Let $S(p)$ be the set of all symmetric matrices of order p, which is considered as a vector space over $\boldsymbol{R}$. The orthogonal group $O(p, \boldsymbol{R})$ acts on $S(p)$ as follows: for any symmetrix matrix B and any orthogonal matrix C, $C(B) = {}^tCBC$. Since the matrix A is invariant under this action, the map f_2 is well-defined and is a positive definite symmetric form of rank $p_1 = n(n + 1)/2$, so that it can be normalized as

$$f_2(X, Y) = \lambda_{n+1}(x^{n+1})^2 + \cdots + \lambda_{n+p_1}(x^{n+p_1})^2 .$$

This means that we can choose a new orthonormal frame $\{e_i, e_{\alpha_1}, e_\beta\}$ at x such that

$$\text{(4.10)} \qquad \begin{aligned} &\omega_{i\alpha_1} \neq 0 , \qquad \omega_{i\beta} = 0 \\ &\qquad \text{for } n + 1 \leqq \alpha_1 \leqq n + p_1,\ n + p_1 + 1 \leqq \beta \leqq n + p . \end{aligned}$$

We take the orthonormal frame $\{e_i, e_{\alpha_1}, e_\beta\}$ satisfying (4.10). Similarly, we consider a matrix $B = (B^\alpha_\beta)$ of order p defined by $B^\alpha_\beta = \sum_{i,j,k} h^\alpha_{ijk} h^\beta_{ijk}$. Then B is also symmetric positive semi-definite, and invariant under $O(p, \boldsymbol{R})$. This implies that a map f_3 of $N_x(M) \times N_x(M)$ into $\boldsymbol{R}$ defined by $f_3(X, Y) = \sum_{\alpha,\beta} B^\alpha_\beta x^\alpha y^\beta$, where $X = \sum_\alpha x^\alpha e_\alpha$ and $Y = \sum_\alpha y^\alpha x_\alpha$, is well-defined and is also a positive semi-definite symmetric form of rank $p_2 = n(n + 4)(n - 1)/6$, because of Lemma 4.3. Therefore we can also choose a new orthonormal frame $\{e_i, e_{\alpha_1}, e_{\alpha_2}, e_\beta\}$ at x such that

$$\omega_{\alpha_1\alpha_2} \neq 0 , \qquad \omega_{\alpha_1\beta} = 0 ,$$

for $n + 1 \leqq \alpha_1 \leqq n + p_1$, $n + p_1 + 1 \leqq \alpha_2 \leqq n + p_1 + p_2$, $\beta \geqq n + p_1 + p_2 + 1$. We restrict our consideration on the subset $F'(\tilde{M}, M)$ of $F(\tilde{M}, M)$ consisting of the orthonormal frames $\{x, e_i, e_{\alpha_1}, e_{\alpha_2}, e_\beta\}$ satisfying

$$\text{(4.11)} \qquad \begin{aligned} &\omega_{i\alpha_1} \neq 0 , \quad \omega_{i\beta} = 0 , && \beta \geqq n + p_1 + 1 , \\ &\omega_{\alpha_1\alpha_2} \neq 0 , \quad \omega_{\alpha_1\beta} = 0 , && \beta \geqq n + p_1 + p_2 + 1 , \\ &\omega_{\alpha_2\beta} = 0 , && \beta \geqq n + p_1 + p_2 + 1 . \end{aligned}$$

In fact, we can choose such an orthonormal frame, because the immersion is with deg $= 3$ and h^{α}_{ijkl} is a polynomial in h^{α}_{ij} and δ_{ij}.

Now, we consider the following Pfaffian equations on $F(\tilde{M})$:

$$\begin{aligned} &\tilde{\omega}_{\beta} = 0\,, \quad \tilde{\omega}_{i\beta} = 0\,, \quad \tilde{\omega}_{\alpha_1\beta} = 0\,, \quad \tilde{\omega}_{\alpha_2\beta} = 0\,, \\ &\beta \geqq n + p_1 + p_2 + 1\,, \qquad n + 1 \leqq \alpha_1 \leqq n + p_1\,, \\ &n + p_1 + 1 \leqq \alpha_2 \leqq n + p_1 + p_2\,. \end{aligned} \tag{4.12}$$

We denote by $(*)$ the set of 1-forms of the left hand sides of (4.12). Then the structure equations of $\tilde{M}$ give

$$\begin{aligned} &d\tilde{\omega}_{\beta} + \textstyle\sum_j \tilde{\omega}_{\beta j} \wedge \tilde{\omega}_j + \sum_{\alpha_1} \tilde{\omega}_{\beta\alpha_1} \wedge \tilde{\omega}_{\alpha_1} + \sum_{\alpha_2} \tilde{\omega}_{\beta\alpha_2} \wedge \tilde{\omega}_{\alpha_2} \\ &\qquad + \textstyle\sum_{\gamma} \tilde{\omega}_{\beta\gamma} \wedge \tilde{\omega}_{\gamma} \equiv 0 \qquad (\mathrm{mod}\ *)\,, \\ &d\tilde{\omega}_{i\beta} + \textstyle\sum_j \tilde{\omega}_{ij} \wedge \tilde{\omega}_{j\beta} + \sum_{\alpha_1} \tilde{\omega}_{i\alpha_1} \wedge \tilde{\omega}_{\alpha_1\beta} + \sum_{\alpha_2} \tilde{\omega}_{i\alpha_2} \wedge \tilde{\omega}_{\alpha_2\beta} \\ &\qquad + \textstyle\sum_{\gamma} \tilde{\omega}_{i\gamma} \wedge \tilde{\omega}_{\gamma\beta} \equiv 0 \qquad (\mathrm{mod}\ *)\,, \\ &d\tilde{\omega}_{\alpha_1\beta} + \textstyle\sum_j \tilde{\omega}_{\alpha_1 j} \wedge \tilde{\omega}_{j\beta} + \sum_{\beta_1} \tilde{\omega}_{\alpha_1\beta_1} \wedge \tilde{\omega}_{\beta_1\beta} + \sum_{\alpha_2} \tilde{\omega}_{\alpha_1\alpha_2} \wedge \tilde{\omega}_{\alpha_2\beta} \\ &\qquad + \textstyle\sum_{\gamma} \tilde{\omega}_{\alpha_1\gamma} \wedge \tilde{\omega}_{\gamma\beta} \equiv 0 \qquad (\mathrm{mod}\ *)\,, \\ &d\tilde{\omega}_{\alpha_2\beta} + \textstyle\sum_j \tilde{\omega}_{\alpha_2 j} \wedge \tilde{\omega}_{j\beta} + \sum_{\alpha_1} \tilde{\omega}_{\alpha_2\alpha_1} \wedge \tilde{\omega}_{\alpha_1\beta} + \sum_{\beta_2} \tilde{\omega}_{\alpha_2\beta_2} \wedge \tilde{\omega}_{\beta_2\beta} \\ &\qquad + \textstyle\sum_{\gamma} \tilde{\omega}_{\alpha_2\gamma} \wedge \tilde{\omega}_{\gamma\beta} \equiv 0 \qquad (\mathrm{mod}\ *)\,, \end{aligned}$$

where $\gamma \geqq n + p_1 + p_2 + 1$, $n + 1 \leqq \beta_1 \leqq n + p_1$ and $n + p_1 + 1 \leqq \beta_2 \leqq n + p_1 + p_2$, so (4.12) is completely integrable. Since $F'(\tilde{M}, M)$ satisfies (4.12), for any point b in $F'(\tilde{M}, M)$ we consider the integral submanifold $Q(b)$ of maximal dimension of the Pfaffian equations (4.12) through b, where $\dim Q(b) = (n + p)(n + p - 1)/2 - (n + p_1 + p_2)(p - p_1 - p_2 - 1)$. $Q(b)$ contains the part of $F'(\tilde{M}, M)$ about b and so it is easily seen that the image $\pi(Q(b))$ of $Q(b)$ under the projection π is an $(n + p_1 + p_2)$-dimensional totally geodesic submanifold $\bar{M}$ of $\tilde{M}$, because of the first and second equations of (4.12). Through each point $\iota(x)$, there exists a piece of a submanifold of this kind. Therefore there exists an $(n + p_1 + p_2)$-dimensional totally geodesic submanifold immersed in $\tilde{M}$ which contains M as a submanifold.

Thus the dimension of the normal space at any point must be equal to $p_1 + p_2 = n(n^2 + 6n - 1)/6$, because the immersion ι is full.

Q.E.D.

Any two $(n + q)$-dimensional totally umbilical submanifolds in $M^{n+p}(\tilde{c})$ with the same constant curvature are congruent to each other and any rotation of a totally umbilical submanifold can be extended to an isometry of the ambient space. By means of Lemmas 4.2 and 4.4, the isotropic immersion ι can be reduced to the full isotropic immersion of M into a totally geodesic submanifold in the ambient space. Furthermore, taking account of the property obtained in the previous section, we see that $\iota(M)$ lies minimally in a totally umbilical submanifold $M^{n+q}(\bar{c})$ of $M^{n+p}(\tilde{c})$ with curvature $\bar{c} = \tilde{c} + H^2 > 0$. Thus we obtain

Theorem 4.5. *Let M be an n-dimensional space form of constant curvature c and $\tilde{M}$ an $(n + p)$-dimensional space form of constant curvature $\tilde{c}$. Let ι be the non-zero constant isotropic immersion of M into $\tilde{M}$. If the mean curvature H satisfies*

$$0 < H^2 \leqq \frac{3(n + 2)}{n} c - \tilde{c} ,$$

then the immersion ι is congruent to the immersion $\iota_0 \circ \iota_r$ of an n-dimensional sphere $S^n(c)$ into $\tilde{M}$, where $r = 1, 2, 3$ and ι_0 is a totally umbilical immersion of $S^{N(r)}(\bar{c})$ into $\tilde{M}$, $N(r) = (n + 2r - 1)(n + r - 2)!/r!\,(n-1)!$, $\bar{c} = \tilde{c} + H^2$, and ι_r is the standard immersion of $S^n(c)$ into $S^{N(r)}(\tilde{c})$ mentioned in the introduction.

Bibliography

[1] M. P. do Carmo and N. R. Wallach, *Minimal immersions of spheres into spheres,* Ann. of Math., **93** (1971), 43–62.

[2] J. Erbacher, *Reduction on the codimension of an isometric immersion,* J. Differential Geometry, **5** (1971), 333–340.

[3] T. Itoh and K. Ogiue, *Isotropic immersions and Veronese manifolds,* Trans. Amer. Math. Soc., **209** (1975), 109–117.

[4] B. O'Neill, *Isotropic and Kaehler immersions,* Canad. J. Math., **17** (1965), 909–915.

[5] B. Smyth, *Submanifolds of constant mean curvature,* Math. Ann., **205** (1973), 265–280.

[6] T. Takahashi, *Minimal immersions of Riemannian manifolds,* J. Math. Soc. Japan, **18** (1966), 380–385.

[7] N. R. Wallach, *Minimal immersions of symmetric spaces into spheres,* Symmetric spaces, Dekker, New York, 1972, 1–40.

TOKYO UNIVERSITY
OF AGRICULTURE AND TECHNOLOGY
FUCHU, TOKYO, 183
JAPAN
UNIVERSITY OF TSUKUBA
IBARAKI, 300-31
JAPAN

Minimal Submanifolds and Geodesics
Kaigai Publications, Tokyo, 1978, 273–282

COMPACT RIEMANNIAN MANIFOLDS WHICH ARE ISOSPECTRAL TO THREE DIMENSIONAL LENS SPACES. I

MINORU TANAKA

It is an interesting problem to study what sort of riemannian structures are determined by the eigenvalues of the Laplacian Δ^M for a compact riemannian manifold M. This problem has been studied by many people, Berger [1], Colin de Verdière [2], Duistermaat-Guillemin [3], Makean-Singer, Sakai, Tanno [11] and so on. For example, two compact riemannian manifolds which are isospectral have the same volume and the same dimension. This fact is an easy conclusion from the asymptotic expansion of Minakshisundaram-Pleijel (see [1]). Here the spectrum of M, Spec (M), is defined by the set of eigenvalues of Δ^M, i.e.

$$\text{Spec}\,(M) = \{0 = \lambda_0 < \lambda_1 \leq \lambda_2 \leq \cdots\}\,,$$

where each λ_i is written a number of times equal to its multiplicity. The manifold M is called *isospectral* to a compact riemannian manifold N when Spec $(M) =$ Spec (N). Throughout this paper, we shall consider only the eigenvalues of the Lapacian defined on smooth (C^∞) functions. It is also known that a compact riemannian manifold is of constant curvature K if the manifold is isospectral to an m-dimensional ($m = 2, 3, 4, 5$) compact riemannian manifold of constant curvature K (see [1], [11]).

Another asymptotic expansion, which generalizes theirs, is given by Colin de Verdière [2]. His result says that the spectrum for a generic compact riemannian manifold determines the set of lengths of the closed geodesics. An easy application of the above fact yields the following theorem.

Main theorem. *If a compact riemannian manifold is isospectral to a* 3*-dimensional lens space with fundamental group of order q, then the manifold is isometric to one of the* 3*-dimensional lens spaces with fundamental group of order q. In particular if the manifold is isospectral to the homogeneous* 3*-dimensional lens space, then the manifold is isometric to the lens space.*

Note. If two compact riemannian manifolds M and N are isometric, then the two manifolds are isospectral. But there exist two flat tori which are not isometric but are isospectral ([1]).

We shall sketch the proof of the theorem. Every 3-dimensional lens space with fundamental group of order q has a closed geodesic of length $2\pi/q$. It follows from Colin de Verdière's asymptotic expansion that M has also a closed geodesic γ of length $2\pi/q$. Since M is of constant curvature 1 and its fundamental group is of order q, γ is a generator of the fundamental group $\pi_1(M)$ of M. This implies that the universal covering space of M is isometric to the 3-dimensional sphere of constant curvature 1 and that the fundamental group of M is cyclic of order q, Z_q. Thus M is isometric to a 3-dimensional lens space with fundamental group of order q. To show the latter claim and that M has a closed geodesic of length $2\pi/q$, we need a study of the closed geodesics of length $2\pi/q$ on the 3-dimensional lens space. Though closed geodesics on a lens space are studied by Sakai [7], we need their more delicate properties such as indices and nullities.

Recently the author obtained a sharper result for a certain class of lens spaces, which is stated as follows.

Theorem. *If a compact riemannian manifold M is isospectral to a 3-dimensional lens space whose fundamental group is of order q or $2q$, where q is prime, then M is isometric to the lens space.*

To prove this, we must compute the spectrum of each 3-dimensional lens space with fundamental group of order q. A proof of the theorem will be given in a forthcoming paper. However it is still hard to conclude that any two lens spaces are isometric if and only if they are isospectral.

Since a riemannian manifold is of constant curvature K if the manifold is isospectral to a 5-dimensional compact riemannian manifold of constant curvature K, it will be possible to get some results about a manifold which is isospectral to a 5-dimensional lens space in the same way.

1. Closed geodesics on lens spaces

Let S^3 be a unit sphere embedded in $\boldsymbol{R}^4$. We shall consider relatively prime integers p and $q(\geq 3)$. Let T denote an orthogonal matrix defined by

$$T = \begin{pmatrix} R(1/q) & 0 \\ 0 & R(p/q) \end{pmatrix}, \quad \text{where} \quad R(\theta) = \begin{pmatrix} \cos 2\pi\theta & -\sin 2\pi\theta \\ \sin 2\pi\theta & \cos 2\pi\theta \end{pmatrix}.$$

Then a lens space with fundamental group of order q, $L(q\,;p)$, is defined by $L(q\,;p) = S^3/\{T, T^2, \cdots, T^q\}$. It is known that the fundamental group of $L(q\,;p)$ is isomorphic to a cyclic group of order q and that $L(q\,;p)$ is homogeneous if and only if $p \equiv \pm 1 \bmod q$ [7]. We should note that two lens spaces $L(q\,;p_1)$ and $L(q\,;p_2)$ are isometric if $p_1 \equiv \pm p_2 \bmod q$. Since

Sakai does not mention the fact explicitely in his paper [7], we prove it here.

Lemma 1. *If* $p_1 \equiv \pm p_2 \bmod q$, *then* $L(q\,;\,p_1)$ *and* $L(q\,;\,p_2)$ *are isometric.*

Proof. In case $p_1 \equiv p_2 \bmod q$, it is trivial from the definition of the lens spaces. Thus we consider the other case. In fact an orthogonal matrix

$$I' = \begin{pmatrix} 1 & & & 0 \\ & 1 & & \\ & & 1 & \\ 0 & & & -1 \end{pmatrix}$$

induces an isometry of $L(q\,;\,p_1)$ onto $L(q\,;\,p_2)$, because

$$I'\begin{pmatrix} R(1/q) & 0 \\ 0 & R(p_1/q) \end{pmatrix} = \begin{pmatrix} R(1/q) & 0 \\ 0 & R(p_2/q) \end{pmatrix} I' .$$

Analogously we get,

Lemma 2. *If* $p_1p_2 \equiv 1 \bmod q$, *then* $L(q\,;\,p_1)$ *and* $L(q\,;\,p_2)$ *are isometric. Hence* $L(q\,;\,p_1)$ *and* $L(q\,;\,p_2)$ *are isometric if*

$$p_1 \equiv \pm p_2 \quad \text{or} \quad p_1p_2 \equiv \pm 1 \bmod q .$$

Proof. An orthogonal matrix

$$J = \begin{pmatrix} & & 1 & 0 \\ & 0 & 0 & 1 \\ 1 & 0 & & \\ 0 & 1 & & 0 \end{pmatrix}$$

induces an isometry, because

$$J\begin{pmatrix} R(1/q) & 0 \\ 0 & R(p_1/q) \end{pmatrix} = \begin{pmatrix} R(p_1/q) & 0 \\ 0 & R(1/q) \end{pmatrix} J$$

and

$$\begin{pmatrix} R(p_1/q) & 0 \\ 0 & R(1/q) \end{pmatrix}^{p_2} = \begin{pmatrix} R(1/q) & 0 \\ 0 & R(p_2/q) \end{pmatrix} .$$

Let π be the natural projection of S^3 onto $L(q\,;\,p)$. If we define two great circles $\tilde{c}$ and $\tilde{c}^*$ by

$$\tilde{c}(t) = \begin{pmatrix} \cos t \\ \sin t \\ 0 \\ 0 \end{pmatrix} \quad \text{and} \quad \tilde{c}^*(t) = \begin{pmatrix} 0 \\ 0 \\ \cos t \\ \sin t \end{pmatrix},$$

then $\pi(\tilde{c}(t))$ and $\pi(\tilde{c}^*(t))$ are closed geodesics of length $2\pi/q$ on $L(q\,;p)$. Now we state two facts on the closed geodesics on lens spaces obtained by Sakai [7].

Lemma 3. *If $p \not\equiv \pm 1 \bmod q$, then the closed geodesics of length $2\pi/q$ on $L(q\,;p)$ are $\pi(\tilde{c}(t))$ and $\pi(\tilde{c}^*(t))$ up to parametrization. If $p \equiv 1 \bmod q$, then for each point of $L(q\,;1)$ there exists a unique closed geodesic of length $2\pi/q$ through that point.*

Note that $T(\tilde{c}(t)) = \tilde{c}(t + 2\pi/q)$ for all $t \in \boldsymbol{R}$. Thus $\tilde{c}$ is a closed geodesic invariant under T (see [4], [10]). Let (u_1, u_2, u_3) be a local coordinate system for S^3 such that

$$x = \begin{pmatrix} x_1 \\ x_2 \\ x_3 \\ x_4 \end{pmatrix} = \begin{pmatrix} \cos u_1 \cos u_2 \cos u_3 \\ \sin u_1 \cos u_2 \cos u_3 \\ \sin u_2 \cos u_3 \\ \sin u_3 \end{pmatrix} \quad \text{for each point } x \in S^3 - \left\{ \begin{pmatrix} 0 \\ 0 \\ 0 \\ \pm 1 \end{pmatrix} \right\}.$$

Then three vector fields along $\tilde{c}(t)$,

$$\left(\frac{\partial}{\partial u_1}\right)_{\tilde{c}(t)}, \left(\frac{\partial}{\partial u_2}\right)_{\tilde{c}(t)}, \left(\frac{\partial}{\partial u_3}\right)_{\tilde{c}(t)},$$

are orthogonal parallel vector fields. By an easy calculation, we have

$$\begin{aligned} T_*\left(\frac{\partial}{\partial u_1}\right)_{\tilde{c}(t)} &= \left(\frac{\partial}{\partial u_1}\right)_{\tilde{c}(t+2\pi/q)} \\ T_*\left(\frac{\partial}{\partial u_2}\right)_{\tilde{c}(t)} &= \cos 2\pi p/q \left(\frac{\partial}{\partial u_2}\right)_{\tilde{c}(t+2\pi/q)} \\ &\quad + \sin 2\pi p/q \left(\frac{\partial}{\partial u_3}\right)_{\tilde{c}(t+2\pi/q)} \\ T_*\left(\frac{\partial}{\partial u_3}\right)_{\tilde{c}(t)} &= -\sin 2\pi p/q \left(\frac{\partial}{\partial u_2}\right)_{\tilde{c}(t+2\pi/q)} \\ &\quad + \cos 2\pi p/q \left(\frac{\partial}{\partial u_3}\right)_{\tilde{c}(t+2\pi/q)} . \end{aligned} \tag{4}$$

Here T_* denotes the differential of T. Any Jacobi field $Y(t)$ which is orthogonal to $\tilde{c}(t)$ is a linear combination of $\sin t(\partial/\partial u_2)_{\tilde{c}(t)}$, $\cos t(\partial/\partial u_2)_{\tilde{c}(t)}$, $\sin t(\partial/\partial u_3)_{\tilde{c}(t)}$ and $\cos t(\partial/\partial u_3)_{\tilde{c}(t)}$. That is, we get

$$\begin{aligned} Y(t) &= (A \cos t + B \sin t)\left(\frac{\partial}{\partial u_2}\right)_{\tilde{c}(t)} \\ &\quad + (C \cos t + D \sin t)\left(\frac{\partial}{\partial u_3}\right)_{\tilde{c}(t)}, \end{aligned} \tag{5}$$

where A, B, C and D are constants.

We shall introduce a path space $\Omega(M, f)$. Here f denotes an isometry of order s on a riemannian manifold $(M, \langle\ ,\ \rangle)$. $\Omega(M, f)$ is the space of all piecewise smooth curves $\sigma: [0, 2\pi/q] \to M$ with $f(\sigma(0)) = \sigma(2\pi/q)$. As usual $E^f: \Omega(M, f) \to \boldsymbol{R}$ is defined by

$$E^f(\sigma) = \int_0^{\theta_0} \langle\dot{\sigma}(t), \dot{\sigma}(t)\rangle dt \qquad \text{for } \sigma \in \Omega(M, f)\ ,$$

which is called the *energy function*. Here $\theta_0 = 2\pi/q$, and $\dot{\sigma}$ denotes the velocity vector field of σ. It is well-known that $c \in \Omega(M, f)$ is critical for E^f if and only if c is a geodesic invariant under f or a point curve in the set of all fixed points by f. Provided that c is critical for E^f, we mean the *index* (resp. the *nullity*) of the critical submanifold

$$S^1 \cdot c = \{(c(t + \alpha))_{(0 \leqslant t \leqslant \theta_0)}\ ;\ 0 \leqslant \alpha \leqslant s\theta_0\}$$

by the index $\lambda(c, f)$ (resp. the nullity $\nu(c, f)$) of the critical point c. In particular in case $f = \mathrm{id}_M$, $\Omega(M, \mathrm{id}_M)$, $\lambda(c, \mathrm{id}_M)$ and $\nu(c, \mathrm{id}_M)$ will be written as $\Omega(M)$, $\lambda(c)$ and $\nu(c)$ respectively.

Now we come back to our case. Note that $\tilde{c} \in \Omega(S^3, T)$ is critical for E^T, and so is $\tilde{c}^* \in \Omega(S^3, T^{p_1})$ for $E^{T^{p_1}}$, where $p_1 \cdot p \equiv 1 \bmod q$.

Lemma 5. *If* $p \not\equiv \pm 1 \bmod q$, *then* $\nu(\tilde{c}, T) = 0$ *and if* $p \equiv \pm 1 \bmod q$, *then* $\nu(\tilde{c}, T) = 2$.

Proof. In [10] we obtained a formula $\nu(\tilde{c}, T) = \dim J(\tilde{c})$, where $J(\tilde{c})$ is a vector space of all Jacobi fields $Y(t)$ which are orthogonal to $\tilde{c}(t)$ and have the property

$$T_*(Y(t)) = Y(t + \theta_0)\ . \tag{6}$$

Using (4) and (5), (6) is equivalent to

$$\begin{aligned} &(A \cos t + B \sin t) \cos p\theta_0 + (C \cos t + D \sin t)(-\sin p\theta_0) \\ &\qquad = A \cos (t + \theta_0) + B \sin (t + \theta_0) \\ &(A \cos t + B \sin t) \sin p\theta_0 + (C \cos t + D \sin t) \cos p\theta_0 \\ &\qquad = C \cos (t + \theta_0) + D \sin (t + \theta_0)\ . \end{aligned} \tag{7}$$

Comparing the coefficients of $\cos t$ and $\sin t$, we obtain four equalities (8) which are equivalent to (7).

$$\begin{aligned} A \cos p\theta_0 - C \sin p\theta_0 &= A \cos \theta_0 + B \sin \theta_0 \\ B \cos p\theta_0 - D \sin p\theta_0 &= -A \sin \theta_0 + B \cos \theta_0 \\ A \sin p\theta_0 + C \cos p\theta_0 &= C \cos \theta_0 + D \sin \theta_0 \\ B \sin p\theta_0 + D \cos p\theta_0 &= -C \sin \theta_0 + D \cos \theta_0\ . \end{aligned} \tag{8}$$

Put $a = \cos(p\theta_0) - \cos\theta_0$, $b = \sin\theta_0$, $c = \sin(p\theta_0)$ and $\lambda = (a^2+b^2+c^2)/2bc$. It follows from (8) that

$$\lambda A = D,\ \lambda D = A,\ -\lambda B = C \text{ and } B = -\lambda C\ . \tag{9}$$

If $p \not\equiv \pm 1 \bmod q$, then $a \neq 0$ and $|\lambda| > 1$. Therefore the solution of (9) is $A = B = C = D = 0$. This implies $\lambda(\tilde{c}, T) = \dim J(\tilde{c}) = 0$, if $p \not\equiv \pm 1 \bmod q$. If $p \equiv 1 \bmod q$, (8) is equivalent to

$$B + C = 0\ , \qquad A - D = 0\ . \tag{10}$$

Hence $\nu(\tilde{c}, T) = \dim J(\tilde{c}) = 2$. Analogously $\nu(\tilde{c}, T) = 2$ if $p \equiv -1 \bmod q$.

The following lemma is proved similarly.

Lemma 11. *If $p \not\equiv \pm 1 \bmod q$, then $\nu(\tilde{c}^*, T^{p_1}) = 0$. Namely $S^1\tilde{c}^*$ is a non-degenerate critical submanifold* [6].

Lemma 12. *If $p \equiv 1 \bmod q$, then a critical submanifold*

$$W^+ = \left\{c\,;\ c(t) = \begin{pmatrix} \cos t & -\sin t & & \\ \sin t & \cos t & & 0 \\ & 0 & \cos t & -\sin t \\ & & \sin t & \cos t \end{pmatrix}\begin{pmatrix} x_1 \\ x_2 \\ x_3 \\ x_4 \end{pmatrix},\ \begin{pmatrix} x_1 \\ x_2 \\ x_3 \\ x_4 \end{pmatrix} \in S^3\right\}$$

in $\Omega(S^3, T)$ is non degenerate.

Proof. It is easy to see that each point of W^+ is critical for E^T. It is sufficient to show that $\nu(c, T) = \dim W^+ - 1 = 2$ for each $c \in W^+$. Introducing an orthogonal matrix

$$Q = \begin{pmatrix} x_1 & -x_2 & -x_3 & -x_4 \\ x_2 & x_1 & x_4 & -x_3 \\ x_3 & -x_4 & x_1 & x_2 \\ x_4 & x_3 & -x_2 & x_1 \end{pmatrix}, \qquad \text{where} \quad \begin{pmatrix} x_1 \\ x_2 \\ x_3 \\ x_4 \end{pmatrix} = c(0)\ ,$$

we see easily $Q(\tilde{c}(t)) = c(t)$ and $TQ = QT$. Therefore Q_* induces a linear isomorphism of $J(\tilde{c})$ onto $J(c) = \{Y\,;\ Y$ is a Jacobi field along c, which is orthogonal to c, with the property $T_*(Y(t)) = Y(t+\theta_0)\}$. This implies that $\nu(c, T) = \nu(\tilde{c}, T) = 2$.

Lemma 13. *$\lambda(\tilde{c}, T) = 0$ and $\lambda(\tilde{c}^*, T^{p_1}) = 0$.*

Proof. Let $V_{\tilde{c}}$ be the linear space of all smooth vector fields along $\tilde{c}$ which are orthogonal to $\tilde{c}$. Let $\tilde{L}$ denote a differential operator on $V_{\tilde{c}}$ defined by

$$\tilde{L}X = -(X''(t) + \tilde{R}(X, \dot{\tilde{c}}(t))\dot{\tilde{c}}(t)) \qquad \text{for } X \in V_{\tilde{c}}\ ,$$

where $\tilde{R}$ denotes the curvature tensor of S^3 and $X'(t)$ denotes the covariant derivative of $X(t)$ along $\tilde{c}$. In [4] or [10], we obtained a formula

$$\lambda(\tilde{c}, T) = \sum_{\mu<0} \dim \{X \in V_{\tilde{c}} ; \tilde{L}X = \mu X \text{ and } X(t+\theta_0) = T_*(X(t))\} .$$

$X \in V_{\tilde{c}}$ satisfies $\tilde{L}X = \mu X$ if and only if

$$\ddot{X}^j(t) + X^j(t) = -\mu X^j(t) , \qquad j = 2, 3 \tag{14}$$

hold, where $X^i(t)$, $i = 1, 2, 3$, denote the coefficients of $X(t)$ with respect to the orthogonal parallel basis $(\partial/\partial u_i)_{\tilde{c}(t)}$, $i = 1, 2, 3$. Suppose that $\mu < -1$. Then $X \in V_{\tilde{c}}$ satisfies $\tilde{L}X = \mu X$ if and only if

$$X^j(t) = A^j e^{\sqrt{-(1+\mu)}t} + B^j e^{-\sqrt{-(1+\mu)}t} , \qquad j = 2, 3 \tag{15}$$

hold for some constants A^j and B^j, $j = 2, 3$. Therefore if $X \in V_{\tilde{c}}$ satisfies $\tilde{L}X = \mu X$ and $T_*(X(t)) = X(t + \theta_0)$, then $X = 0$. Note that $X(t) = X(t + 2\pi)$. Suppose that $\mu = -1$. Then if $X \in V_{\tilde{c}}$ satisfies $\tilde{L}X = \mu X$, then

$$X(t) = \sum_{j=2}^{3} (a_j t + b_j)\left(\frac{\partial}{\partial u_j}\right)_{\tilde{c}(t)}$$

for some constants a_j and b_j ($j = 2, 3$). Furthermore if $T_*(X(t)) = X(t + \theta_0)$, then all the constants a_j and b_j are equal to zero. Hence

$$\dim \{X \in V_{\tilde{c}} ; \tilde{L}X = \mu X \text{ and } X(t + \theta_0) = T_*(X(t))\} = 0$$

for any $\mu \leqslant -1$. Suppose $-1 < \mu < 0$. Then $X \in V_{\tilde{c}}$ satisfies $\tilde{L}X = \mu X$ if and only if

$$\begin{aligned} X(t) &= (A \cos \sqrt{1+\mu}t + B \sin \sqrt{\mu+1}t)\left(\frac{\partial}{\partial u_2}\right)_{\tilde{c}(t)} \\ &\quad + (C \cos \sqrt{1+\mu}t + D \sin \sqrt{1+\mu}t)\left(\frac{\partial}{\partial u_3}\right)_{\tilde{c}(t)} \end{aligned} \tag{16}$$

holds for some constants A, B, C and D. Thus the vector field $X(t)$ satisfies $T_*(X(t)) = X(t + \theta_0)$ if and only if the four equations

$$\begin{aligned} A \cos p\theta_0 - C \sin p\theta_0 &= A \cos \theta_0\sqrt{1+\mu} + B \sin \theta_0\sqrt{1+\mu} \\ B \sin p\theta_0 - D \sin p\theta_0 &= -A \sin \theta_0\sqrt{1+\mu} + B \cos \theta_0\sqrt{1+\mu} \\ A \sin p\theta_0 + C \cos p\theta_0 &= C \cos \theta_0\sqrt{1+\mu} + D \sin \theta_0\sqrt{1+\mu} \\ B \sin p\theta_0 + D \cos p\theta_0 &= -C \sin \theta_0\sqrt{1+\mu} + D \cos \theta_0\sqrt{1+\mu} \end{aligned} \tag{17}$$

hold. Put $a = \cos p\theta_0 - \cos \theta_0\sqrt{1+\mu}$, $b = \sin \theta_0\sqrt{1+\mu}$, $c = \sin p\theta_0$

and $\lambda = (a^2 + b^2 + c^2)/2bc$. Then (17) is equivalent to

$$(18) \qquad A = \lambda D,\ \lambda A = D,\ B = -\lambda C \quad \text{and} \quad C = -\lambda B\ .$$

Because $a \neq 0$, $|\lambda|$ is greater than 1. Hence (18) is equivalent to $A = B = C = D = 0$. Finally we obtain that for each $\mu < 0$,

$$\dim \{X \in V_{\tilde{c}}\,;\ \tilde{L}X = \mu X \text{ and } X(t + \theta_0) = T_*(X(t))\} = 0\ .$$

This means $\lambda(\tilde{c}, T) = 0$. We can prove that $\lambda(\tilde{c}^*, T^{p_1}) = 0$ analogously.

Lemma 19. *If we put $c = \pi \circ \tilde{c}$ and $c^* = \pi \circ \tilde{c}^*$, then c and c^* are closed geodesics of length $2\pi/q$ on $L(q\,;p)$ and $\lambda(c) = \lambda(c^*) = 0$. If $p \not\equiv \pm 1 \bmod q$, then $\nu(c) = \nu(c^*) = 0$, and if $p \equiv 1 \bmod q$, then $\bar{\pi}(W^+) = \{\pi \circ \gamma\,;\ \gamma \in W^+\}$ is non degenerate.*

Proof. Let V_c be the vector space of all smooth vector fields along the closed geodesic c which are orthogonal to c. Then we define a differential operator L on V_c by

$$LX = -X''(t) - R(X, \dot{c})\dot{c}\ ,$$

where $X'(t)$ denotes the covariant derivative of $X(t)$ along c and R denotes the curvature tensor of the lens space $L(q\,;p)$. Since $\pi : S^3 \to L(q\,;p)$ is a riemannian covering,

$$\begin{aligned}\pi_* : \{\tilde{X} \in V_{\tilde{c}}\,;\ \tilde{L}\tilde{X} = \mu\tilde{X} \text{ and } \tilde{X}(t + \theta_0) = T_*(\tilde{X}(t))\} \\ \to \{X \in V_c\,;\ LX = \mu X \text{ and } X(t + \theta_0) = X(t)\}\end{aligned}$$

is a linear isomorphism for each real number μ. Hence

$$\begin{aligned}\lambda(c) &= \sum_{\mu<0} \dim \{X \in V_c\,;\ LX = \mu X \text{ and } X(t + \theta_0) = X(t)\} \\ &= \sum_{\mu<0} \dim \{\tilde{X} \in V_{\tilde{c}}\,;\ \tilde{L}\tilde{X} = \mu\tilde{X} \text{ and } \tilde{X}(t + \theta_0) = T_*(\tilde{X}(t))\} \\ &= \lambda(\tilde{c}, T) = 0\end{aligned}$$

and $\nu(c) = \nu(\tilde{c}, T)$. We can prove that $\lambda(c^*) = 0$ similarly. Since W^+ is non degenerate, so also is $\bar{\pi}(W^+)$. Note that $2 = \nu(\gamma, T) = \nu(\pi \circ \gamma)$ for each $\gamma \in W^+$ and $\dim \bar{\pi}(W^+) = 3$.

Lemma 20. *If a compact riemannian manifold M is isospectral to $L(q\,;p)$, then M has a closed geodesic of length $2\pi/q$.*

Proof. The claim is obvious from Lemma 19 and Colin de Verdière's asymptotic expansion [2]. Note that $\bar{\pi}(W^+)$ is the set of all closed geodesics of length $2\pi/q$ in $L(q\,;1)$ up to parametrization.

Lemma 21. *If a compact riemannian manifold M is isospectral to $L(q\,;p)$, then a closed geodesic γ of length $2\pi/q$ on M becomes a generator* *of the fundamental group* $\pi_1(M)$ *of* M, *which is cyclic of order* q.

Proof. Let $\tilde{M}$ be the universal covering space of M and introduce the induced metric to $\tilde{M}$. Since $\tilde{M}$ is compact and of constant curvature 1, $\tilde{M}$ is isometric to S^3. By Corollaire A. III. 2 [1, p. 16], we have

$$\mathrm{Vol}\,(S^3) = q\,\mathrm{Vol}\,(L(q\,;p)) \quad \text{and} \quad \mathrm{Vol}\,(\tilde{M}) = {}^{\#}\pi_1(M)\,\mathrm{Vol}\,(M)\;,$$

where ${}^{\#}\pi_1(M)$ denotes the order of the group $\pi_1(M)$ and $\mathrm{Vol}\,(\cdot)$ means its volume. Therefore ${}^{\#}\pi_1(M) = q$, since $\mathrm{Vol}\,(\tilde{M}) = \mathrm{Vol}\,(S^3)$ and $\mathrm{Vol}\,(M) = \mathrm{Vol}\,(L(q\,;p))$. Next we shall see that a closed geodesic γ of length $2\pi/q$ on M becomes a generator of $\pi_1(M)$. Without loss of generality we may assume that the closed geodesic γ is defined on $\boldsymbol{R}$ and its fundamental period is equal to 1. Let $\{\gamma\}$ denotes the homotopy class of $\gamma\,|\,[0,1]$ in $\pi_1(M,\gamma(0))$, the fundamental group of M at $\gamma(0)$. Suppose that $\{\gamma\}^i = 0$ for some i, $1 \leqslant i < q$. Define $\gamma^i(t) = \gamma(it)$ for $0 \leqslant t \leqslant 1$. Since $\{\gamma^i\} = \{\gamma\}^i = 0$, γ^i is homotopic to zero. Now take a point $\tilde{m} \in \tilde{M} = S^3$ with $\varphi(\tilde{m}) = \gamma(0)$, where $\varphi\colon \tilde{M} \to M$ is the covering projection. Let $\tilde{\gamma}^i$ be the closed geodesic on $\tilde{M} = S^3$ with $\tilde{\gamma}^i(0) = \tilde{m}$ and $\varphi\circ\tilde{\gamma}^i = \gamma^i$. The length of the geodesic segment $\tilde{\gamma}^i\,|\,[0,1]$ is less than 2π. Hence the point $\tilde{\gamma}^i(1)$ is not $\tilde{m}$. On the other hand $\tilde{\gamma}^i(1)$ is $\tilde{m}$ since γ^i is homotopic to zero (see Lemma III 15.2 in [5]). This is a contradiction. Therefore $\{\{\gamma\}, \{\gamma\}^2, \cdots, \{\gamma\}^q\}$ is a subgroup of $\pi_1(M,\gamma(0))$ and is of order q. This implies that $\{\gamma\}$ is a generator of the group.

2. Proof of the main theorem

It follows from [1] that M is of constant curvature 1. Thus we may take S^3 as the universal covering space of M. Let $\varphi\colon S^3 \to M$ be the riemannian covering projection. From Lemma 21 M has a closed geodesic γ of length $2\pi/q$, which is a generator of $\pi_1(M,\gamma(0))$. We identify $\alpha \in \pi_1(M,\gamma(0))$ and the covering transformation on S^3 induced from it. Then α acts freely on S^3 and $\varphi(\alpha(x)) = \varphi(x)$ for each $x \in S^3$ (see Theorem III, 16.6 in [5]). Since $\alpha \in \pi_1(M,\gamma(0))$ acts on S^3 isometrically, there exists a unique isometry T_α on $\boldsymbol{R}^4$ with $T_\alpha\,|\,S^3 = \alpha$ and $T_\alpha(0) = 0$, where 0 is the origin of $\boldsymbol{R}^4$. Since T_α is an orthogonal (linear) map, there exists an orthonormal basis $\{e_i,\ i = 1,2,3,4\}$ and real numbers θ and η such that

$$T_\alpha(e_1) = \cos\theta\cdot e_1 + \sin\theta\cdot e_2\,, \qquad T_\alpha(e_2) = -\sin\theta\cdot e_1 + \cos\theta\cdot e_2\,,$$
$$T_\alpha(e_3) = \cos\eta\cdot e_3 + \sin\eta\cdot e_4 \quad \text{and} \quad T_\alpha(e_4) = -\sin\eta\cdot e_3 + \cos\eta\cdot e_4\,.$$

Since $(T_\alpha)^q = \mathrm{id}_{\boldsymbol{R}^4}$, there exist integers l_1 and l_2 such that $\theta = 2\pi l_1/q$ and $\eta = 2\pi l_2/q$. Suppose that α is a generator of $\pi_1(M,\gamma(0))$. Then each l_i and q are relatively prime, because $\pi_1(M,\gamma(0))$ acts freely on S^3. Therefore M is isometric to the lens space $S^3/\{T_\alpha ; \alpha \in \pi_1(M, \gamma(0))\}$. In case M is isospectral to $L(q ; 1)$, it follows from Colin de Verdière's expansion that M has infinitely many closed geodesics of length $2\pi/q$. From Lemmas 2 and 3, M is isometric to $L(q ; 1)$.

Combining Lemma 2 and the main theorem, we obtain

Corollary. *Suppose* $q = 12, 14, 18$ *or* $3 \leqslant q \leqslant 10$. *Then the manifold* M *in the theorem is isometric to* $L(q ; p)$.

Note that for each q in Corollary, there exist at most only two non isometric lens spaces among the lens spaces with fundamental group of order q.

References

[1] M. Berger, P. Gauduchon and E. Mazet, Le Spectre d'une Varieté Riemannienne, Lecture Notes in Math. vol. 194, Springer Verlag, 1971.

[2] Y. Colin de Verdière, *Spectre du Laplacian et longueurs des géodésiques périodiques II,* Compositio Math., **27** (1973), 159–184.

[3] J. J. Duistermaat and V. W. Guillemin, *The spectrum of positive elliptic operators and periodic bicharacteristics.* Inventiones Math. **29**, (1975), 39–79.

[4] K. Grove and M. Tanaka, *On the number of invariant closed geodesics,* Acta. Math., to appear.

[5] S. T. Hu, Homotopy theory, Academic press, 1959.

[6] W. Meyer, *Kritishe Mannigfaltigkeiten in Hilbert mannigfaltigkeiten,* Math. Ann. **170** (1967), 45–66.

[7] T. Sakai, *On closed geodesics of lens spaces,* Tôhoku Math. J. **23** (1971), 403–411.

[8] T. Sakai, *On the spectrum of lens spaces,* Kōdai Math. Sem. Rep., **27** (1975), 249–257.

[10] M. Tanaka, *On invariant closed geodesics under isometries,* Kōdai Math. Sem. Rep. **28** (1977), 262–277.

[11] S. Tanno, *Eigenvalues of the Laplacian of riemannian manifolds,* Tôhoku Math. J. **25** (1973), 391–403.

TOKAI UNIVERSITY
HIRATSUKA, 259-12
JAPAN

Minimal Submanifolds and Geodesics
Kaigai Publications, Tokyo, 1978, 283–292

GEODESIC FLOWS ON C_L-MANIFOLDS AND EINSTEIN METRICS ON $S^3 \times S^2$

SHÛKICHI TANNO

1. Introduction

An m-dimensional complete Riemannian manifold (M, g) is called a *C_L-manifold*, if every complete geodesic is closed and of the same length $2\pi L$. The set of all oriented closed geodesics in a C_L-manifold (M, g) is denoted by Geod (M, g). Geod (M, g) is a $(2m - 2)$-dimensional compact manifold. A natural metric on Geod (M, g) is constructed as follows. Let (T_1M, G^S) be the unit tangent sphere bundle of M with the Sasaki metric G^S. The orbit space (i.e., the space of all trajectories) of the geodesic flow vector field ξ is identified with Geod (M, g). So, our metric G on Geod (M, g) is defined by averaging G^S along each orbit and projecting the result.

If (M, g) is a unit m-sphere, ξ is a Killing vector field with respect to G^S and G is the natural projection of G^S to Geod (M, g).

A sufficient condition for curves in Geod (M, g) to be geodesic with respect to G is given by Proposition 3.6. Since we do not have applications of this at present, we give only statements of lemmas without proofs.

In § 4, as a special case of C_L-manifolds we consider a unit m-sphere (S^m, g_0). Then the contact metric structure (η, G^S) on (T_1S^m, G^S) is Sasakian, and (Geod (M, g), G) is identified with a real Grassmann manifold $G(2, m - 1)$ with the canonical metric.

Since Grassmann manifolds are symmetric spaces of compact type and are Einstein spaces, (T_1S^m, G^S) is seen to be η-Einstein and T_1S^m admits a homogeneous and irreducible Einstein metric (Proposition 5.1).

Corollary. $S^3 \times S^2$ *admits a homogeneous and irreducible Einstein metric as a Sasakian manifold.*

In particular, this corollary says that, contrary to the case of dimension 4, there are two different types of homogeneous Einstein metrics on a simply connected manifold $S^3 \times S^2$.

2. Preliminaries

Let (M, g) be an m-dimensional C_L-manifold of class C^∞. We consider the following five Riemannian manifolds.

(i) (M, g, ∇) denotes the given C_L-manifold, where ∇ denotes the Riemannian connection defined by g. By (U, x^i) we mean a local coordinate neighborhood with coordinate system $(x^i; i = 1, \cdots, m)$. By $(\Gamma_j{}^i{}_k)$ we denote Christoffel's symbols and by R the Riemannian curvature tensor; $R(X, Y)Z = \nabla_X\nabla_Y Z - \nabla_Y\nabla_X Z - \nabla_{[X,Y]}Z$.

(ii) (TM, g^S, ∇^S) is the tangent bundle of (M, g), where g^S denotes the Sasaki metric and ∇^S denotes its Riemannian connection. By $\pi: TM \to M$ we denote the natural projection. Then, for a local coordinate neighborhood (U, x^i) in (M, g), $(\pi^{-1}U, x^i, y^i)$ is a local coordinate neighborhood in TM with local coordinate system (x^i, y^i) where $y^i\partial/\partial x^i$ denotes a tangent vector at (x^i).

We denote a vector field $Z = Z^i\partial/\partial x^i + Z^{m+i}\partial/\partial y^i$ on TM by

$$Z = (x^i, y^i; Z^i, Z^{m+i}) .$$

For a vector field $X = (X^i)$ on M,

$$X^H = (x^i, y^i; X^i, -\Gamma_r{}^i{}_s y^r X^s) , \tag{2.1}$$

$$X^C = (x^i, y^i; X^i, y^r\partial_r X^i) , \tag{2.2}$$

$$X^V = (x^i, y^i; 0, X^i) \tag{2.3}$$

are vector fields on TM called the *horizontal* lift of X, the *complete* lift of X, and the *vertical* lift of X, respectively. As is seen from (2.1) and (2.3), X^H and X^V are defined also for a tangent vector X at a point (x^i) of M and for (x^i, y^i) of TM. The Sasaki metric g^S is characterized by

$$g^S(X^H, Y^H) = g(X, Y) \circ \pi , \tag{2.4}$$

$$g^S(X^H, Y^V) = 0 , \tag{2.5}$$

$$g^S(X^V, Y^V) = g(X, Y) \circ \pi , \tag{2.6}$$

where X and Y are tangent vectors at a point or vector fields on M (cf. S. Sasaki [8], P. Dombrowski [1]). The meaning of horizontal or vertical is now clear from (2.4) $\sim$ (2.6).

A vector field F on TM defined by

$$F = (x^i, y^i; y^i, -\Gamma_r{}^i{}_s y^r y^s) \tag{2.7}$$

is called the *geodesic flow vector field* of (TM, g^S). The one parameter group of transformations $\{\phi_s\}$ generated by F is the geodesic flow of (TM, g^S).

For a (1, 1)-tensor fiels $A = (A^i_j)$ on M,

(2.8) $$\iota A = (x^i, y^i; 0, A^i_j y^j)$$

is a vertical vector field on TM. If A is a (1, 1)-tensor at a point x of M, then ιA is a tangent vector at (x, y) of TM. For a vector field Y on M, Y^C is expressed by

(2.9) $$Y^C = Y^H + \iota(\nabla Y) .$$

(iii) (T_1M, G^S, D^S) denotes the unit tangent sphere bundle. This is a submanifold of (TM, g^S) defined by $g(y, y) = 1$. We denote the projection also by the same letter π. By G^S we denote the induced metric from g^S to T_1M. D^S is its Riemannian connection. F is tangent to T_1M, and we denote the restriction of F to T_1M by ξ. ξ is a unit vector field on (T_1M, G^S). The 1-form dual to ξ with respect to G^S is denoted by η, which is called the contact form of T_1M. (η, ξ, G^S) defines a contact metric structure on T_1M (S. Sasaki [8]). By $\{\phi_s; -\infty < s < \infty, \phi_{s+2\pi L} = \phi_s\}$ we denote also the restriction of $\{\phi_s\}$ on TM to T_1M. This is the geodesic flow of (T_1M, G^S). Namely for a point (x_0, y_0) of T_1M,

$$\{\pi\phi_s(x_0, y_0), 0 \leq s \leq 2\pi L\}$$

is a closed geodesic in (M, g), and conversely every closed geodesic is expressed in this form for some (x_0, y_0).

Remark 2.1. For a Riemannian manifold (M, g), ξ is a Killing vector field on (T_1M, G^S) if and only if (M, g) is of constant curvature 1 (Y. Tashiro [14], S. Tanno [12]).

(iv) (T_1M, G^*, D^*). We define a new Riemannian metric G^* on T_1M from G^S by averaging by the geodesic flow:

(2.10) $$G^* = (1/2\pi L) \int_0^{2\pi L} \phi_s^* G^S ds .$$

Then ξ is a Killing vector field with respect to G^*. By D^* we denote the Riemannian connection defined by G^*.

(v) Geod (M, g) = (Geod $(M, g), G)$. A closed geodesic $\{c(s)\}$ with arclength parameter s in (M, g) is lifted as a closed geodesic $\{\phi_s\dot{c}(0)\}$ in (T_1M, G^S), where $\dot{c}(s) = dc(s)/ds$. $\{\phi_s\dot{c}(0)\}$ is a trajectory of ξ in T_1M and the orbit space T_1M/ξ = Geod (M, g) is the space of all oriented closed geodesics in (M, g). We denote this projection by $\pi_1: T_1M \to$ Geod (M, g). Since G^* is invariant by ϕ_s, G^* defines a Riemannian metric G on Geod (M, g) and we get (Geod $(M, g), G)$.

If one considers (T_1M, G^*) as a principal S^1-bundle over Geod (M, g), then η is an infinitesimal connection form. The subspace of the tangent space $(T_1M)_p$ defined by $\eta = 0$ is called the *η-horizontal* subspace at p

of T_1M. The η-horizontal distribution is invariant by ϕ_s, because $L_\xi \eta = 0$, where L_ξ denotes the Lie derivation by ξ.

For a vector field Z along a submanifold B of Geod (M, g), its η-horizontal lift Z^* is a vector field along $\pi_1^{-1}B$ such that

$$\pi_1 Z^* = Z \quad \text{and} \quad \eta(Z^*) = 0 .$$

The following diagram may be helpful to understand the situation;

$$(TM, g^S, \nabla^S) \supset (T_1M, G^S, D^S) \cdots (T_1M, G^*, D^*)$$

$$\pi \downarrow \qquad \swarrow \pi \qquad \qquad \qquad \qquad \downarrow \pi_1$$

$$(M, g, \nabla) \longleftarrow \qquad\qquad (\text{Geod}\,(M, g), G) .$$

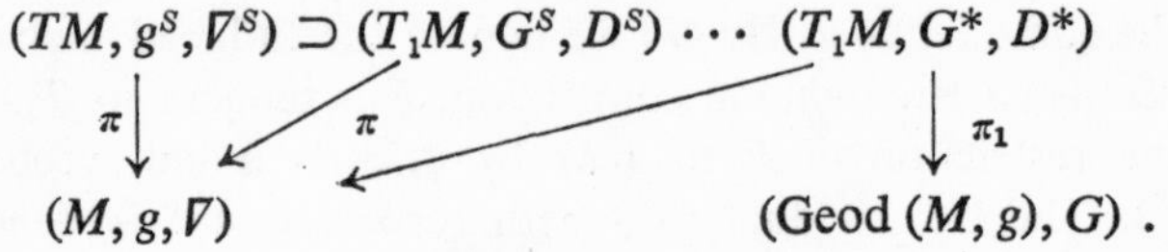

3. A curve in Geod (M, g)

We fix a point $c_0 = \{c_0(s)\}$ in Geod (M, g). A C^∞-curve $\{c_u\}$ passing through c_0 in Geod (M, g) is a one parameter family $\{c_u ; -\varepsilon < u < \varepsilon\}$ of closed geodesics

$$\{c_u(s) ; 0 \le s \le 2\pi L, s: \text{arclength parameter}\}$$

in (M, g) depending differentiably on u, and $d(c_u)/du \ne 0$ for each u; $-\varepsilon < u < \varepsilon$.

Let $[P] = \pi_1^{-1}\{c_u\}$. Then $[P]$ is a 2-dimensional immersed submanifold of T_1M. $[P]$ is invariant by ϕ_s.

Let $P = \pi[P]$. Then $P - Q$ is a 2-dimensional immersed submanifold of M for some singular point set Q. P is identified with the point set defined by

$$P = \{c_u(s) ; -\varepsilon < u < \varepsilon, 0 \le s \le 2\pi L\} .$$

Then $[P] = \{\dot{c}_u(s)\}$, where $\dot{c}_u(s) = \partial(c_u(s))/\partial s$.

Now let $Y^\# = d(c_u)/du$ be the tangent vector field to $\{c_u\}$ along $\{c_u\}$. Let Y^* be the η-horizontal lift of $Y^\#$. Take a point w_0 in $\pi_1^{-1}c_0$ such that $\pi w_0 \in P - Q$, and let $\{w_u\}$ be the η-horizontal curve in T_1M covering $\{c_u\}$, which passes through w_0. $\{w_u\}$ is nothing but the integral curve of Y^* passing through w_0. $\pi\{w_u\}$ is a non-trivial curve in $P - Q$. Here, if necessary we change ε to sufficiently small one. Then we can assume that the arclength parameter s of each closed geodesic c_u is chosen so that s starts at πw_u; that is,

$$c_u(0) = \pi w_u \quad \text{for each } u; -\varepsilon < u < \varepsilon .$$

Then $[P]$ has a coordinate system (u, s) such that $\dot{c}_u(s)$ has its coordinates (u, s). With respect to this coordinate system, Y^* is given by

$$Y^* = \partial(\dot{c}_u(s))/\partial u \ . \tag{3.1}$$

Furthermore, $P - Q$ has a coordinate system (u, s) such that $c_u(s)$ has its coordinates (u, s). Let (U, x^i) be a local coordinate neighborhood such that $U \cap (P - Q) \neq \emptyset$. Then, on $U \cap (P - Q)$, $P - Q$ is expressed by

$$c_u(s) = (x^i(u, s)) \ . \tag{3.2}$$

Let Y be the vector field along P defined by

$$Y = \partial(c_u(s))/\partial u \ .$$

Y vanishes on Q and Y on $P - Q$ is written as

$$Y = (\partial x^i(u, s)/\partial u) \ .$$

If we restrict Y to each closed geodesic c_u, then Y_u is a Jacobi field along c_u.

Lemma 3.1. *The vector field Y along P is lifted as the complete lift Y^C along $[P]$ in T_1M, and $Y^* = Y^C$.*

Let $\dot{c}$ be the vector field along $P - Q$ defined by

$$(\dot{c}) \text{ at } c_u(s) = \dot{c}_u(s) \ .$$

Lemma 3.2. *Y and $\dot{c}$ are orthogonal on $P - Q$, i.e., each Jacobi field Y_u along c_u satisfies $g(Y_u, \dot{c}_u) = 0$ and $g(\nabla_{\dot{c}} Y_u, \dot{c}_u) = 0$.*

Let $Z^\sharp$ be a tangent vector at c_0, and let Z^* be its η-horizontal lift and put $\pi Z^* = Z$. Then Z is a Jacobi field along c_0 such that $g(Z, \dot{c}_0) = g(\nabla_{\dot{c}} Z, \dot{c}_0) = 0$.

Lemma 3.3. *For two tangent vectors $Y^\sharp$ and $Z^\sharp$ at c_0, $G(Y^\sharp, Z^\sharp)$ is given by*

$$G(Y^\sharp, Z^\sharp) = \frac{1}{2\pi L} \int_{c_0} [g(Y, Z) + g(\nabla_{\dot{c}} Y, \nabla_{\dot{c}} Z)] ds \ . \tag{3.3}$$

This gives the geometric meaning of G in $(\mathrm{Geod}\,(M, g), G)$ and one sees that (3.3) is familiar in the theory of path spaces.

Since ξ is a Killing vector field with respect to G^*, in the fibering $\pi_1 \colon (T_1M, \eta, \xi, G^*) \to (\mathrm{Geod}\,(M, g), G)$, we have

$$D^*_{Y^*} Y^* = (D_{Y^\sharp} Y^\sharp)^* \ . \tag{3.4}$$

Therefore $\{c_u\}$ is a geodesic in $(\mathrm{Geod}\,(M, g), G)$ if and only if $D^*_{Y^*} Y^*$ is proportional to Y^*. Since $Y^* = Y^C$, our next step is to give the relation relation between $D^*_{Y^*}Y^*$ and $D^S_{Y^*}Y^* = D^S_{Y^C}Y^C$.

Lemma 3.4. *Let $Z^\sharp$ be a tangent vector at c_0, and let Z^* be its η-horizontal lift and $\pi Z^* = Z$ be the Jacobi field along c_0. Then, for $p \in \pi_1^{-1}c_0$ we have*

$$G^*_p(D^*_{Y^*}Y^*, Z^*) = \frac{1}{2\pi L}\int G^S_{\phi_s p}(D^S_{Y^*}Y^*, Z^*)ds \,. \tag{3.5}$$

In the next place we study

$$G^S(D^S_{Y^*}Y^*, Z^*) = G^S(D^S_{Y^C}Y^C, Z*) \tag{3.6}$$

and for this purpose, we calculate $\nabla^S_{Y^C}Y^C$.

Lemma 3.5. *For Y^C at c_0 we have*

$$(\nabla^S_{Y^C}Y^C)_{\dot{c}_0} = ((\nabla_Y Y)^C)_{\dot{c}_0} + (R(\dot{c}_0, \nabla_{\dot{c}}Y)Y)^H - (R(\dot{c}_0, Y)Y)^V \,. \tag{3.7}$$

Applying these lemmas we obtain the following.

Proposition 3.6. *Let $\{c_u\}$ be a curve in* (Geod $(M, g), G$). *Let $Y^\sharp$ be its canonical tangent vector field and let Y^* be its η-horizontal lift in T_1M. Then $Y = \pi Y^*$ is a vector field along $P = \{c_u(s)\,;\, -\varepsilon < u < \varepsilon, 0 \le s \le 2\pi L\}$ in the given C_L-manifold (M, g), such that the restriction Y_u of Y to each closed geodesic c_u is a Jacobi field satisfying $g(\dot{c}_u, Y_u) = 0$. $\{c_u\}$ is a geodesic in* (Geod $(M, g), G$) *if and only if, at each point c_u of $\{c_u\}$, Y along P satisfies*

$$\int_{c_u} [g(\nabla_Y Y + R(\dot{c}, \nabla_{\dot{c}}Y)Y, Z) + g(\nabla_{\dot{c}}\nabla_{\dot{c}}Y - R(\dot{c}, Y)Y, \nabla_{\dot{c}}Z)]ds = 0 \tag{3.8}$$

for any Jacobi field Z along c_u such that $g(\dot{c}_u, Z) = 0$ and

$$\int_{c_u} [g(Y, Z) + g(\nabla_{\dot{c}}Y, \nabla_{\dot{c}}Z)]ds = 0 \,. \tag{3.9}$$

Remark 3.7. Let $\{c_u\}$ be a curve in (Geod $(M, g), G$) and let P and Y be as in Proposition 3.6. Assume that

(1) $\nabla_Y(g(Y, Y))/g(Y, Y)$ is constant along each c_u in $P - Q$,
(2) $R(\dot{c}, Y)Y =$ proportional to $\dot{c}$,
(3) $R(\dot{c}, \nabla_{\dot{c}}Y)Y =$ proportional to $\dot{c}$,
(4) $P - Q$ is minimal in (M, g).

Then, $\{c_u\}$ is a geodesic in (Geod $(M, g), G$), by Proposition 3.6.

Example 3.8. Let (S^m, g_0) be a unit m-sphere. Let S^2 be a totally geodesic submanifold imbedded in (S^m, g_0). As a one parameter family of closed geodesics in S^2 we consider ones $\{c_u\}$ which pass the north pole and the south pole of S^2. By the standard parameterization u of $\{c_u\}$ we see that the vector field Y canonically related to $\{c_u\}$ is a constant speed rotation vector field and satisfies (1) of Remark 3.7. Hence, $\{c_u\,;\ -\pi < u \leq \pi\}$ is a closed geodesic in $(\mathrm{Geod}\,(M, g), G)$.

Remark 3.9. The author is not sure up to today if some differential geometry on a C_L-manifold (M, g) could be carried out on $(\mathrm{Geod}\,(M, g), G)$.

4. Grassmann manifolds $G(2, m-1)$

In this section as a C_L-manifold we consider a unit m-sphere (S^m, g_0). Then ξ on (T_1S^m, G^S) is a Killing vector field and the contact metric structure (η, ξ, G^S) is Sasakian (in the normalization such that the sectional curvature $K(X, \xi)$ for each 2-plane which contains ξ is equal to $1/4$).

Remark 4.1. The usual Sasakian structure (in the normalization such that $K(X, \xi) = 1$) is defined by $(\eta/2, 2\xi, G^S/4)$.

If we consider an oriented great circle of S^m as an oriented 2-plane in the Euclidean $(m+1)$-space E^{m+1}, then $\mathrm{Geod}\,(S^m, g_0)$ is identified with a real Grassmann manifold

$$G(2, m-1) = SO(m+1)/SO(2) \times SO(m-1)\ .$$

The Sasakian manifold $(T_1S^m, \eta, \xi, G^S = G^*)$ is a circle bundle over $(\mathrm{Geod}\,(S^m, g_0), G)$. We see that G on $\mathrm{Geod}\,(S^m, g_0) = G(2, m-1)$ is the canonical metric by the following two observations.

(i) $SO(m+1)$ acts on (S^m, g_0) as its isometry group. For each $f \in SO(m+1)$ the natural extension $f_*: T_1S^m \to T_1S^m$ is an isometry with respect to G^S and f_* leaves ξ invariant. So, $SO(m+1)$ is a group of automorphisms of $(T_1S^m, \eta, \xi, G^S = G^*)$. Since f_* leaves ξ invariant, it induces an isometry of $(G(2, m-1), G)$ and $SO(m+1)$ is a group of isometries of $(G(2, m-1), G)$. Since the canonical metric on $G(2, m-1)$ is the unique $SO(m+1)$-invariant metric (up to a constant factor) except the case $m = 3$ (cf. K. Leichtweiss [5]), G is canonical (even the case $m = 3$).

(ii) Since each $(G(r, s), G)$ is isometrically imbedded in a bigger $(G(u, v), G)$ as a totally geodesic submanifold, consider G in the case $m = 2$. For (S^2, constant curvature 1), we have

$$(T_1S^2, G^S) = (RP^3,\ \text{constant curvature } 1/4)$$

(W. Klingenberg and S. Sasaki [4], S. Tanno [13]), and

$$(\mathrm{Geod}\,(S^2, g_0), G) = (CP^1 = S^2,\ \text{constant curvature } 1)\ .$$

Thus G of $G(1, m)$ (i.e., an m-sphere) has constant curvature 1.

Now, it is well known that the Ricci tensor Ric (G) of $(G(r, s), G)$ is given by (cf. K. Leichtweiss [5], p. 351); Ric $(G) = (r + s - 2)G$. Therefore, on $(G(2, m-1), G)$ we obtain

$$\text{(4.1)} \qquad \operatorname{Ric}(G) = (m-1)G \,.$$

Remark 4.2. If the Ricci tensor of a Sasakian manifold (M, η, ξ, g) is of the form; Ric $= ag + b\eta \otimes \eta$ for some scalar fields a and b, it is called an *η-Einstein space*. If $\dim M \geq 5$, a and b are necessarily constant.

Remark 4.3. From now on by a Sasakian structure we mean one under the normalization $K(X, \xi) = 1$.

Lemma 4.4. *Let (M, η, ξ, g) be an η-Einstein Sasakian manifold*; Ric $= ag + b\eta \otimes \eta$. *If $a > -2$, then for $\alpha = (a + 2)/(2u + 2)$ where $2u + 1 = \dim M$,*

$$(M, {}^*\eta = \alpha\eta, {}^*\xi = (1/\alpha)\xi, {}^*g = \alpha g + (\alpha^2 - \alpha)\eta \otimes \eta)$$

is an Einstein Sasakian manifold; ${}^*\text{Ric} = ((\alpha + 2 - 2\alpha)/\alpha){}^*g$.

Proof. By (2.15) of [10] we get

$$\begin{aligned} {}^*\text{Ric} &= ag + b\eta \otimes \eta - 2(\alpha - 1)g + (\alpha - 1)(2u + 2 + 2u\alpha)\eta \otimes \eta \\ &= ((a + 2 - 2\alpha)/\alpha)(\alpha g + (\alpha^2 - \alpha)\eta \otimes \eta) \,, \end{aligned}$$

where we have used $a + b = 2u = \dim M - 1$. Q.E.D.

$(T_1M, \eta/2, 2\xi, G^*/4)$ is an η-Einstein space, if and only if $(G(2, m-1), G/4)$ is an Einstein space (cf. [11], § 5). The relation is given by

$$\text{(4.2)} \qquad \operatorname{Ric}(G^*/4) = aG^* + b(1/4)\eta \otimes \eta \,,$$

with $a + b = \dim T_1M - 1$, and

$$\text{(4.3)} \qquad \operatorname{Ric}(G/4) = (a + 2)(G/4) \,.$$

By Ric (G) = Ric $(G/4)$, (4.1), (4.2), and (4.3) we get

$$\text{(4.4)} \quad \operatorname{Ric}(G^*/4) = 2(2m - 3)(G^*/4) + 2(2 - m)(\eta/2) \otimes (\eta/2) \,.$$

Theorem 4.5. *The unit tangent sphere bundle (T_1M, G^S) of a unit m-sphere has a Sasakian structure $(\eta/2, 2\xi, G^S/4)$ for which $(G^S/4) = (G^*/4)$ is an η-Einstein metric* (4.4). *This structure can be deformed to a new Sasakian structure*;

$$\begin{aligned} {}^*\eta = (\alpha/2)\eta \,, \qquad {}^*\xi = (2/\alpha)\xi \,, \\ {}^*G^S = (\alpha/4)G^S + ((\alpha^2 - \alpha)/4)\eta \otimes \eta \,, \end{aligned}$$

so that $^*G^S$ *is an Einstein metric*; $^*\mathrm{Ric} = 2(m-1)^*G^S$, *where* $\alpha = 2(m-1)/m$. $SO(m+1)$ *acts on* T_1M *as a (transitive) group of automorphisms of both Sasakian structures.*

5. Einstein metrics

Since the metric of every Sasakian structure is irreducible (cf. S. Tanno [9]), from Theorem 4.5 we obtain

Proposition 5.1. *The unit tangent sphere bundle of a unit m-sphere admits an Einstein metric which is homogeneous and irreducible.*

Since T_1S^3 is topologically $S^3 \times S^2$, we obtain

Proposition 5.2. *On* $S^3 \times S^2$ *there are two different Einstein metrics*;

(i) *a Riemannian product*, $(S^3, 2g_0) \times (S^2, g_0)$,

(ii) *the metric given by* Proposition 5.1.

Remark 5.3. The second example in Proposition 5.2 is not locally symmetric, because every locally symmetric Sasakian manifold is of constant curvature 1 (cf. [9]).

Homogeneous, simply connected Einstein spaces of dimension 4 are classified by G. R. Jensen [2]. They are all symmetric spaces. So the possible lowest dimensional example of homogeneous simply connected Einstein spaces which are not locally symmetric is given by Proposition 5.2 (ii) on $S^3 \times S^2$.

Remark 5.4. For $\dim M = 4$ there are three types of homogeneous simply connected compact Einstein spaces; (S^4, g_0), $(S^2, g_0) \times (S^2, g_0)$ and (CP^2, g_0) [complex projective 2-space]. Topological types of these spaces are all different. Contrary to this, for the case of dimension 5 the same $S^3 \times S^2$ admits different homogeneous Einstein metrics.

Remark 5.5. Under the normalization such that $(G(1, m), G)$ is of constant curvature 1, the volume of $(G(2, m-1), G)$ is given by

$$\mathrm{Vol}\,(G(2, m-1), G) = c_m \cdot c_{m-1}/2\pi ,$$

where c_m denotes the volume of a unit m-sphere $(S^m, g_0) = (G(1, m), G)$.

Remark 5.6. Theorem 4.5 or Proposition 5.1 gives a geometric construction of the homogeneous Einstein metric on $(T_1S^m, {}^*G^S) = SO(m+1)/SO(m-1)$ using the Sasaki metric. For more general case $SO(p+q)/SO(q)$, see [6] by A. Sagle, and for general cases of principal fiber bundles over symmetric spaces, see [3] by G. R. Jensen.

References

[1] Dombrowski, P., *On the geometry of the tangent bundle*, J. reine angew. Math., **210** (1962), 73–88.

[2] Jensen, G. R., *Homogeneous Einstein spaces of dimension four*, J. Differ-

ential Geometry, **3** (1969), 309–349.

[3] Jensen, G. R., *Einstein metrics on principal fibre bundles,* J. Differential Geometry, **8** (1973), 599–614.

[4] Klingenberg, W. and S. Sasaki, *On the tangent sphere bundle of a 2-sphere,* Tôhoku Math. J., **27** (1975), 49–56.

[5] Leichtweiss, K., *Zur Riemannschen Geometrie in Grassmannschen Mannigfaltigkeiten,* Math. Zeit., **76** (1961), 334–366.

[6] Sagle, A., *Some homogeneous Einstein manifolds,* Nagoya Math. J., **39** (1970), 81–106.

[7] Sakai, T., *On the riemannian structure all of whose geodesics are closed and of the same length,* J. Math. Soc. Japan, **27** (1975), 239–247.

[8] Sasaki, S., *On the differential geometry of tangent bundles of Riemannian manifolds,* Tôhoku Math. J., **10** (1958), 338–354; Part II, **14** (1962), 146–155.

[9] Tanno, S., *Locally symmetric K-contact Riemannian manifolds,* Proc. Japan Acad., **43** (1968), 581–583.

[10] Tanno, S., *The topology of contact Riemannian manifolds,* Illinois J. Math., **12** (1968), 700–717.

[11] Tanno, S., *Harmonic forms and Betti numbers of certain contact Riemannian manifolds,* J. Math. Soc. Japan, **19** (1967), 308–316.

[12] Tanno, S., *Killing vectors and geodesic flow vectors on tangent bundles,* J. reine angew. Math. **282** (1976), 162–171.

[13] Tanno, S., *Orthonormal frames on 3-dimensional Riemannian manifolds,* J. Differential Geometry, **11** (1976), 467–474.

[14] Tashiro, Y., *On contact structures of tangent sphere bundles,* Tôhoku Math. J., **21** (1969), 117–143.

Tôhoku University
Sendai, 980
Japan